Bayesian Models for Categorical Data

A complete list of the titles in this series appears at the end of this volume.

Bayesian Models for Categorical Data

PETER CONGDON
Queen Mary, University of London, UK

John Wiley & Sons, Ltd

Chichester • New York • Weinheim • Brisbane • Singapore • Toronto

Other Wiley Editorial Offices

John Wiley & Sons Inc., 111 River Street, Hoboken, NJ 07030, USA

Jossey-Bass, 989 Market Street, San Francisco, CA 94103-1741, USA

Wiley-VCH Verlag GmbH, Boschstr. 12, D-69469 Weinheim, Germany

John Wiley & Sons Australia Ltd, 33 Park Road, Milton, Queensland 4064, Australia

John Wiley & Sons (Asia) Pte Ltd, 2 Clementi Loop #02-01, Jin Xing Distripark, Singapore 129809

John Wiley & Sons Canada Ltd, 22 Worcester Road, Etobicoke, Ontario, Canada M9W 1L1

Wiley also publishes its books in a variety of electronic formats. Some content that appears in
print may not be available in electronic books.

Library of Congress Cataloging-in-Publication Data

Congdon, P.
   Bayesian models for categorical data/Peter Congdon.
      p. cm. – (Wiley series in probability and statistics)
   Includes bibliographical references and index.
   ISBN 0-470-09237-8 (cloth : alk. paper)
      1. Bayesian statistical decision theory. 2. Monte Carlo method. 3. Markov processes.
   4. Multivariate analysis. I. Title. II. Series.
   QA279.5.C6495 2005
   519.5'42–dc22   2005005158

British Library Cataloguing in Publication Data

A catalogue record for this book is available from the British Library

   ISBN-13 978-0-470-09237-8 (HB)
   ISBN-10 0-470-09237-8 (HB)

Typeset in 11/13pt Times by Thomson Press (India) Limited, New Delhi, India.

This book is printed on acid-free paper responsibly manufactured from sustainable forestry
in which at least two trees are planted for each one used for paper production.

Contents

Preface

This book continues the themes in my two earlier Wiley books in seeking to make modern Bayesian methods accessible via a practically oriented exposition, with statistical computing and applied data analysis at the forefront. As before, I have focused on the WINBUGS package, which has now reached a wide degree of acceptance in application to the ever expanding corpus of methodological work and applications based on Monte Carlo Markov Chain techniques. I hope that the applied focus will help students and researchers alike, including those with primary disciplinary interests outside statistics (e.g. psychology, geography and epidemiology). Nevertheless a wide range of simple or more advanced modelling techniques are discussed and I hope the approach is also helpful for courses in Bayesian data analysis and statistical computing. I have sought to review recent Bayesian methodology for categorical outcomes (binary, count and multinomial data) but my take on this will obviously emphasize some themes more than others: particular aspects that I have focused on include non-parametric and non-linear regression models, model choice, time series and spatio-temporal models. Missing data models are also considered. As with my earlier Wiley books, a set of worked examples with documented code forms part of the book's presentation and can be downloaded from the publisher's website ftp:// www.wiley.co.uk/pub/books/congdon, the WINBUGS site or STATLIB.

Peter Congdon

CHAPTER 1

Principles of Bayesian Inference

1.1 BAYESIAN UPDATING

Bayesian inference differs from classical inference in treating parameters as random variables and using the data to update prior knowledge about parameters and functionals of those parameters. We are also likely to need model predictions and these are provided as part of the updating process. Prior knowledge about parameters and updated (or posterior) knowledge about them, as well as implications for functionals and predictions, are expressed in terms of densities. One of the benefits of modern Monte Carlo Markov Chain (MCMC) sampling methods (e.g. Chib and Greenberg, 1995; Tierney, 1994; Gelfand and Smith, 1990; Gilks *et al.*, 1996a; Smith and Roberts, 1993) is the ease with which full marginal densities of parameters may be obtained. In a regression model the parameters would be regression coefficients and possible variance parameters, and functionals of parameters might include elasticities (in econometrics) or effective dose (in biometrics).

The new Bayesian sampling-based estimation techniques obtain samples from the posterior density, either of parameters themselves, or functionals of parameters. They improve considerably on multiple integration or analytical approximation methods that are infeasible with large numbers of parameters. Nevertheless many issues remain in the application of sampling-based techniques, such as obtaining convergence, and choice of efficient sampling method. There are also more general

Bayesian Models for Categorical Data   P. Congdon
© 2005 John Wiley & Sons, Ltd

problems in Bayesian methods such as choice of priors (and possible sensitivity of inferences to alternative choices).

The basis for Bayesian inference may be derived from simple probability theory. Thus the conditional probability theorem for events A and B is that

$$Pr(A|B) = Pr(A, B)/Pr(B) = Pr(B|A)Pr(A)/Pr(B)$$

Replacing B by observations y, A by a parameter set θ and probabilities by densities results in the relation

$$p(\theta|y) = p(\theta, y)/p(y) = p(y|\theta)p(\theta)/p(y) \qquad (1.1)$$

where $p(y|\theta)$ is the likelihood of y under a model and $p(\theta)$ is the prior density, or the density of θ before y is observed. This density expresses accumulated knowledge about θ, or, viewed another way, the degree of uncertainty about θ. It may also include working model assumptions (e.g. assumptions about the nature of error structures); for example, one model might assume uncorrelated errors over time or space and another model assume correlated errors.

Classical analysis via maximum likelihood focuses on the likelihood $p(y|\theta)$ without introducing a prior, whereas fully Bayesian analysis updates the prior information about θ with the information contained in the data. The denominator $p(y) = \int p(y|\theta)p(\theta)d\theta$ in (1.1) defines the 'marginal likelihood' or 'prior predictive density' of the data and may be set to be an unknown constant c. So posterior inferences about θ under (1.1) can be equivalently stated in the relation

$$p(\theta|y) = p(y|\theta)p(\theta)/c$$

or

$$p(\theta|y) \propto p(y|\theta)p(\theta)$$

This can be stated as 'posterior is proportional to likelihood times prior'.

1.2 MCMC TECHNIQUES

The basis of modern Bayesian inference regarding $p(\theta|y)$ is the use of iterative MCMC methods that involve repeated sampling from the posterior distribution, using long, possibly multiple, chains of parameter samples. One is then interested in posterior summaries of parameters or functionals from the MCMC output in the form of expectations, densities or probabilities. These summaries typically include posterior means and variances of the parameters themselves, or of functions $\Delta = \Delta(\theta)$ of the parameters, which analytically are

$$E(\theta_k|y) = \int \theta_k p(\theta|y) d\theta \qquad (1.2)$$

$$Var(\theta_k|y) = \int \theta_k^2 p(\theta|y) d\theta - [E(\theta_k|y)]^2$$

$$= E(\theta_k^2|y) - [E(\theta_k|y)]^2 \qquad (1.3)$$

$$E[\Delta(\theta)|y] = \int \Delta(\theta) p(\theta|y) d\theta$$

$$Var[\Delta(\theta)|y] = \int \Delta^2 p(\theta|y) d\theta - [E(\Delta|y)]^2$$

$$= E(\Delta^2|y) - [E(\Delta|y)]^2$$

Often the major interest is in marginal densities of the parameters themselves. Let the model dimension be d, so that $\theta = (\theta_1, \ldots, \theta_d)$. Then the marginal density of the jth parameter θ_j is obtained by integrating out all other parameters

$$p(\theta_j|y) = \int p(\theta|y) d\theta_1 d\theta_2 \ldots \theta_{j-1} \theta_{j+1} \ldots \theta_d$$

The predictive density for new or replicate data useful in model checking and comparison is

$$p(y_{new}|y) = \int p(y_{new}, \theta|y) d\theta = \int p(y_{new}|y, \theta) p(\theta|y) d\theta$$

Posterior probabilities might relate to the probability that θ_j exceeds a threshold b, and involve integrals of the form

$$\Pr(\theta_j > b|y) = \int_b^\infty p(\theta_j|y) d\theta_j \qquad (1.4)$$

Such expectations, densities or probabilities may be obtained analytically for conjugate analyses, such as a binomial likelihood where the probability has a beta prior. Results can be obtained under asymptotic approximations (Bernardo and Smith, 1994), similar to those used in classical statistics, or by analytic approximations (e.g. Laplace) based on expanding the relevant integral (Kass *et al.*, 1988). Such approximations tend to be less good for posteriors that are not approximately normal or where there is multimodality. An alternative strategy facilitated by contemporary computer technology is to use sampling-based approximations based on the Monte Carlo principle. One such sampling method is importance sampling (Geweke, 1989; McFadden, 1989), and other precursors of modern Bayesian sampling include data augmentation for Bayes inference in missing-data problems (Tanner and Wong, 1987).

1.3   THE BASIS FOR MCMC

The canonical Monte Carlo method assumes a sample of independent d-dimensional simulations $u^{(1)}, u^{(2)}, \ldots, u^{(T)}$ from a target density $\pi(u)$ whereby $E[g(u)] = \int g(u)\pi(u)du$ is estimated as

$$\bar{g}_T = \sum_{t=1}^{T} g(u^{(t)})$$

With probability 1, $\bar{g}_T$ tends to $E_\pi[g(u)]$ as $T \to \infty$. However, independent sampling from the posterior density $p(\theta|y)$ is not feasible in general. It is valid, however, to use dependent samples $\theta^{(t)}$ provided the sampling satisfactorily covers the support of $p(\theta|y)$ (Gilks *et al.*, 1996b). In order to sample approximately from $p(\theta|y)$, MCMC methods generate pseudorandom-dependent draws via Markov chains. Specifically let $\theta^{(0)}, \theta^{(1)}, \ldots$ be a sequence of random variables. Then $p(\theta^{(0)}, \theta^{(1)}, \ldots, \theta^{(T)})$ is a Markov chain if

$$p(\theta^{(t)}|\theta^{(0)}, \theta^{(1)}, \ldots, \theta^{(t-1)}) = p(\theta^{(t)}|\theta^{(t-1)})$$

so that only the preceding state is relevant to the future state. Suppose $\theta^{(t)}$ is defined on a discrete state space $S = \{s_1, s_2, \ldots\}$; generalisation to continuous state spaces is described by Tierney (1996). Assume $p(\theta^{(t)}|\theta^{(t-1)})$ is defined by a constant one-step transition matrix

$$Q_{i,j} = \Pr(\theta^{(t)} = s_j|\theta^{(t-1)} = s_i)$$

with t-step transition matrix $Q_{i,j}(t) = \Pr(\theta^{(t)} = s_j|\theta^{(0)} = s_i)$. Sampling from a constant one-step Markov chain converges to a stationary distribution $\pi(\theta) = p(\theta|y)$ if additional requirements[1] on the chain are satisfied (irreducibility, aperiodicity and positive recurrence) – see Roberts (1996, p 46) and Norris (1997). Sampling chains meeting these

---

[1]Suppose a chain is defined on a space S. A chain is irreducible if for any pair of states $(s_i, s_j) \in S$ there is a non-zero probability that the chain can move from s_i to s_j in a finite number of steps. A state is positive recurrent if the number of steps the chain needs to revisit the state has a finite mean. If all the states in a chain are positive recurrent then the chain itself is positive recurrent. A state has period k if it can only be revisited after a number of steps that is a multiple of k. Otherwise the state is aperiodic. If all its states are aperiodic then the chain itself is aperiodic. Positive recurrence and aperiodicity together constitute ergodicity.

requirements have a unique stationary distribution $\lim_{t \to \infty} Q_{i,j}(t) = \pi_{(j)}$ satisfying the full balance condition $\pi_{(j)} = \sum_i \pi_{(i)} Q_{i,j}$. Many Markov chain methods are additionally reversible, meaning $\pi_{(i)} Q_{i,j} = \pi_{(j)} Q_{j,i}$.

With this type of sampling mechanism, the ergodic average $\bar{g}_T$ tends to $E\pi[g(u)]$ with probability 1 as $T \to \infty$ despite dependent sampling. Remaining practical questions include establishing an MCMC sampling scheme and establishing that convergence to a steady state has been obtained for practical purposes (Cowles and Carlin, 1996).

Estimates of quantities such as (1.2) and (1.3) are routinely obtained from sampling output along with 2.5% and 97.5% percentiles that provide credible intervals for the value of the parameter. A full posterior density estimate may be derived also (e.g. by kernel smoothing of the MCMC output of a parameter). For $\Delta(\theta)$ its posterior mean is obtained by calculating $\Delta^{(t)}$ at every MCMC iteration from the sampled values $\theta^{(t)}$. The theoretical justification for this is provided by the MCMC version of the law of large numbers (Tierney, 1994), namely that

$$\sum_{t=1}^{T} \Delta(\theta^{(t)})/T \to E_\pi[\Delta(\theta)]$$

provided that the expectation of $\Delta(\theta)$ under $\pi(\theta) = p(\theta|y)$, denoted $E_\pi[\Delta(\theta)]$, exists.

The probability (1.4) would be estimated by the proportion of iterations where $\theta_j^{(t)}$ exceeded b, namely $\sum_{t=1}^{T} 1(\theta_j^{(t)} > b)/T$, where $1(A)$ is an indicator function which takes value 1 when A is true, 0 otherwise. Thus one might in a disease mapping application wish to obtain the probability that an area's smoothed relative mortality risk θ_k exceeds zero, and so count iterations where this condition holds, avoiding the need to evaluate the integral

$$\Pr(\theta_k > 0) = \int_0^\infty p(\theta_k|y)\mathrm{d}\theta_k$$

This principle extends to empirical estimates of the distribution function, $F()$ of parameters or functions of parameters. Thus the estimated probability that $\Delta < d$ for values of d within the support of Δ is

$$\hat{F}(d) = \sum_{t=1}^{T} 1(\Delta^{(t)} \le d)/T$$

The sampling output also often includes predictive replicates $y_{\text{new}}^{(t)}$ that can be used in posterior predictive checks to assess whether a model's predictions are consistent with the observed data. Predictive replicates are obtained by sampling $\theta^{(t)}$ and then sampling y_{new} from the likelihood model $p(y_{\text{new}}|\theta^{(t)})$. The posterior predictive density can also be used for model choice and residual analysis (Gelfand, 1996, sections 9.4–9.6).

1.4   MCMC SAMPLING ALGORITHMS

The Metropolis–Hastings (M–H) algorithm is the baseline for MCMC sampling schemes and is based on a binary transition kernel. Following Hastings (1970), the chain is updated from $\theta^{(t)}$ to θ^* with probability

$$\alpha(\theta^*|\theta^{(t)}) = \min\left(1, \frac{p(\theta^*|y)f(\theta^{(t)}|\theta^*)}{p(\theta^{(t)}|y)f(\theta^*|\theta^{(t)})}\right) \qquad (1.5)$$

with transition kernel $\alpha(\theta^*|\theta^{(t)})f(\theta^*|\theta^{(t)})$, where f is known as a proposal or jumping density (Chib and Greenberg, 1995). If the proposed update is rejected the next state is the same as the current state. The algorithm works most successfully when the proposal density matches, at least approximately, the shape of the target density $p(\theta|y)$. The rate at which a proposal generated by f is accepted (the acceptance rate) depends on how close θ^* is to $\theta^{(t)}$, and this depends on the variance σ^2 assumed in the proposal density. For a normal proposal density a higher acceptance rate follows from reducing σ^2, but with the risk that the posterior density will take longer to explore. Performance also tends to be improved if parameters are transformed to take the full range of positive and negative values $(-\infty, \infty)$ so lessening the occurrence of skewed parameter densities.

If the proposal density is symmetric, with $f(\theta^*|\theta^{(t)}) = f(\theta^{(t)}|\theta^*)$, then the Hastings algorithm reduces to an algorithm used by Metropolis *et al.* (1953) for indirect simulation of energy distributions, whereby

$$\alpha(\theta^*|\theta^{(t)}) = \min\left[1, \frac{p(\theta^*|y)}{p(\theta^{(t)}|y)}\right] \qquad (1.6)$$

A particular symmetric density in which $f(\theta^*|\theta^{(t)}) = f(|\theta^{(t)} - \theta^*|)$ leads to the random walk Metropolis (Gelman *et al.*, 1996). While it is possible for the proposal density to relate to the entire parameter set, it is often computationally simpler to divide θ into blocks or components, and use componentwise updating, where updating is used in a generic sense allowing for possible non-acceptance of proposed values.

Thus let $\theta_{[j]} = (\theta_1, \theta_2, \ldots, \theta_{j-1}, \theta_{j+1}, \ldots, \theta_d)$ denote the parameter set omitting θ_j and $\theta_j^{(t)}$ be the value of θ_j after iteration t. At step j of iteration $t+1$ the preceding $j-1$ parameters are already updated via the M–H algorithm while $\theta_{j+1}, \ldots, \theta_d$ are still at their iteration t values (Chib and Greenberg, 1995). Let the vector of partially updated parameters be denoted

$$\theta_{[j]}^{(t,t+1)} = (\theta_1^{(t+1)}, \theta_2^{(t+1)}, \ldots, \theta_{j-1}^{(t+1)}, \theta_{j+1}^{(t)}, \ldots, \theta_d^{(t)})$$

The proposed value θ_j^* for $\theta_j^{(t+1)}$ is generated from the jth proposal density, denoted $f(\theta_j^*|\theta_j^{(t)}, \theta_{[j]}^{(t,t+1)})$. Also governing the acceptance of a proposal are full conditional densities $p(\theta_j^{(t)}|\theta_{[j]}^{(t,t+1)})$ specifying the density of θ_j conditional on other parameters $\theta_{[j]}$. The candidate θ_j^* is accepted with probability

$$\alpha(\theta_j^{(t)}, \theta_{[j]}^{(t,t+1)}, \theta_j^*) = \min\left[1, \frac{p(\theta_j^*|\theta_{[j]}^{(t,t+1)})f(\theta_j^{(t)}|\theta_j^*, \theta_{[j]}^{(t,t+1)})}{p(\theta_j^{(t)}|\theta_{[j]}^{(t,t+1)})f(\theta_j^*|(\theta_j^{(t)}, \theta_{[j]}^{(t,t+1)})}\right]$$

The Gibbs sampler (Gelfand and Smith, 1990; Gilks *et al.*, 1996a; Casella and George, 1992) is a special componentwise M–H algorithm whereby the proposal density for updating θ_j is the full conditional $p(\theta_j^*|\theta_{[j]})$ so that proposals are accepted with probability 1. This sampler was originally developed by Geman and Geman (1984) for Bayesian image reconstruction, with its full potential for simulating marginal distributions by repeated draws recognised by Gelfand and Smith (1990). The Gibbs sampler involves parameter-by-parameter updating which when completed forms the transition from $\theta^{(t)}$ to $\theta^{(t+1)}$:

1. $\theta_1^{(t+1)} \sim f_1(\theta_1|\theta_2^{(t)}, \theta_3^{(t)}, \ldots, \theta_d^{(t)})$;
2. $\theta_2^{(t+1)} \sim f_2(\theta_2|\theta_1^{(t+1)}, \theta_3^{(t)}, \ldots, \theta_d^{(t)})$;
   -
   -
   -
d. $\theta_d^{(t+1)} \sim f_d(\theta_d|\theta_1^{(t+1)}, \theta_3^{(t+1)}, \ldots, \theta_{d-1}^{(t+1)})$.

Repeated sampling from M–H samplers such as the Gibbs sampler generates an autocorrelated sequence of numbers that, subject to regularity conditions (ergodicity etc.), eventually 'forgets' the starting values $\theta^{(0)} = (\theta_1^{(0)}, \theta_2^{(0)}, \ldots, \theta_d^{(0)})$ used to initialize the chain and converges to a stationary sampling distribution $p(\theta|y)$.

The full conditional densities may be obtained from the joint density $p(\theta, y) = p(y|\theta)p(\theta)$ and in many cases reduce to standard densities (normal, exponential, gamma, etc.) from which sampling is straightforward. Full conditional densities can be obtained by abstracting out from the full model density (likelihood times prior) those elements including θ_j and treating other components as constants (Gilks, 1996). Consider a conjugate model for Poisson count data y_i with means μ_i that are themselves gamma distributed; this is a model appropriate for overdispersed count data with actual variability $\text{Var}(y)$ exceeding that under the Poisson model. Suppose $\mu_i \sim \text{Ga}(\alpha, \beta)$, namely

$$f(\mu_i|\alpha, \beta) = \mu_i^{\alpha-1}\exp(-\beta\mu_i)\beta^\alpha/\Gamma(\alpha)$$

and further that $\alpha \sim E(a)$, and $\beta \sim G(b, c)$, where a, b and c are preset constants; this prior structure is used by George *et al.* (1993). So the posterior density of $\theta = (\mu_1, \ldots, \mu_n, \alpha, \beta)$ given y is proportional to

$$e^{(-a\alpha)}\beta^{b-1}e^{(-c\beta)}\prod_{i=1}^{n}\exp(-\mu_i)\mu_i^{y_i}\left\{\prod_{i=1}^{n}\mu_i^{\alpha-1}\exp(-\beta\lambda_i)\right\}[\beta^\alpha/\Gamma(\alpha)]^n$$

where all constants (such as the denominator $\prod y_i!$ in the Poisson likelihood) are combined in the proportionality constant. It is apparent that the conditional densities of μ_i and β are $\text{Ga}(y_i + \alpha, \beta + 1)$ and $\text{Ga}(b + n\alpha, c + \sum \mu_i)$ respectively. The full conditional density of α is

$$f(\alpha|y, \beta, \underset{\sim}{\mu}) \propto \exp(-a\alpha)[\beta^\alpha/\Gamma(\alpha)]^n\left(\prod_{i=1}^{n}\mu_i\right)^{\alpha-1}$$

This density is non-standard but log-concave and cannot be sampled directly (as can the gamma densities for μ_i and β). However, adaptive rejection sampling (Gilks and Wild, 1992) may be used.

As examples of how M–H sampling might be carried out in practice, consider a Poisson density with unknown mean μ and data $y = (y_1, \ldots, y_n)$. One possible reference prior for μ is $p(\mu) = 1/\mu$ so the posterior density on μ can be written

$$p(\mu|\mathbf{y}) \propto \prod_{i=1}^{n}[\exp(-\mu)\mu^{y_i}]/\mu$$

Suppose μ is transformed to $\theta = \log(\mu)$. Then the posterior in θ includes a Jacobian adjustment, $\partial\mu/\partial\theta = \exp(\theta) = \mu$, which cancels with $p(\mu)$, and so

$$p(\theta|\mathbf{y}) \propto \prod_{i=1}^{n}\exp(-e^\theta + y_i\theta) = \exp\left(-ne^\theta + \sum_{i=1}^{n}y_i\theta\right)$$

If the prior is placed directly on θ rather than μ, e.g. $\theta \sim N(a, b)$ where a and b are known, a Jacobian adjustment is not needed. Then

$$p(\theta|y) \propto \exp\left(-ne^\theta + \sum_{i=1}^n y_i\theta\right) \exp\left[\frac{-(\theta - a)^2}{2b}\right]$$

In either case one might use a symmetric normal proposal density to generate potential new values θ^* via (1.6). This density is centred on θ_t with variance σ_p^2 that might be set at a default such as $\sigma_p^2 = 1$ and increased or reduced in order to obtain better acceptance rates. Alternatively it might be based on $\text{Var}(\theta)$ from a maximum likelihood (ML) analysis but with variance increased, namely $\sigma_p^2 = K\,\text{Var}(\theta_{ML})$, $K > 1$.

If the variance is too low, acceptance rates will be high but the chain will mix slowly (i.e. move slowly through the parameter space) and converge slowly to $p(\theta|y)$. If the variance is too large, the acceptance rate will be low because the proposed new parameter values will have low values of the ratios $p(\theta^*|y)/p(\theta_t|y)$. For example, if the required acceptance rate is ρ (e.g. $\rho = 0.45$), then one might run the sampler for a certain number N of iterations and compare the actual number of proposals accepted N_1 with required number $N_2 = \rho N$ and revise the scale according as $N_1 \leq N_2$ or $N_1 > N_2$. If $N_1 \leq N_2$,

$$\sigma_{p,\text{new}} = \sigma_{p,\text{old}}/(2 - N_1/N_2)$$

while if $N_1 > N_2$,

$$\sigma_{p,\text{new}} = \sigma_{p,\text{old}}/[2 - (N - N_1)/(N - N_2)]$$

Multiparameter updating will involve a multivariate density such as a multivariate normal with dispersion matrix Σ_p. For example, consider the mean μ and standard deviation σ in a univariate Student t density with known degrees of freedom ν. Taking

$$p(\mu, \sigma) = 1/\sigma$$

as one among several possible reference priors, the posterior density is

$$p(\mu, \sigma|y) \propto \frac{1}{\sigma}\prod_{i=1}^n\left[1 + \frac{(y_i - \mu)^2}{\nu\sigma^2}\right]^{-(\nu+1)/2}$$

Transforming to $\theta = \log \sigma$, the Jacobian is $\partial\sigma/\partial\theta = \sigma$ and the posterior density in μ and θ is

$$p(\mu, \theta|y) \propto \frac{1}{e^\theta}\prod_{i=1}^n\left[1 + \frac{(y_i - \mu)^2}{\nu e^{2\theta}}\right]^{-(\nu+1)/2}$$

This is the comparator density in (1.5) or (1.6), with a possible proposal density, when $\theta = \log \sigma$, being a bivariate normal centred at $(\mu^{(t)}, \theta^{(t)})$.

1.5 MCMC CONVERGENCE

There are many unresolved questions around the assessment of convergence of MCMC sampling procedures (Cowles and Carlin, 1996). It is generally accepted to be preferable to use two or more parallel chains with diverse starting values to ensure full coverage of the sample space of the parameters, and so diminish the chance that the sampling will become trapped in a small part of the space (Gelman and Rubin, 1992; Gelman, 1996). Single long runs may be adequate for straightforward problems, or as a preliminary to obtain inputs to multiple chains. Convergence for multiple chains may be assessed using Gelman–Rubin scale reduction factors that compare variation in the sampled parameter values within and between chains. Parameter samples from poorly identified models will show wide divergence in the sample paths between different chains and the variability of sampled parameter values between chains will considerably exceed the variability within any one chain. To measure variability of samples $\theta_j^{(t)}$ within the jth chain ($j = 1, \ldots, J$) define

$$V_j = \sum_{t=s+1}^{s+T} (\theta_j^{(t)} - \bar{\theta}_j)^2 / (T - 1)$$

over T iterations after an initial burn-in of s iterations. Ideally the burn-in period is a short initial set of samples where the effect of the initial parameter values tails off; during the burn-in the parameter trace plots will show clear monotonic trends as they reach the region of the posterior. Convergence is therefore assessed from iterations $s + 1$ to $s + T$.

Variability within chains V_W is then the average of the V_j. Between-chain variance is measured by

$$V_B = \frac{T}{J - 1} \sum_{j=1}^{J} (\bar{\theta}_j - \bar{\theta})^2$$

where $\bar{\theta}$ is the average of the $\bar{\theta}_j$. The scale reduction factor (SRF) compares a pooled estimator of $\text{Var}(\theta)$, given by $V_P = V_B/T + TV_W/(T - 1)$, with the within-sample estimate V_W. Specifically the SRF is $(V_P/V_W)^{0.5}$ with values under 1.2 indicating convergence.

Parameter samples obtained by MCMC methods are correlated, which means extra samples are needed to convey the same information.

Additionally, as in any iterative estimation, there may be a delay in seeking the region of the posterior density where the modal value is located. The extent of correlation, and the convergence towards the modal region, will depend on a number of factors including the form of parameterization, the sample size, the complexity of the model and the form of sampling (e.g. block or univariate sampling of parameters).

A more recently proposed convergence statistic is that due to Brooks and Gelman (1998) and known as the Brooks–Gelman–Rubin (BGR) statistic. This is a ratio of parameter interval lengths, where for chain j the length of the $100(1 - \alpha)\%$ interval for parameter θ is obtained, i.e. the gap between 0.5α and $(1 - 0.5\alpha)$ points from T simulated values. This provides J within-chain interval lengths, with mean I_U. For the pooled output of T_J samples, the same $100(1 - \alpha)\%$ interval I_P is also obtained. Then the ratio I_P/I_U should converge to one if there is convergent mixing over different chains.

Analysis of sequences of samples from an MCMC chain amounts to an application of time series methods, in regard to problems such as assessing stationarity in an autocorrelated sequence. Autocorrelation at lags 1, 2 and so on may be assessed from the full set of sampled values $\theta^{(t)}$, $\theta^{(t+1)}$, $\theta^{(t+2)}$,..., or from subsamples K steps apart, $\theta^{(t)}$, $\theta^{(t+K)}$, $\theta^{(t+2K)}$,..., etc. If the chains are mixing satisfactorily then the auto-correlations in the one-step apart iterates $\theta^{(t)}$ will fade to zero as the lag increases (e.g. at lag 10 or 20). Non-vanishing autocorrelations at high lags mean that less information about the posterior distribution is provided by each iterate and a higher sample size T is necessary to cover the parameter space. Slow convergence will show in trace plots that wander, and that exhibit short-term trends rather than fluctuating rapidly around a stable mean.

Problems of convergence in MCMC sampling may reflect problems in model identifiability due to overfitting or redundant parameters. Running multiple chains often assists in diagnosing poor identifiability of models. This is illustrated most clearly when identifiability constraints are missing from a model, such as in discrete mixture models that are subject to 'label switching' during MCMC updating (Frühwirth-Schnatter, 2001). One chain may have a different 'label' to others so that obtaining a G–R statistic for some parameters is not sensible. Choice of diffuse priors tends to increase the chance of poorly identified models, especially in complex hierarchical models or small samples (Gelfand and Sahu, 1999). Elicitation of more informative priors or application of parameter con-straints may assist identification and convergence. Correlation between parameters within the parameter set $\theta = (\theta_1, \theta_2, \ldots, \theta_d)$ tends to delay convergence and increase the dependence between successive iterations. Reparameterisation to reduce correlation – such as centring predictor variables in regression – usually improves convergence (Gelfand *et al.*, 1995; Zuur *et al.*, 2002).

1.6   COMPETING MODELS

Generally there are several possible competing models for the data, differing in likelihood or prior specifications. It is necessary either to choose between them, or to have some way of averaging inferences over them in terms of their relative probability or likelihood. From the conditional probability rule one can obtain posterior model probabilities. Let M_j be one among several possible alternative models $\{M_1, \ldots, M_J\}$. Then with $B = y$ and $A = M_j$, we have that

$$p(M_j|y) = p(y|M_j)p(M_j)/p(y)$$

where the denominator is equivalent to $\sum_k p(y|M_k)p(M_k)$. The formal Bayes model choice is based on posterior model probabilities $p(M_j|y)$ and associated quantities, such as the Bayes factor $B_{jk} = p(y|M_j)/p(y|M_k)$ comparing models j and k. This and other methods for comparing and checking models are considered in Chapter 2.

1.7   SETTING PRIORS

Priors may be chosen to encapsulate existing knowledge (e.g. based on results or parameter estimates from historical data) or to impart relatively little information in relation to that provided by the data. In the latter category are 'reference priors' that are constructed automatically without needing to choose tuning parameters. The latter are also sometimes known as objective priors (e.g. Casella and Moreno, 2002), whereas priors based to some degree on existing knowledge are known as subjective or elicited priors (Garthwaite *et al.*, 2004). While objective priors have the benefit of being 'off the shelf' and of 'letting the data speak for themselves', they may also create problems in sampling-based estimation. Improper priors, for example, may lead to an improper joint posterior distribution even when all the full conditional posteriors are proper (Casella and George, 1992; Hobert and Casella, 1996).

In practice other principles may guide prior specification: for example, formal model choice (section 1.6 and Chapter 2) may be adversely affected by using diffuse or just proper priors. Complex models for relatively small samples may require relatively informative priors in order to be identifiable from the data. Conjugate priors in which the posterior has the same form as the prior have advantages in tractability, and also in interpretation, since the prior can be interpreted in terms of a prior sample size or as pseudo data.

In many applications involving small or modest sample sizes posterior inferences about parameters or functionals of parameters may be affected by the prior used. Formal model choice via the Bayes factor may remain sensitive to prior specification for all sample sizes. In fact, the Bayes factor B_{21} for model 2 vs. model 1 (where model 2 is more parameterized) tends to zero as the sample size n increases. Also when model 2 contains an additional parameter and the diffuseness of the prior on that parameter is increased, B_{21} tends to zero (Bartlett, 1957; Lindley, 1957). On the other hand, problems with Bayes factor stability may be overstated. Although the Bayes factor may change with a change in the prior, it is pertinent to ask whether the ranking of the leading models changes when the prior is changed in reasonable ways. Possible alternatives to the formal Bayes factor include Bayes factors using minimal training samples from $y = (y_1, \ldots, y_n)$ to provide proper priors, examples being the fractional Bayes factor (O'Hagan, 1995) and the intrinsic Bayes factor (Berger and Pericchi, 1996a; 1996b).

A good principle for any application is to carry out a sensitivity analysis with a variety of prior specifications. This is especially so for parameters that are known to have sensitivity implications, e.g. the variances in hierarchical random effects models, or where the model is only weakly identified (e.g. see Chapter 8). A practical procedure implicit in the WINBUGS package and with some support in the literature (e.g. Besag *et al.*, 1995) is the use of just proper, minimally informative priors. However, some argue that improper or just proper priors can be avoided by some subject matter reasoning. Thus Kadane and Lazar (2003, p 281) state that 'if statisticians were to think about the reality underlying the parameter, they should always be able to describe it reasonably well using a prior distribution'.

1.8   THE NORMAL LINEAR MODEL AND GENERALIZED LINEAR MODELS

In this book the focus is on discrete outcome models $p(y|\theta)$ that generalize the tools used for continuous data, e.g. regression models, time series models and panel analysis. Such generalization usually starts with non-hierarchical linear regression. The mainstay of regression modelling for continuous data is the normal linear model whereby a response variable y_i, measured for subjects $i = 1, \ldots, n$, is related linearly to regressor variables (or predictors) $x_{i1}, \ldots, x_{ip}$. Assuming that predictors are measured without error, model discrepancies or measurement errors in the y_i are expressed via errors $\varepsilon = (\varepsilon_1, \ldots, \varepsilon_n)$ assumed to beindependently normally distributed with a common variance σ^2. Thus

$$y_i = \beta_0 + \beta_1 x_{i1} + \beta_2 x_{i2} + \cdots + \beta_p x_{ip} + \varepsilon_i$$
$$\varepsilon_i \sim N(0, \sigma^2)$$

Equivalently

$$Y = X\beta + \varepsilon$$

where Y is an $n \times 1$ column vector, $\beta = (\beta_0, \ldots, \beta_p)'$ is a $(p+1) \times 1$ column vector of regression parameters and X is an $n \times (p+1)$ matrix of predictor variables including an intercept $x_{i0} = 1$. Letting $\eta_i = \beta_0 + \beta_1 x_{i1} + \beta_2 x_{i2} + \ldots + \beta_p x_{ip} = X_i\beta$, where X_i is the ith row of X, the normal linear model may also be written in the form

$$y_i \sim N(\eta_i, \sigma^2)$$

However, for discrete responses (e.g. binary or count data) the distribution of y_i will not usually be normal. Also, to ensure that predictions $\hat{y}_i$ are of the same form as y_i itself (e.g. if y_i is binary it is necessary that $\hat{y}_i$ be between 0 and 1), the mean μ_i of the dependent variable may need to be a transformation of the linear predictor η_i rather than equalling it. In the class of generalized linear models (Nelder and Wedderburn, 1972), y_i may follow one of a wide set of distributions within the linear exponential family instead of being assumed normal. Thus

$$y_i \sim p(y_i | \theta_i, \phi)$$
$$l(\theta_i, \phi | y) = \log[p(y_i | \theta_i, \phi] = [y_i \theta_i - b(\theta_i)] / a_i(\phi) + c(y_i, \phi) \qquad (1.7)$$

where a, b and c are known functions that define the particular distribution, $\theta_i = h(\eta_i) = h(X_i\beta)$ are unknown location parameters, and the scale parameter ϕ and hence $a_i(\phi)$ may be known or unknown. Canonical models are where $\theta_i = \eta_i$. By using the relations $E(\partial l/\partial \theta) = 0$ and $E(\partial^2 l/\partial \theta^2) + E(\partial l/\partial \theta)^2 = 0$, one may show that $E(y_i) = b'(\theta_i)$ and $\text{Var}(y_i) = a_i(\phi) b''(\theta_i)$.

To ensure predictions of the right form, a transform or link function $g(\mu_i)$ is usually needed to link the mean $\mu_i = E(y_i)$ to the regression term in the predictors X_i. So assuming a linear regression term, one obtains

$$g(\mu_i) = X_i\beta = \eta_i$$

with the inverse link being

$$\mu_i = g^{-1}(X_i\beta) = g^{-1}(\eta_i)$$

Different types of link may be used for any particular distribution (e.g. for the binomial distribution, commonly used links are the logit, probit or extreme value). The normal linear model with an identity link is in fact a special case of the linear exponential family, in which $\mu_i = \theta_i = X_i\beta$, and the relevant functions are $a_i(\phi) = \sigma^2$, $b(\theta_i) = \theta_i^2/2$, $c(y_i, \phi) = -0.5[y_i^2/\sigma^2 + \log(2\pi\sigma^2)]$.

The Poisson with log link between η_i and μ_i leads to $\theta_i = \log(\mu_i) = X_i\beta$, with relevant functions being $a_i(\phi) = 1$, $b(\theta_i) = \exp(\theta_i)$, $c(y_i, \phi) = -\log y_i!$. Thus

$$y_i \sim \text{Po}(y_i|\mu_i)$$

where

$$\text{Po}(y_i|\mu) = \exp[y_i \log(\mu_i) - \mu_i - \log y_i!]$$

The variance = mean relation of the Poisson is obtained by differentiating $b(\theta_i)$. For the binomial with mean π_i and n_i subjects at risk in the ith group or trial the likelihood may be written as

$$\binom{n_i}{y_i} [\pi_i/(1 - \pi_i)]^{y_i} (1 - \pi_i)^{n_i}$$

This is obtained from the exponential family with $\theta_i = X_i\beta = \log[\pi_i/(1 - \pi_i)]$, $a_i(\phi) = 1$, $b(\theta_i) = n_i \log[1 + \exp(\theta_i)]$, and

$$c(y_i, \phi) = \log\binom{n_i}{y_i}$$

Thus the logit function is the canonical link between π_i and $\eta_i = \theta_i$.

Random variation in subject means (μ_i in the Poisson, π_i in the binomial) may be introduced. If modelled in the regression link, this might take the form

$$g(\mu_i) = X_i\beta + \alpha_i$$

where α_i might be normal, or a mixture of normals. Alternatively conjugate mixing may be considered, as when Poisson means μ_i are assumed to be gamma distributed over subjects, and when binomial probabilities are taken to be beta distributed (Nelder and Lee, 2000). The motivations for such extensions often lie in heterogeneity greater than expected under the standard exponential family models or to obtain improved estimates of varying means for subjects that allow 'pooling of strength'.

Random variation between subjects may be unstructured and exchangeable (the features of the density are unaffected by permutation of the subject identifiers). However, structured variation is likely for observations arranged in space or time (e.g. spatially correlated variation if the observation units are areas). Consider the mixed model in spatial epidemiology (Besag et al., 1991) for Poisson distributed event totals y_i with means $\mu_i = \lambda_i E_i$, where E_i are expected events and λ_i are relative disease risks (see Chapter 8). Then

$$\log(\lambda_i) = X_i\beta + \alpha_{i1} + \alpha_{i2} \qquad (1.8)$$

where α_{i1} and α_{i2} are white–noise and spatial errors respectively.[2] The impact of unstructured random effects α_{i1} is to pool strength over all areas towards the overall mean. With structured errors, the configuration of areas affects the nature of the pooling of strength: smoothing is towards local rather than global averages.

Classical ML analysis of models following particular forms of the exponential family density (1.7) involves Newton–Raphson iteration or iteratively reweighted least squares, and usually relies on asymptotic normality to provide parameter densities or hypothesis tests. However, the asymptotic approximations may not hold for small or moderate sample sizes or for non-linear models (e.g. see Zellner and Rossi, 1984). Inferences based on asymptotic normality of the ML parameters can be affected, perhaps distorted, by choice of parameterization, by sample size, by the form of regression (non-linear vs. linear) and so on. An illustration by Pastor-Barriuso et al. (2003) is for a logit model subject to a change point in the impact of regressors. Bayesian analysis of discrete data follows the generalized linear model (GLM) structure but is not constrained to asymptotic normality to obtain posterior inferences.

---

[2]E_i denotes the expected deaths, taking account of the size of the population exposed to risk and its age structure but assuming a standard death rate schedule (e.g. national death rates).

Other distributions for discrete data not encompassed by (1.7) may sometimes be obtained as a special form of GLM providing certain parameters are fixed. Thus the negative binomial with dispersion parameter $1/r$ is often appropriate when count data are overdispersed. This has form

$$y_i \sim NB(\mu_i, r)$$

$$NB(\mu_i, r) = \frac{\Gamma(y_i + r)}{\Gamma(r)\Gamma(y_i + 1)} \left(\frac{r}{\mu_i + r}\right)^r \left(\frac{y_i}{\mu_i + r}\right)^{y_i}$$

$$E(y_i) = \mu_i = \exp(X_i\beta)$$

$$Var(y_i) = \mu_i + \mu_i^2/r$$

If r is fixed, then the negative binomial can be expressed in the linear exponential form (see McCullagh and Nelder, 1989, chapter 11).

The GLM framework also extends to cases where subjects are clustered (e.g. in firms, schools) or are measured repeatedly, sometimes known as general linear mixed models (GLMMs). For clustered data, let $i = 1, \ldots, I$ denote the level 1 units (clusters) and $j = 1, \ldots, J_i$ denote the level 1 units (observations) nested within each cluster. Then random effects may be defined at both cluster and observation level. Let W_{ij} be a predictor of dimension r and γ_{ij} be a vector random effect at observation level, so that for y_{ij} Poisson or binomial with means ν_{ij} and link g

$$g(\nu_{ij}) = X_{ij}\beta + Z_{ij}\alpha_i + W_{ij}\gamma_{ij}$$

Bayesian analysis of clustered GLMMs is exemplified by applications in longitudinal data (Chib and Carlin, 1999) and health services research (Normand *et al.*, 1997; Daniels and Gatsonis, 1999).

1.9   DATA AUGMENTATION

Data augmentation via latent or auxiliary variables is a strategy used to convert the likelihood to a form (such as the normal linear regression) for which simple Gibbs sampling can be used. For example, data augmentation is used to ensure conjugacy of prior and likelihood such that the posterior takes a standard form (Damien *et al.*, 1999; van Dyk and Meng, 2001; van Dyk, 2002). It is used to simplify sampling in discrete mixture models (Chapter 5), and also forms the basis for missing-data models (Chapter 11). Bayesian developments in factor analysis and item response analysis can also be cast in terms of augmented data for the factor scores (Aitkin and Aitkin, 2005).

Consider for example a binary regression for responses $y_i = 1$ or 0 with logit link to regressors x_i. Suppose a mixture of regression components is proposed (e.g. distinct regression models for a small number of subpopulations, $j = 1, \ldots, J$ as in market segmentation studies), with prior probabilities π_j that subject i belongs to a group j. In practice the group indicators $D_i \in (1, \ldots, J)$ are unknown but are considered as augmented data to be sampled at each iteration in an MCMC chain. So conditional on D_i

$$y_i | D_i = j \sim \text{Bern}(\pi_{ij})$$
$$g(\pi_{ij}) = x_i \beta_j$$

where g is typically the probit or logit link. Given a set of sampled $D = (D_1, \ldots, D_n)$ the likelihood in y_i has the form

$$L_{ij} = \pi_{ij}^{y_i} (1 - \pi_{ij})^{(1-y_i)}$$

and the total likelihood is

$$\prod_{i=1}^{n} \sum_{j=1}^{J} \pi_j L_{ij}$$

The full conditionals for binary and count regression in such situations have been presented by several authors (e.g. Robert, 1996) but are not necessarily of standard form; this necessitates the M–H algorithm or adaptive rejection sampling for the regression parameters β_j.

Instead a further augmentation produces Gibbs sampling throughout and also provides continuous quantities W_i with substantive relevance (e.g. consumer utilities). These are the unobserved responses in a linear regression

$$W_i = x_i \beta_j + u_i \tag{1.9}$$

with $u_i \sim N(0, \sigma^2)$. For equivalence to a probit link, such augmentation is unproblematic. The W_i are generated by truncated sampling according to whether $y_i = 1$ or 0 as

$$W_i | D_i = j, \quad \beta_j \sim N(x_i \beta_j, \sigma^2) I(0,) \quad y_i = 1$$
$$W_i | D_i = j, \quad \beta_j \sim N(x_i \beta_j, \sigma^2) I(, 0) \quad y_i = 0$$

The form of the information provided by y does not allow all the regression parameters for the metric model to be identified unless the variance σ^2 is fixed, typically $\sigma^2 = 1$.

A logit link is consistent with a heavier tail density for the underlying W_i, and here augmentation relies on the close approximation to a logit link by a Student t density. Specifically the W_i are taken to be Student t with eight degrees of freedom (Albert and Chib, 1993) and scale $\sigma^2 = 2.49$. In terms of the linear regression (1.9), the u_i are now Student t errors, and following Andrews and Mallows (1974), the Student t with ν degrees of freedom is obtainable by mixing the scale in a normal density, i.e. $u_i \sim \mathrm{N}(0, \sigma^2/\lambda_i)$, with $\lambda_i \sim \mathrm{Ga}(0.5\nu, 0.5\nu)$.

The latent metric scale may be envisaged to result from an underlying comparison of scales depending on the application (e.g. patient severity in the medical application, consumer utility in the economic application) with value U_{i1} for subject i if the response is yes and U_{i0} if the response is no. Define $W_i = U_{i1} - U_{i0}$; then according as $y_i = 1$ or 0 the metric scale has values $W_i \geq 0$ or $W_i < 0$. Modelling on the latent scale with W rather than with the original (e.g. Bernoulli) likelihood will entail no loss of information provided $p(\theta|W, Y) = p(\theta|W)$. One can see that this holds for the example just cited since $\{W_i \geq 0, W_i < 0\}$ is equivalent to $\{y_i = 1, \ y_i = 0\}$. The sampling for the metric scale model will then consist of sampling from conditional densities $p(\theta_j|W, \theta_{[j]})$ and $p(W|Y, \theta)$ where $\theta_{[j]}$ consists of all regression model parameters apart from θ_j. The form of the conditional $p(W|Y, \theta)$, namely estimating proxy metric scale data corresponding to the actual discrete data, follows from the density assumed for u (see Chapters 4 et seq).

1.10   IDENTIFIABILITY

Devices such as data augmentation, together with other features of MCMC sampling–based estimation, facilitate the fitting of highly complex models. However, highly parameterized models raise questions about practical identification of the model's assumed structure with the data at hand, for instance in hierarchical random effects models (e.g. Gelfand and Sahu, 1999; Vines *et al.*, 1996), and in specific applications in phylogeny (Rannala, 2002) and factor analysis (Fokoue, 2004).

Model identifiability has several possible meanings: it may relate to the ability to estimate the entirety of the parameters postulated by the model. It may also relate to the fact that several distinct models have closely similar fits to the same data; or it may relate to certain parameters in a model that cannot be effectively updated by the data. Among relevant factors are the size of sample, the complexity of the model, and the level of information provided by the prior. Also relevant is the type of model: for example, the parameters of non-linear regressions and random effects models are more subject to identification problems than those of linear regression.

The first sort of situation occurs in multiple random effects models where the level of several random effects is confounded with the intercept. In a model with a single random effect as in two–way analysis of variance, one might combine the random effect and the intercept, so that

$$y_{ij} = \mu + \alpha_i + e_{ij}$$

where the α_i have prior mean zero, is replaced by

$$y_{ij} = \beta_i + e_{ij}$$

where β_i have mean μ. Another possibility is a centring operation at each MCMC iteration, or a constraint such as $\alpha_1 = 0$ or $\alpha_1 = \alpha_2$. Identifiability is compromised by random effects priors that do not specify an average level (e.g. $\alpha_i \sim N(0, \tau_\alpha)$) but instead how successive values are related, e.g. in a first-order random walk with $\alpha_i \sim N(\alpha_{i-1}, \tau_\alpha)$. Identifiability is also complicated in some types of models over space or time when there are several random effects; in spatial applications involving the mixed model of Besag et al. (1991) the identifiability problems focus on separating structured and unstructured errors since only their sum is identified. In disease models with age and time as dimensions an additional cohort effect may be identified and constraints on random effects have to be imposed to ensure identification.

Similar fits from models with different substantive implications may occur suggesting perhaps a model averaging strategy (see Hoeting, 1999; also Chapter 2). Structural equation models provide examples when statistical fit is similar but the character of the model might differ substantially, e.g. in one model factors are uncorrelated and in another they are correlated, but similar fits are obtained (Muthen, 2003). Formal observational equivalence has been defined by econometricians: two models that generate identical outcomes are observationally equivalent and data alone cannot distinguish between them (Hendry et al., 2001).

An example of the third kind occurs in the setting of higher stage priors (which, suitably applied, has value for assessing sensitivity of inferences to priors). For example, the inverse variance $\tau = \sigma^{-2}$ in a linear regression with errors u may be assigned a gamma prior $Ga(\alpha_1, \alpha_2)$ and α_1 and

α_2 are typically taken as known (e.g. $\alpha_1 = \alpha_2 = 0.001$). However, they can be extra parameters, e.g. $\alpha_1 \sim \text{Exp}(\gamma)$, $\alpha_2 \sim \text{Exp}(\gamma)$. If γ is preset then the data may effectively update the priors on α_1 and α_2, but if γ itself is taken as unknown[3] the data are unlikely to identify α_1, α_2 and γ.

1.11 ROBUSTNESS AND SENSITIVITY

Model comparison is subject to sensitivity when inferences about para-meters θ are affected by choice of sampling model or choice of prior (Geweke, 1998), and especially when standard assumptions (e.g. normal-ity of errors, unimodal densities) are questionable. Model elaborations better to reflect data features may involve robust alternatives in the prior or in the likelihood components of a model, or in both (Weiss *et al.*, 1998; Carlin, 1992).

If alternative models have close fit but differ in their implications then one strategy is model averaging rather than trying to choose one model as best (Chapter 2). Another strategy is to recognize that inferences are not certain and report the results of a sensitivity analysis using perturbations in likelihood or prior assumptions. Non-parametric classes of prior that may provide greater robustness in the face of prior uncertainty have been suggested, such as density ratio priors (DeRobertis and Hartigan, 1981), Dirichlet process priors (Ferguson, 1973; Dey *et al.*, 1998), and ε-contamination classes (Berger, 1994) where

$$p(\theta) = (1 - \varepsilon)p_b(\theta) + \varepsilon\zeta \qquad \zeta \in Z$$

where $p_b(\theta)$ is the base prior, ε is the postulated error in the base prior, and Z is the class of contaminations considered.

---

[3]Consider a Poisson regression where the error u represents overdispersion in relation to the Poisson assumption that variance = mean. To illustrate, 100 points were generated from a model

$$y_i \sim \text{Po}(\mu_i)$$
$$\log(\mu_i) = 2 + x_i + u_i$$
$$x_i \sim \text{U}(-1, 1)$$
$$u_i \sim \text{N}(0, 2)$$

(see Program 1.1 Simulated Overdispersed Poisson). The first prior on $\tau = 1/\text{Var}(u)$ using preset parameters ($\alpha_1 = 1, \alpha_2 = 0.001$) produces a posterior mean (standard deviation), $\tau = 0.56(0.10)$. Taking $\alpha_1 \sim \text{E}(\gamma)$, $\alpha_2 \sim \text{E}(\gamma)$ with $\gamma = 1$ also gives a well-identified model with $\tau = 0.54(0.10)$, $\alpha_1 = 0.97(0.69)$, $\alpha_2 = 1.29(1.00)$. However, taking γ as additional unknown with an E(1) prior leads to poorly identified α_1 and α_2, with inflated standard deviations.

Inference sensitivity tends to be related to factors such as sample size or features of the data: inferences for small samples or samples with multiple modes or unusual cases are more sensitive to changes in prior settings. Hence diagnostics for outliers and other checks comparing model predictions to the data (see Chapter 2 on posterior predictive checks) are important in assessing robustness. Often a heavier tailed sampling density $p(y|\theta)$ or heavier tailed priors $p(\theta)$ such as t_ν (ν under five) for means or regression coefficients instead of normal priors reduce the impact of outliers (Angers, 1992). A univariate or multivariate Student t density instead of a normal sampling model $p(y|\theta)$ provides outlier weights if the scale mixture version of the t density is used (West, 1984); this of course adds to the parameter count of the model. Other generalizations of standard densities allow for skewness (Branco and Dey, 2001; Azzalini and Capitanio, 1999). Discrete mixture models also provide great flexibility in modelling unusual densities or clusters of unusual cases (Roberts, 1996). Overdispersed alternatives to standard discrete densities (e.g. negative binomial instead of Poisson) can also be seen as robustified versions of these densities (Gelman et al., 2003).

Sensitivity to prior assumptions also depends on the type of parameter under consideration. For example, fixed effect regression parameters are typically less problematic than hyperparameters governing random effects models. A further type of sensitivity is when inferences are affected by the type of population model, whether parametric random effects or non-parametric mixtures. The latter are often proposed as a way of avoiding parametric assumptions (Ishwaran, 2000).

Much analysis has occurred in relation to the specification of variances for random effects in hierarchical models. In applications of the spatial mixed model, non-informative gamma $Ga(\varepsilon, \varepsilon)$ priors (with ε small such as $\varepsilon = 0.001$) on the inverse variances compound the identifiability problems inherent in separating structured and unstructured variance components. A $Ga(0.001, 0.001)$ prior on $\tau = \sigma^{-2}$ is incompatible with small levels of variation and priors such as $Ga(0.5, 0.0005)$ (Wakefield and Morris, 1999) or $Ga(0.1, 0.1)$ (MacNab, 2003) have been suggested instead; these priors are compatible with stronger prior belief in the possibility of small variability.

Gelman (2005) points out problems with this form of prior in multi-level models with repetitions $j = 1, \ldots, n_i$ within clusters $i = 1 \ldots I$, e.g.

$$y_{ij} \sim N(\mu + \alpha_i, \sigma^2)$$

where the cluster effects α_i are often assumed normal with mean 0 and variance σ_α^2. The choice $1/\sigma_\alpha^2 \sim Ga(\varepsilon, \varepsilon)$ with $\varepsilon = 0.001$ can be problematic when I is small. Gelman instead recommends a uniform prior, e.g.

U(0, 100) on the standard deviation σ_α, or a half Student t prior, where σ_α is the absolute value of a draw from a Student t with mean 0, known scale and small degrees of freedom (e.g. $\nu = 1$ for a half Cauchy).

1.12 CHAPTER THEMES

In the following chapters the themes mentioned above will recur within an overriding focus on categorical data models. Chapter 2 considers in detail the question of model comparison, checking and choice. Increasingly model averaging rather than choice is being emphasized and this is in practice what happens in regression selection models whether for metric data (Chapter 3), or for count and binomial data (Chapter 4). Chapters 3, 4 and 5 also consider issues such as robust alternatives to the standard densities such as the Poisson and binomial, e.g. non-parametric regression modelling from a Bayesian perspective, the modelling of overdispersion and the accommodation of outliers or population subgroups. Chapter 6 considers multinomial data generalizations of common links such as the multinomial logit and probit models, while Chapter 7 concerns ordinal data models. Chapter 8 considers a particular form of discrete data analysis, namely spatial data modelling, where issues of both robustness and parameter identification figure large. Chapters 9 and 10 consider respectively time series analysis for categorical data and clustered data (multilevel and panel models). With clustered data, correlation over time periods or within clusters becomes an issue. Chapter 11 considers possible models for missing categorical data from a Bayesian perspective.

REFERENCES

Aitkin, M. and Aitkin, I. (2005) Bayesian inference for factor scores. In *Contemporary Advances in Psychometrics*. Erlbaum (in press).

Albert, J. and Chib, S. (1993) Bayesian analysis of binary and polychotomous response data. *Journal of the American Statistical Association*, **88**, 669–679.

Andrews, D. and Mallows, C. (1974) Scale mixtures of normal distributions. *Journal of the Royal Statistical. Society, Series B*, **36**, 99–102.

Angers, J.-F. (1992) Use of Student-t prior for the estimation of normal means: a computational approach. In *Bayesian Statistics 4*, Bernardo, J., Berger, J., Dawid, A. and Smith, A. (eds). Clarendon Press: Oxford, 567–575.

Azzalini, A. and Capitanio, A. (1999) Statistical applications of the multivariate skew-normal distribution. *Journal of the Royal Statistical Society, Series B*, **61**, 579–602.

Bartlett, M. (1957) A comment on D. V. Lindley's statistical paradox. *Biometrika*, **44**, 533–534.

Berger, J. (1994) An overview of robust Bayesian analysis. *Test*, **3**, 5–124.

Berger, J. and Pericchi, L. (1996a) The intrinsic Bayes factor for model selection and prediction. *Journal of the American Statistical Association*, **91**, 109–122.

Berger, J. and Pericchi, L. (1996b) The intrinsic Bayes factor for linear models. In *Bayesian Statistics 5*, Bernardo, J., Berger, J., Dawid, A. and Smith, A. (eds.). Clarendon Press: Oxford, 23–42.

Bernardo, J. and Smith, A. (1994) *Bayesian Theory. Chichester John Wiley & Sons.*

Besag, J. York, J. and Mollié, A. (1991) Bayesian image restoration with two applications in spatial statistics. *Annals of the Institute of Statistics and Mathematics*, **43**, 1–59.

Besag, J., Green, P., Higdon, D. and Mengersen, K. (1995) Bayesian computation and stochastic systems. *Statistical Science*, **10**, 3–66.

Branco, M. and Dey, D. (2001) A general class of multivariate skew-elliptical distributions. *Journal of Multivariate Analysis*, **79**, 99–113.

Brooks, S. and Gelman, A. (1998) Alternative methods for monitoring convergence of iterative simulations. *Journal of Computational and Graphical Statistics*, **7**, 434–455.

Carlin, B. (1992) State space modeling of non-standard actuarial time series. *Insurance Mathematics and Economics*, **11**, 209–222.

Casella, G. and George, E. (1992) Explaining the Gibbs sampler. *American Statistician*, **46**, 167–174.

Casella, G. and Moreno, E. (2002) Objective Bayesian variable selection. Dept. of Statistics, University of Florida.

Chib, S. and Carlin, B. (1999) On MCMC sampling in hierarchical longitudinal models. *Statistics and Computing*, **9**, 17–26.

Chib S. and Greenberg, E. (1995) Understanding the Metropolis-Hastings algorithm. *American Statistician*, **49**, 327–335.

Cowles, M. and Carlin, B. (1996) Markov Chain Monte Carlo convergence diagnostics: a comparative review. *Journal of the American Statistical Association*, **91**, 883–904.

Damien, P., Wakefield, J. and Walker, S. (1999) Gibbs sampling for Bayesian non-conjugate and hierarchical models by using auxiliary variables. *Journal of the Royal Statistical Society, Series B*, **61**, 331–344.

Daniels, M. and Gatsonis, C. (1999) Hierarchical generalized linear models in the analysis of variations in health care utilization. *Journal of the American Statistical Association*, **94**, 29–42.

DeRobertis, L. and Hartigan, J. (1981) Bayesian inference using intervals of measures. *Annals of Statistics*, **9**, 235–244.

Dey, D., Muller, P. and Sinha, D. (eds) (1998) *Practical Nonparametric and Semiparametric Bayesian Statistics*. Springer: New York.

Ferguson, T. (1973) A Bayesian analysis of some nonparametric problems. *Annals of Statistics*, **1**, 209–230.

Fokoue, E. (2004) Stochastic determination of the intrinsic structure in Bayesian factor analysis. *Technical Report* #2004-17, Statistical and Applied Mathematical Sciences Institute.

Frühwirth-Schnatter, S. (2001) MCMC estimation of classical and dynamic switching and mixture models. *Journal of the American Statistical Association*, **96**, 194–209.

Garthwaite, P., Kadane, J. and O'Hagan, A. (2004) Elicitation. *Journal of the American Statistical Association*, to appear.

Gelfand, A. (1996) Model determination using sampling-based methods. In *Markov Chain Monte Carlo in Practice*, Gilks, W., Richardson, S. and Spiegelhalter, D. (eds). Chapman and Hall: London, 145–162.

Gelfand, A. and Sahu, S. (1999) Identifiability, improper priors, and Gibbs sampling for generalized linear models. *Journal of the American Statistical Association*, **94**, 247–253.

Gelfand, A. and Smith, A. (1990) Sampling-based approaches to calculating marginal densities. *Journal of the American Statistical Association*, **85**, 398–409.

Gelfand, A., Sahu, S. and Carlin, B. (1995) Efficient parameterization for normal linear mixed models. *Biometrika*, **82**, 479–488.

Gelman, A. (1996) Inference and monitoring convergence. In *Markov Chain Monte Carlo in Practice*, Gilks, W., Richardson, S. and Spiegelhalter, D. (eds). Chapman and Hall: London, 131–143.

Gelman, A. (2005) Analysis of variance: why it is more important than ever (with discussion). *Annals of Statistics*, to appear.

Gelman, A. and Rubin, D. (1992) Inference from iterative simulation using multiple sequences. *Statistical Science*, **7**, 457–511.

Gelman, A., Roberts, G. and Gilks, W. (1996) Efficient Metropolis jumping rules. In *Bayesian Statistics 5*, Bernardo, J., Berger, J., Dawid, A. and Smith, A. (eds). Oxford University Press: Oxford, 599–607.

Gelman, A., Carlin, J., Stern, H. and Rubin, D. (2003) *Bayesian Data Analysis*, 2nd Edition. Boca Raton, FL: CRC Press.

Geman, S. and Geman, D. (1984) Stochastic relaxation, Gibbs distributions, and the Bayesian restoration of images. *Transactions on Pattern Analysis and Machine Intelligence*, **6**, 721–741.

George, E., Makov, U., and Smith, A. (1993) Conjugate likelihood distributions. *Scandinavian Journal of Statistics*, **20**, 147–156.

Geweke, J. (1989) Bayesian inference in econometrics models using Monte Carlo integration. *Econometrica*, **57**, 1317–1340.

Geweke, J. (1998) Simulation methods for model criticism and robustness analysis. In *Bayesian Statistics 6*, Bernardo, J. (ed.). Oxford University Press: Oxford.

Gilks, W. (1996) Full conditional distributions. In *Markov Chain Monte Carlo in Practice*, Gilks, W., Richardson, S. and Spiegelhalter, D. (eds). Chapman and Hall: London, 75–88.

Gilks, W. and Wild, P. (1992) Adaptive rejection sampling for Gibbs sampling. *Applied Statistics*, **41**, 337–348.

Gilks, W., Richardson, S. and Spiegelhalter, D. (eds) (1996a) *Markov Chain Monte Carlo in Practice*. Chapman and Hall: London.

Gilks, W., Richardson, S. and Spiegelhalter, D. (1996b) Introducing Markov chain Monte Carlo. In *Markov Chain Monte Carlo in Practice*, Gilks, W., Richardson, S. and Spiegelhalter, D. (eds). Chapman and Hall: London, 1–20.

Hastings, W. (1970) Monte Carlo sampling methods using Markov chains and their applications. *Biometrika*, **57**, 97–109.

Hendry, D., Lu, M. and Mizon, G. (2001) Model identification and non-unique structure. *Nuffield College, Oxford Working Paper Series* 2002-W10.

Hobert, J. and Casella, G. (1996) The effect of improper priors on Gibbs sampling in hierarchical linear mixed models. *Journal of the American Statistical Association*, **91**, 1461–1473.

Hoeting, J., Madigan, D., Raftery, A. and Volinsky, C. (1999) Bayesian model averaging: a tutorial. *Statistical Science*, **14**, 382–401.

Ishwaran, H. (2000) Inference for the random effects in Bayesian generalized linear mixed models. *ASA Proceedings of the Bayesian Statistical Science Section*, 1–10.

Kadane, J. and Lazar, N. (2003) Methods and criteria for model selection. *Journal of the American Statistical Association*, **99**, 279–290.

Kass, R., Tierney, L. and Kadane, J. (1988) Asymptotics in Bayesian computation. In *Bayesian Statistics 3*, Bernardo, J., DeGroot, M., Lindley, D. and Smith, A. (eds). Oxford University Press: Oxford, 261–278.

Lindley, D. (1957) A statistical paradox. *Biometrika*, **44**, 187–192.

MacNab, Y. (2003) Hierarchical Bayesian modeling of spatially correlated health service outcome and utilization rates. *Biometrics*, **59**, 305–316.

McCullagh, P. and Nelder, J. (1989) *Generalized Linear Models*, 2nd Edition. Chapman and Hall: London.

McFadden, D. (1989) A method of simulated moments for estimation of discrete response models without numerical integration. *Econometrica*, **7**, 995–1026.

Metropolis, N., Rosenbluth, A., Rosenbluth, M., Teller, A. and Teller, E. (1953) Equations of state calculations by fast computing machines. *Journal of Chemical Physics*, **21**, 1087–1092.

Muthen, B. (2003) Statistical and substantive checking in growth mixture modeling. *Psychological Methods*, **8**, 369–377.

Nelder, J. and Lee, Y. (2000) Two ways of modelling overdispersion in non-normal data. *Applied Statistics*, **49**, 591–598.

Nelder, J. and Wedderburn, R. (1972) Generalized linear models. *Journal of the Royal Statistical Society, Series A*, **135**, 370–384.

Normand, S., Glickman, M. and Gatsonis, C. (1997) Statistical methods for profiling providers of medical care: issues and applications. *Journal of the American Statistical Association*, **92**, 803–814.

Norris, J. (1997) *Markov Chains*. Cambridge University Press: Cambridge.

O'Hagan, A. (1995) Fractional Bayes factors for model comparison. *Journal of the Royal Statistical Society*, **57**B, 99–138.

Pastor-Barriuso, R., Guallar, E. and Coresh, J. (2003) Transition models for change-point estimation in logistic regression. *Statistics in Medicine*, **22**, 1141–1162.

Rannala, B. (2002) Identifiability of parameters in MCMC Bayesian inference of phylogeny. *Systematic Biology*, **51**, 754–760.

Robert, C. (1996) Mixtures of distributions: inference and estimation. In *Markov Chain Monte Carlo in Practice*, Gilks, W., Richardson, S. and Spiegelhalter, D. (eds). Chapman and Hall: London, 441–464.

Roberts, C. (1996) Markov chain concepts related to sampling algorithms. In *Markov Chain Monte Carlo in Practice*, Gilks, W., Richardson, S. and Spiegelhalter, D, (eds). Chapman and Hall: London, 45–59.

Smith, A. and Roberts, G. (1993) Bayesian computation via the Gibbs sampler and related Markov chain Monte Carlo methods. *Journal of the Royal Statistical Society, Series B*, **55**, 3–23.

Tanner, M. and Wong, W. (1987) The calculation of posterior distributions by data augmentation (with discussion). *Journal of the American Statistical Association*, **82**, 528–550.

Tierney, L. (1994) Markov chains for exploring posterior distributions. *Annals of Statistics*, **22**, 1701–1762.

Tierney, L. (1996) Introduction to general state-space Markov chain theory. In *Markov Chain Monte Carlo in Practice*, Gilks, W., Richardson, S. and Spiegelhalter, D. (eds). Chapman and Hall: London, 59–74.

van Dyk, D. (2002) Hierarchical models, data augmentation, and MCMC. In *Statistical Challenges in Modern Astronomy III*, Babu, G. and Feigelson, E. (eds). Springer: New York, 41–56.

van Dyk, D. and Meng, X. (2001) The art of data augmentation. *Journal of Computational and Graphical Statistics*, **10**, 1–111.

Vines, S., Gilks, W. and Wild, P. (1996) Fitting Bayesian multiple random effects models. *Statistics and Computing*, **6**, 337–346.

Wakefield, J. and Morris, S. (1999) Spatial dependence and error-in-variables in environmental epidemiology. In *Bayesian Statistics 6*, Bernardo, J., Berger, J., Dawid A. and Smith A. (eds). Clarendon Press: Oxford, 657–684.

Weiss, R., Cho, M. and Yanuzzi, M. (1998) On Bayesian calculations for mixture likelihoods and priors. Dept. of Biostatistics, UCLA School of Public Health.

West, M. (1984) Outlier models and prior distributions in Bayesian linear regression. *Journal of the Royal Statistical Society, Series B*, **46**, 431–439.

Zellner, A. and Rossi, P. (1984) Bayesian analysis of dichotomous quantal response models. *Journal of Econometrics*, **25**, 365–393.

Zuur, G., Garthwaite, P., and Fryer, R. (2002) Practical use of MCMC methods: lessons from a case study. *Biometrical Journal*, **44**, 433–455.

CHAPTER 2

Model Comparison and Choice

2.1 INTRODUCTION: FORMAL METHODS, PREDICTIVE METHODS AND PENALIZED DEVIANCE CRITERIA

A range of methods have been proposed for model choice and diagnosis based on Bayesian principles. For instance, among the questions that regression model choice might include are choice between subsets of regressor variables, whether response and regressor variables should be transformed, and whether a linear sum of regression effects should be used or various non-linear forms including general additive models (Chapters 3 and 4). The error structure, or more generally the specifications of possibly multiple random effects (e.g. see Chapters 6 and 10), is an additional question. One may wish to assess the gains from adopting heavy-tailed densities instead of the default normal errors assumption of multiple linear regression (West, 1984), or whether to adopt error forms adapted to the possibility of a small number of outlier points (Vernardinelli and Wasserman, 1991).

For models $j = 1, \ldots, J$, let m be a multinomial model indicator, and θ_j be the parameters under each model. Then formal Bayesian model assessment (sections 2.2–2.3) is based on prior model probabilities $P(m = j)$ and posterior model probabilities $P(m = j|Y)$ after observing data. Instead of choosing one model as best, the goal may be to allow for model uncertainty by model averaging (Hoeting *et al.*, 1999). Problems with the formal choice method occur with diffuse or improper priors $P(\theta_j|m = j)$; for example, if the prior is improper then so is $P(Y|m = j)$ (Gelfand and Dey, 1994). A variety of Bayesian predictive approaches to

Bayesian Models for Categorical Data   P. Congdon
© 2005 John Wiley & Sons, Ltd

model choice and diagnosis (sections 2.4–2.5) shift the focus onto
observables away from parameters (Geisser and Eddy, 1979) and seek
to alleviate the impact of factors such as specification of priors. In
particular, repeated sampling of replicate data Z_j from each model's
parameters provides estimates of the posterior predictive densities

$$P(Z_j|Y, m = j) = \int P(Z_j|Y, \theta_j, m = j)P(\theta_j|Y, m = j)\mathrm{d}\theta_j$$

and model choice may be based on comparing such 'predictive replicates'
with the observations (Laud and Ibrahim, 1995). Predictive methods may
also be used in a variety of model checks, with posterior predictive
checks (Gelman *et al.*, 1995) the simplest to apply, though more
sophisticated predictive check methods have been suggested that are
less conservative.

Analogies to classical methods are illustrated by formal out-of-sample
validation methods (section 2.6); for example, this may involve leaving
some of the data out of the analysis (the 'validation sample') and making
predictions of the validation data from a model using only the remaining data (the 'training sample'). Similarly, Bayesian methods for estimating parametric complexity lead to penalized deviance measures similar to
those in classical methods (section 2.7). Of relevance both to formal
choice based on posterior model probabilities and to deviance penaliza-
tion are methods based on parallel sampling (sections 2.8–2.9), as
discussed by Congdon (2005a, 2005b, 2005d).

2.2 FORMAL BAYES MODEL CHOICE

Formal Bayesian model assessment is based on updating prior model
probabilities $P(m = j)$ to posterior model probabilities $P(m = j|Y)$ after
observing the data. Thus if J models are being compared,

$$P(m = j|Y) = [P(m = j)P(Y|m = j)]/\sum_{j=1}^{J}[P(m = j)P(Y|m = j)] \qquad (2.1)$$

Evaluation of (2.1) involves multidimensional integration over each θ_j,

$$P(m = j|Y) = P(m = j)\int P(Y|\theta_j)P(\theta_j)\,\mathrm{d}\theta_j/\sum_{j=1}^{J}[P(m = j)$$

$$\times \int P(Y|\theta_j)P(\theta_j)\,\mathrm{d}\theta_j]$$

The marginal likelihoods in (2.1)

$$P(Y|m=j) = \int P(Y|\theta_j)P(\theta_j)\,d\theta_j$$

give the probability of the data conditional on model j and are often difficult to estimate. Approximation methods for $P(Y|m=j)$ include those presented by Gelfand and Dey (1994), Newton and Raftery (1994) and Chib (1995) (section 2.3). The Bayes factor is used for comparing one model against another under the formal approach and is the ratio of marginal likelihoods

$$B_{12} = P(Y|m=1)/P(Y|m=2)$$

From (2.1) it can be seen that the Bayes factor converts the ratio of prior model probabilities $P(m=2)/P(m=1)$ to the posterior ratio of such probabilities

$$P(m=2|Y)/P(m=1|Y)$$

Often equal prior model probabilities, namely $P(m=j) = 1/J$, are assumed. However, if alternate models are being compared in an MCMC analysis (rather than single models being estimated one at a time) one may use the prior model probabilities to penalize more complex models.

Assuming equal prior model probabilities, the Bayes factor is used for model choice under the formal approach. While there are no preset values at which one model can be chosen as correct, values of $\log_{10}(B_{12})$ above 2 decisively support model 1 (Jeffreys, 1961; Kass and Raftery, 1995). Values between 0.5 and 2 provide weaker support for model 1 and values in 0–0.5 are inconclusive. Taking $\log_e(B_{12})$ instead, values above 3 are conclusive in support of model 1.

Whereas data analysis is often based on selecting a single best model and making inference as if that model were true, such an approach neglects uncertainty about the model itself, as expressed in the posterior model probabilities $P(m=j|Y)$ (Hoeting et al., 1999; Draper, 1995; Wasserman, 2000). Consider a quantity $\Delta(\theta_j, Y)$ possibly depending on both the data and the parameters of each model $j = 1, \ldots, J$, though it might just be a function of the parameters. Averaging over each model's posterior density $\Delta(\theta_j, Y)$ on this quantity implies the model averaged estimate

$$E[\Delta(\theta, Y)] = \sum_j P(m=j|Y)\Delta(\theta_j|Y) \qquad (2.2)$$

Assuming a single model true means that uncertainty about such quantities can be understated. Moreover, several studies have shown better predictive performance via model averaging as opposed to choosing a single model (Fernandez *et al.*, 2001; Hoeting *et al.*, 2002); formal reasons for this are mentioned by Raftery *et al.* (1997).

2.3 MARGINAL LIKELIHOOD AND BAYES FACTOR APPROXIMATIONS

The formal model choice process has a probabilistic basis, but the evaluation of marginal likelihoods may become unstable when vague (non-informative) priors $P(\theta_j)$ are used on parameters. Arguably this is not a problem with the formal model approach so much as with adopting flat priors that are not Bayesian in spirit (e.g. see the discussion in Kadane and Lazar, 2004, p 281). Several approximations to the marginal likelihood and hence to the Bayes factor have been suggested. Analytic approximations include the Laplace approximation (Tierney and Kadane, 1986). Thus using the Laplace method for integrals one can show for a model of dimension d that

$$\log[P(Y)] \approx 0.5d \log(2\pi) + 0.5 \log |G^*| + \log[P(Y|\theta^*)P(\theta^*)]$$

where

(a) θ^* might be a componentwise vector of posterior means, the maximum likelihood estimate of θ, or the multivariate median (Lewis and Raftery, 1997).
(b) $P(\theta^*)$ is the set of prior densities evaluated at $P(\theta^*)$, providing what is known as the prior ordinate.
(c) G^* is minus the inverse of the Hessian matrix $\partial^2 h(\theta, y)/(\partial\theta\, \partial\theta')$ of $h(\theta, y) = \log[P(Y|\theta)P(\theta)]$ evaluated at θ^*. Asymptotically this equals the posterior covariance matrix, and can be estimated by the covariance between parameter samples from MCMC output, though more robust estimators are available (Rousseeuw, 1985).

As another example of an approximation using MCMC output, consider the relation (with the conditioning on model m omitted)

$$1 = \int g(\theta)\, d\theta = \int g(\theta) \frac{P(Y)P(\theta|Y)}{P(Y|\theta)P(\theta)}\, d\theta$$

Since $P(Y)$ is a constant one can move it to the left hand side, so that the marginal likelihood is given by

$$P(Y) = \left[\int \frac{g(\theta)}{P(Y|\theta)P(\theta)} P(\theta|Y)\, d\theta \right]^{-1} \tag{2.3a}$$

Gelfand and Dey (1994) recommend that g be an importance density approximation for the posterior $P(\theta|Y)$, such as a multivariate normal, derived as a moment estimator[1] from a run of sample values of the components of θ, namely $\theta_1^{(t)}, \theta_2^{(t)}, \ldots, \theta_d^{(t)}$. Once this estimated density is obtained the probability $g^{(t)}$ under the importance density of each individual sampled parameter $\theta^{(t)}$ can be evaluated (so sampled values far from the average of the density will have a low probability under g). The values of $g^{(t)}, L^{(t)} = P(Y|\theta^{(t)})$, and $\pi^{(t)} = P(\theta^{(t)})$, namely the importance density, likelihood and prior ordinate, are evaluated at each of the sampled values $\theta^{(t)}$. From (2.3a), the marginal likelihood is then approximated by

$$\hat{P}(Y) = 1/\left[T^{-1}\sum_t g^{(t)}/\{L^{(t)}\pi^{(t)}\}\right]$$

$$= T/\left[\sum_t g^{(t)}/\{L^{(t)}\pi^{(t)}\}\right] \tag{2.3b}$$

Note that if g is taken as the prior $P(\theta)$, one obtains the harmonic mean of the likelihoods as an estimator for $P(Y)$, namely

$$\hat{P}(Y) = T/\left[\sum_t \{1/P(Y|\theta^{(t)})\}\right] \tag{2.4}$$

Newton and Raftery (1994) propose methods of stabilizing this estimator which will be affected by the degree of informativeness in the prior.

Another approach to derive Bayes factors from MCMC output was developed by Chib (1995) and is based on approximating the marginal likelihood via a restatement of (1.1). Thus consider

$$P(Y) = P(Y|\theta)P(\theta)/P(\theta|Y)$$

or equivalently

$$\log[P(Y)] = \log[P(Y|\theta)] + \log[P(\theta)] - \log[P(\theta|Y)]$$

at a high density point for θ such as the posterior mean $\bar{\theta}$. The likelihood and prior ordinate at $\bar{\theta}$ are usually readily obtained and the problem lies in estimating $P(\bar{\theta}|Y)$. While approximation methods for the posterior den-

---

[1]For example, if $d = 2$ and the samples of parameters θ_1 and θ_2 were approximately normal then a bivariate normal density g might be estimated with mean $\{\mu_1, \mu_2\}$ given by sample averages from a long MCMC run and covariance matrix Σ estimated from the sample standard deviations and correlations.

sities $P(\theta|Y)$ might be used, such as the Gelfand–Dey approach, Chib (1995) presents a method for estimating this ordinate in terms of a marginal/conditional decomposition of the blocks of θ used in the MCMC estimation procedure. For example, if $\theta = (\theta_1, \theta_2)$ are the blocks with which the MCMC sampling is conducted, the posterior ordinate is expressed as $p(\theta_1|y)p(\theta_2|y, \theta_1)$. The first ordinate in this decomposition is estimated from the output of the full run whereas the second ordinate is available directly. When there are the three blocks, the first ordinate is estimated as above, but the second ordinate, $p(\theta_2|y, \theta_1)$ is estimated from the output of a subsidiary MCMC simulation with blocks θ_2 and θ_3 free but block θ_1 held fixed. The conditional density of the third block given the first two blocks is found directly. This method, which works in this manner for any number of blocks, will be illustrated for binary responses in Chapter 4.

Some approximation methods to formal Bayesian model choice produce posterior model probabilities or Bayes factors without seeking to produce marginal likelihoods. For example, path sampling (Gelman and Meng, 1998; Song and Lee, 2002) involves constructing a path to link two models being compared. Instead of approximating the marginal likelihood of each model the goal is to estimate the Bayes factor as a ratio of normalizing constants. Denote the alternative models as $m = 0$ and $m = 1$. From the relation

$$P(\theta|Y) = P(Y, \theta)/P(Y)$$

where $P(Y, \theta) = P(Y|\theta)P(\theta)$, one may obtain for a metric variable s

$$P(\theta|Y, s) = P(Y, \theta|s)/P(Y|s) \tag{2.5}$$

In particular, define s as a path parameter $s \in [0, 1]$ linking the two models. For example, suppose the alternative models were

$$\text{Model 0}: \quad y_i = \alpha + x_i\beta_1 + \varepsilon_i \tag{2.6}$$
$$\text{Model 1}: \quad y_i = \alpha + w_i\beta_2 + \varepsilon_i \tag{2.7}$$

The models $m = 0$ and $m = 1$ at $s = 0$ and $s = 1$ are linked by the models $m = s$ (s taking values between 0 and 1) defined by

$$\text{Model } s: \quad y_i = \alpha + (1 - s)x_i\beta_1 + sw_i\beta_2 + \varepsilon_i \tag{2.8}$$

Let $Z(s) = P(Y|m = s)$, and in particular $Z(1) = P(Y|m = 1)$ and $Z(0) = P(Y|m = 0)$. Taking logarithms in (2.5) gives

$$\log[P(\theta|Y, s)] = \log[P(Y, \theta|s)] - \log[Z(s)]$$

and differentiating gives

$$\frac{d}{ds}\log[Z(s)] = \int \frac{1}{Z(s)} \frac{d}{ds} P(Y,\theta|s) \, d\theta = E\left[\frac{d}{ds}\log P(Y,\theta|s)\right]$$

where the expectation is with respect to $P(\theta|Y,s)$. If the prior density $P(\theta)$ is independent of s, then

$$\frac{d}{ds}\log[P(Y,\theta|s)] = \frac{d}{ds}\log[P(Y|\theta,s)]$$

Denote

$$R(\theta,s) = \frac{d}{ds}\log[P(Y|\theta,s)]$$

Then the logarithm of the Bayes factor is obtained as

$$\log B_{10} = \log[Z(1)/Z(0)] = \int_0^1 R(\theta,u) \, du$$

Notionally s is continuous between 0 and 1, but to estimate the integral in practice, define an ordered grid between 0 and 1, $s_0 = 0, s_1 < s_2 < s_3 < \ldots < s_G < s_{G+1} = 1$. Then by the trapezoid rule

$$\log \hat{B}_{10} = 0.5 \sum_{j=0}^{G} [\bar{R}_{j+1} + \bar{R}_j][s_{j+1} - s_j]$$

where $\bar{R}_j = \sum_{t=1}^{T} R(\theta^{(t)}, s_j)/T$ is an average over T iterations from an MCMC chain of parameters $\theta^{(t)}$ sampled from $P(\theta|Y, s_j)$. In the above linear regression example

$$\log[P(Y|\theta,s)] = -0.5[\log(2\pi) + n \log(\sigma^2)]$$
$$- 0.5 \sum_{i=1}^{n} [y_i - \alpha - (1-s)\beta_1 x_i - s\beta_2 w_i]^2/\sigma^2$$

and

$$R(\theta,s) = -\sum_{i=1}^{n} [y_i - \alpha - (1-s)\beta_1 x_i - s\beta_2 w_i][\beta_1 x_i - \beta_2 w_i]/\sigma^2$$

2.4 PREDICTIVE MODEL CHOICE AND CHECKING

Another set of model choice and diagnostic procedures is based on adapting the classical principle of predictive cross-validation. This approach reflects a belief that model choice should be based on accurate predictions within and/or beyond the sample, with Geisser and Eddy (1979), Laud and Ibrahim (1995) and Gelfand and Ghosh (1998) being among those providing a Bayesian basis for predictive model selection. Thus suppose the posterior for a parameter θ^m under model m and observations $\{y_i; i = 1, \ldots, n\}$ is given by

$$P(\theta^m|Y) \propto P(\theta^m)P(Y|\theta^m)$$

In choice of predictor applications, with X_m as the predictor set in model m, θ_m in a normal linear regression might be a vector of regression coefficients and a conditional variance, $\theta^m = (\beta^m, \phi)$, and in a Poisson or binomial regression simply a vector of regression coefficients. Let $\eta_{im} = X_{im}\beta_m$, and g be the relevant link function to model means μ_{im}. To assess the suitability of the model one may sample replicate data $Z_m = (z_{1m}, \ldots, z_{nm})$ from the same model assumed to produce Y. These are drawn from the posterior predictive density

$$P(Z_m|Y) = \int P(Z_m, \theta_m|Y)\, d\theta_m$$

$$= \int P(Z_m|Y, \theta_m)P(\theta_m|Y)d\theta_m \qquad (2.9)$$

$$= \int P(Z_m|\theta_m)P(\theta_m|Y)\, d\theta_m$$

Predictions Z_m may be used both in model selection and in predictive checks (Gelman *et al.*, 1995). In practice, predictions Z are obtained by repeated sampling from $P(Z_m^{(t)}|\theta_m^{(t)})$; for example, $z_i \sim N(\mu_{im}, \phi_m)$ in a linear regression with $\mu_{im} = \eta_{im}$, or $z_{im} \sim \text{Poi}(\mu_{im})$ in a log-link Poisson regression with $\log(\mu_{im}) = \eta_{im}$.

Bayesian model selection or diagnostics may be based on predictions based on comparing the 'new data' Z with the observations. For example, Laud and Ibrahim (1995) and Meyer and Laud (2002) propose as a basis for model choice minimization of the criterion

$$C_m = E[c(Z_m, Y)|Y] = \sum_{i=1}^{n}\{\text{Var}(z_{im}) + [y_i - E(z_{im})]^2\} \qquad (2.10)$$

where $c(Z_m, Y)$ is the predictive error sum of squares under model m

$$c(Z_m, Y) = (Z_m - Y)'(Z_m - Y)$$

The criterion (2.10) can be obtained from the posterior means and variances of sampled $z_{im}^{(t)}$ or from the posterior average of $\sum_{i=1}^{n}(z_{im}^{(t)} - y_i)^2$. Meyer and Laud (2002) suggest a forward selection algorithm based on this criterion, or more specifically its square root $C_m^{0.5}$. Carlin and Louis (1996) and Buck and Sahu (2000) propose analogous criteria appropriate to both metric and discrete outcomes. Thus the expected predictive deviance (EPD) is

$$D_m = E[d(Z_m, Y)|Y] \tag{2.11}$$

where $d(Z_m, Y)$ is the relevant deviance. For Y Poisson, for example,

$$d(Z_m, Y) = 2 \sum_{i=1}^{n} \{y_i \log(y_i/z_{im}) - (y_i - z_{im})\}$$

Buck and Sahu (2000) propose the loss function

$$U(Z_m, Y) = 2 \sum_{i=1}^{n} y_{im} \log(y_{im}/z_{im})$$

with an expectation (over the posterior predictive density) that can be written as the sum of the usual likelihood ratio statistic and a complexity penalty.

The predictive loss criteria of Gelfand and Ghosh (1998), Ibrahim *et al.* (2001) and others allow varying trade-offs in the balance between bias in predictions and their precision. Then for k positive, one possible criterion has the form

$$PrL_m(k) = \sum_{i=1}^{n} \left\{ \text{Var}(z_{im}) + \left(\frac{k}{k+1}\right)[y_i - E(z_{im})]^2 \right\} \tag{2.12}$$

This criterion would be compared between models at selected values of k, typical values being $k = 1, k = 10$ and $k = 10\,000$. Higher values of k imply a greater stress on accuracy in predictions (i.e. bias) and less stress on precision. Laud and Ibrahim (1995), Ibrahim *et al.* (2001) and others have considered calibration of such measures, i.e. expressing the uncertainty of C_m, D_m or $PrL_m(k)$ in a variance measure. Then one might move on to make probabilistic statements that model k is inferior to the best fitting model j with a certain probability based on $\{C_k, C_j, s_j\}$ where $s_j = [\text{Var}(C_j)]^{0.5}$. An alternative approach to assessing uncertainty in predictive criteria is considered by Congdon (2005a); see section 2.8.

2.5 POSTERIOR PREDICTIVE CHECKS

Gelman *et al.* (1995) propose a diagnostic procedure known as a posterior predictive checking using predictive replicates Z. Various forms of checking function may be calculated for both new data and actual observations to assess whether the model satisfactorily reproduces certain important aspects of the actual data. Thus such checks go beyond bias and precision.

For example, if continuous data are skewed, or count data are over-dispersed, then the model should reproduce such features in the replicates Z which are sampled from the model. Suppose $C(Y; \theta)$ is the observed criterion (e.g. a ratio of observed variance to mean, or a skewness measure). Let the same criterion based on new data be denoted $C(Z; \theta)$. A reference distribution is obtained from the joint distribution of Z and θ,

$$P_R(Z, \theta) = P(Z|\theta)P(\theta|Y)$$

and the actual value obtained by sampling set against this distribution.

In practice, at each iteration $C(Z^{(t)}, \theta^{(t)})$ and $C(Y, \theta^{(t)})$ are obtained and the proportion of iterations where $C(Z^{(t)}, \theta^{(t)})$ exceeds the other is also obtained, namely

$$\hat{P}_c = \sum_{t=1}^{T} 1(C(Z^{(t)}, \theta^{(t)})) > C(Y, \theta^{(t)})]/T \qquad (2.13)$$

Sometimes $C(Y, \theta^{(t)})$ may in fact not depend on $\theta^{(t)}$ but represent a fixed aspect of the observations (e.g. the proportion of zero counts in a Poisson regression application). So it does not need to be sampled at each iteration, though $C(Z^{(t)}, \theta^{(t)})$ still does.

Values of $\hat{P}_c$ near 0 or 1 (above 0.9 or below 0.1) indicate discrepancy between the observations and the model. Values relatively close to 0.5 mean that the actual data and the new data sampled from the model are closely comparable in terms of the feature that the checking function summarises. Reviews of the posterior predictive check procedure argue that it makes double use of the data and suggest that it may be conservative as a test (Bayarri and Berger, 2000), since the observations y_i can have a strong influence on the replicate z_i. Sinharay and Stern (2003) present simulations showing that posterior predictive checking may not detect departures from normality in random effects models; they consider the educational testing data also analysed by Gelman *et al.* (1995).

Marshall and Spiegelhalter (2003) develop on Gelman *et al.* (1996) by suggesting a mixed predictive scheme where the model for μ_{im} includes random effects. Thus one first simulates new random effects, and then simulates predictions based on the sampled random effects. For example, in the disease mapping model of (1.8) with means $\mu_i = \lambda_i E_i$, and $\log(\lambda_i) = X_i\beta + \alpha_{i1} + \alpha_{i2}$, replicates $\alpha_{i1,\text{new}}$ and $\alpha_{i2,\text{new}}$ would be sampled and replicate disease counts Z^* would be sampled using means $\lambda_{i,\text{new}} = \exp(X_i\beta + \alpha_{i1,\text{new}} + \alpha_{i2,\text{new}})$, whereas usually replicates Z are sampled from means $\lambda_i = \exp(X_i\beta + \alpha_{i1} + \alpha_{i2})$. This procedure considerably reduces the conservatism of posterior predictive checks (and other predictive assessments).

2.6 OUT-OF-SAMPLE CROSS-VALIDATION

Cross-validation is more explicit when predictions of a subset Y_s of the observed data Y are made from parameters estimated using only the remaining part of the data, namely the complement of Y_s, denoted $Y_{[s]}$ (Gelfand and Dey, 1994). The data on which the model is estimated are often called the training data or training set and the data whose fit to the model is to be assessed the prediction or validation set (e.g. Raftery *et al.*, 1997, p 185).

An example of such methods is in time series analysis where observations $y_1, \ldots, y_T$ are available, but only part of the history, say $\{y_1, \ldots, y_{T-m}\}$, is used to predict the remainder of the series $\{y_{T-m+1}, \ldots, y_T\}$. Cross-validatory predictions may, like those of new data, be based on repeated sampling from the posterior of θ in an MCMC sampling chain and have the form

$$P(Y_s|Y_{[s]}) = \int P(Y_s|\theta, Y_{[s]})P(\theta|Y_{[s]}) \, d\theta$$

Even if the prior for θ is improper, this predictive density is proper because the posterior based on using $Y_{[s]}$ to estimate θ, namely $P(\theta|Y_{[s]})$, is proper (Gelfand, 1996). By contrast, the formal Bayes factor is not defined for improper priors.

If $Y_s \equiv y_i$ consists of just one observation, with $Y_{[i]}$ denoting the remaining $n-1$ cases, then the density

$$P(y_i|Y_{[i]}) = P(Y)/P(y_i) = \int P(y_i|\theta, Y_{[i]})P(\theta|Y_{[i]}) \, d\theta \qquad (2.14)$$

is known as the conditional predictive ordinate (CPO) for case i (Geisser and Eddy, 1979). A Monte Carlo estimate of the CPO obtained without actually omitting case i from the estimation is provided by the harmonic mean of the likelihoods for case i (e.g. Sinha *et al.*, 2003):

$$CPO_i = T / \left[\sum_t \{1/P(y_i|\theta^{(t)})\} \right] \qquad (2.15)$$

Thus the full cross-validation based on N separate estimations (the first omitting case 1, the second omitting case 2, etc.) may be approximated by a single estimation run based on the full sample. The CPO values may be scaled (dividing by their maximum) and low values for particular observations (e.g. under 0.01) will then show outlying observations where the model is not fitting well (Weiss, 1994). If there are no scaled CPOs under say 0.01 then a relatively good fit to all data points is suggested. The range in scaled CPOs is useful as an indicator of a good fitting model (Congdon, 2005c).

An improved estimate of the CPO may be obtained by weighted resampling from $P(\theta|Y)$ (Smith and Gelfand, 1992). Thus samples $\theta^{(t)}$ from $P(\theta|Y)$ can be converted to samples from $P(\theta|Y_{[i]})$ by resampling the $\theta^{(t)}$ with weights

$$w_i^{(t)} = H(y_i|\theta^{(t)})/\sum_i H(y_i|\theta^{(t)})$$

where $H(y_i|\theta^{(t)}) = 1/P(y_i|\theta^{(t)})$, i.e. the inverse likelihoods of case i at iteration t. Once θ has been resampled, a sample of replicate data Y can be obtained. From (2.14), $P(y_i|Y_{[i]}) = \int P(y_i|\theta, Y_{[i]})P(\theta|Y_{[i]}) \, d\theta$, so if $\theta_*^{(t)}$ is sampled from $P(\theta|Y_{[i]})$ and if replicate data are obtained as $y_{i*}^{(t)} \sim P(y_i|\theta_*^{(t)})$, then the marginal density of y_{i*} is $P(y_i|Y_{[i]})$. Note that both estimation techniques, the simple one via (2.15) or the resampling one, become difficult when individual cases have low likelihoods (e.g. log likelihoods of -10 or -20).

One may obtain a range of summary measures of the closeness of fit between the observed y_i and the predictions made via (2.14). Gelfand et al. (1992) propose discrepancy functions for comparing the observations with predictions from $P(y_i|Y_{[i]})$. Among the simplest is the cross-validation residual (Carlin and Louis, 2000, p 205)

$$[y_i - E(y_i|Y_{[i]})]/[\text{Var}(y_i|Y_{[i]})]^{0.5}$$

The posterior mean $E(y_i|Y_{[i]}) = \int E(y_i|\theta)P(\theta|Y[i]) \, d\theta$ may be approximated by sampling new data Z_i and setting $E(y_i|Y_{[i]}) = E(Z_i|Y)$. Similarly $\text{Var}(y_i|Y_{[i]})$ may be estimated by the posterior variance of Z_i or by sampling Z_i^2 and setting $\text{Var}(y_i|Y_{[i]}) \approx E(Z_i^2|Y) - [E(Z_i|Y)]^2$.

As discussed by Gelfand (1996), cross-validation based on single case omission provides an alternative to formal marginal likelihood estimation. Thus if y_i consists of only one observation and the remaining $n-1$ observations are used to estimate θ, then a proxy for the marginal likelihood $P(Y)$, known as the pseudo marginal likelihood, is defined by the product of these terms

$$\text{PsML} = \hat{P}(Y) = \prod_{i=1}^{n} P(y_i|Y_{[i]}) \tag{2.16}$$

with $P(y_i|Y_{[i]})$ estimated as in (2.15), or via resampling. The ratio of $P_1(y)$ and $P_2(y)$ for two models 1 and 2 is sometimes known as the pseudo Bayes factor (PsBF), though a Jeffreys' type scale for these is not available. An estimate of the log pseudo marginal likelihood in (2.16) is obtained by totalling over the log CPOs as estimated by (2.15).

Model assessment by cross-validation includes comparison of models via random splitting of data sets, e.g. k-fold validation (Kuo and Peng, 1999). Here the data are split into a small number of groups (e.g. $k = 5$) of roughly equal size and cross-validation is applied k times by leaving out the kth group and using the $k - 1$ remaining groups as the training data.

2.7 PENALIZED DEVIANCES FROM A BAYES PERSPECTIVE

Several model choice techniques proposed for use in Bayesian analysis include features of classical model assessment, such as penalizing excess complexity and the requirement for accurate out-of-sample predictions (or cross-validation). The resulting criteria for model selection may be known as informal if they do not produce a model probability $P(m = j|Y)$ conditional on the data. As an example of a Bayesian adaptation of frequentist model choice by comparing penalized deviances, Spiegelhalter *et al.* (2002) propose an estimate for the effective total number of parameters or model dimension, denoted d_e. This total is generally less than the nominal number of parameters d_n in complex hierarchical random effects models where there is no way to count parameters, and where the number of effects being fitted may exceed the number of observations.

For instance, under the mixed model in spatial epidemiology with p regressors (equation 1.8), the nominal number of parameters is $d_n = (p + 1) + 2n + 2$, allowing for the n parameters in each of α_{i1} and α_{i2}. However, such sets of parameters are not independent of each other because they are generated from a joint density and their effective dimension is lower. Let $L^{(t)} = \log[P(Y|\theta^{(t)})]$ denote the log likelihood obtained at the tth iteration in a long sampling chain, and $D^{(t)} = -2L^{(t)}$ be the deviance. Another definition of the deviance is provided by McCullagh and Nelder (1989) and involves comparing the model and saturated likelihoods (this is termed the GLM deviance below): both definitions may be used to derive the effective parameters.

Then d_e may be approximated by the difference between the expected deviance $E(D|y, \theta)$, given by the mean $\bar{D}$ of the sampled deviances $D^{(t)}$, and the deviance $D(\bar{\theta}|y)$ at the posterior mean $\bar{\theta}$ of the parameters. The latter deviance may also be estimated at the posterior means, e.g. $\bar{\mu}_i = \bar{\lambda}_i E_i$ in the model of (1.8), and this estimate may be denoted $D(\bar{\mu}|y)$.

The analogue of the classical Akaike information criterion (AIC) is then either

$$D(\bar{\theta}|y) + 2d_e \tag{2.17a}$$

or

$$D(\bar{\mu}|y) + 2d_e \tag{2.17b}$$

and termed the deviance information criterion (DIC) by Spiegelhalter et al. (2002). $D(\bar{\mu}|y)$ may be more easily obtainable than $D(\bar{\theta}|y)$ in certain complex (e.g. discrete mixture) models.

Another effective parameter estimate relies on the asymptotic chi-square distribution of $D(\theta|y) - D(\theta_{\min}|Y)$ where $\theta_{\min}$ is the value of θ minimizing the deviance for a given model (Gelman et al., 2003). Then from the properties of the chi-square density, $d_e^* = 0.5 \text{Var}(D^{(t)})$. Both effective parameter estimates in practice include aspects of a model such as the precision of its parameters and predictions.

2.8 MULTIMODEL PERSPECTIVES VIA PARALLEL SAMPLING

As in Congdon (2005a), consider data $Y = (y_1, \ldots, y_n)$ and a set of potential models $j = 1, \ldots, J$ of dimensions $d = (d_1, d_2, \ldots, d_J)$, with parameter sets $\theta = (\theta_1, \theta_2, \ldots, \theta_J)$, prior model probabilities $P(m = j)$, and priors on parameters $P(\theta_j | m = j)$. Likelihood and prior specification is independent between models.

The log-likelihood under model j is $\log[P(Y|\theta_j)] = L(\theta_j|Y)$, with deviance obtained as $D(\theta_j|Y) = -2L(\theta_j|Y)$ or as $D(\theta_j|Y) = -2[L(\theta_j|Y) - L(Y|Y)]$. Consider the output from sampling stream $\theta^{(t)} = (\theta_1^{(t)}, \ldots, \theta_j^{(t)}, \ldots, \theta_J^{(t)})$ of length T. This may be obtained from parallel or sequential sampling. As noted above, posterior means of the log likelihood or deviance (respectively $\bar{L}_j = \sum_{t=1}^{T} L(\theta_j^{(t)}|Y)/T$ and $\bar{D}_j$) have been suggested as measures of model performance by Spiegelhalter et al. (2002) and Dempster (1997). One might also use deviances or log likelihoods with a penalty for dimension; for instance, take the AIC of model j as $\text{AIC}_j = \bar{D}_j + 2d_j$, or the BIC (Bayesian Information Criterion) of model j as $\text{BIC}_j = \bar{D}_j + d_j \log(n)$. An alternative might be to penalize the deviance at the posterior parameter mean (namely, $D(\bar{\theta}_j)$), or posterior

mode. Sometimes inferences are based on maximum a posteriori estimation maximizing the joint density of parameters, data and model (the 'full
model likelihood' for short), namely

$$F_j = P(Y, \theta_j, m = j) = P(Y|\theta_j)P(\theta_j|m = j)P(m = j)$$

Consider a fit criterion such as the difference in AIC between models j
and k,

$$\Delta \text{AIC}_{jk} = \text{AIC}_j - AIC_k$$

where $\text{AIC}_j = D_j + d_j$, or equivalently the penalized likelihood ratio or
evidence ratio (Burnham and Anderson, 2002)

$$E_{jk} = (L_j/L_k) \exp(d_k/d_j) \qquad (2.18)$$

Similarly relevant to model comparisons are Akaike or AIC weights
(Brooks, 2002) obtained by comparing AIC_j to the minimum AIC for
model m^*, giving differences $\Delta \text{AIC}_j = \text{AIC}_j - \text{AIC}_{m^*}$. These are rescaled
to give weights

$$\omega_j = \exp(-0.5\Delta\text{AIC}_j)/ \sum_m [\exp(-0.5\Delta\text{AIC}_m)] \qquad (2.19)$$

One may also consider likelihood weights, based on comparison with the
highest log-likelihood model m^{**}; defining $\Delta L_j = L_j - L_{m^{**}}$, likelihood
weights are obtained as

$$\xi_j = \exp(\Delta L_j)/ \sum_m [\exp(\Delta L_m)] \qquad (2.20)$$

Another option, based on Schwartz (1978), are BIC weights (Wintle
et al., 2003). Comparison with the minimum BIC model m^{***}, gives
$\Delta \text{BIC}_j = \text{BIC}_j - \text{BIC}_{m^{***}}$, and weights

$$\zeta_j = \exp(-0.5\Delta\text{BIC}_j)/ \sum_m [\exp(-0.5\Delta\text{BIC}_m)] \qquad (2.21)$$

Analysing parallel output (or pooled output from models run separately)
enables analysis of the densities of intermodel fit criteria, or of associated
model weights. Posterior averages and densities of such criteria are
obtained and the densities may be used as evidence of support for
models. Monte Carlo estimates of formal model probabilities $P(m =
j|Y)$ are also obtainable using samples of the iteration-specific model
probabilities $P[m = j|Y, \theta^{(t)}]$ (see section 2.9). Iteration-specific values of
model weights or model probabilities provide a basis for model averaging
that takes into account both model and parameter uncertainty; this
approach constrasts with that of Hoeting *et al.* (1999) who base model

averaging on posterior means and variances of parameters in different models.

From a parallel sampling analysis one may consider posterior probabilities that model j provide better predictions than model k (perhaps after adjusting for parameter complexity). Thus the $J(J-1)$ predictive comparison probabilities

$$Q(j,k) = \Pr[D(Z_j, Y) < D(Z_k, Y)] \tag{2.22}$$

provide evidence on the relative predictive success of models j and k. For example, if

$$\Pr[D(Z_j, Y) < D(Z_k, Y)] > 0.5$$

and $d_j < d_k$ also (i.e. model j has a higher probability of providing better predictions and also has a lower dimension), this would be evidence, using Occam's razor, in support of model j. One may also consider the probabilities that model j provides the best predictions among all models:

$$B_j^{(\text{new})} = \Pr[D(Z_j, Y) < D(Z_k, Y), \text{ all } k \neq j] \tag{2.23a}$$

or

$$B_j^{(\text{new})} = \Pr[C(Z_j, Y) < C(Z_k, Y), \text{ all } k \neq j] \tag{2.23b}$$

estimated by the proportion of iterations where

$$D(Z_j^{(t)}, Y) < D(Z_k^{(t)}, Y), \text{ all } k \neq j$$

Optimality may also be defined in terms of the model with the best likelihood or penalized likelihood criterion; for example,

$$B_j^{(\text{AIC})} = \Pr[\text{AIC}(\theta_j, Y) < \text{AIC}(\theta_k, Y), \text{ all } k \neq j] \tag{2.24}$$

estimated by the proportion of iterations where

$$L(\theta_j^{(t)}|y) + 2p_j < L(Z_k^{(t)}, y) + 2p_k, \quad \forall k \neq j$$

As mentioned above, a problem in model selection is in random effects models where the model dimension is unknown. One can, however, substitute an effective parameter count into conventional fit criteria such as the AIC and BIC. Using the effective parameter count proposed by Spiegelhalter *et al.* (2002) leads to a strategy for model choice based on minimum DIC_j where

$$\text{DIC}_j = D(\bar{\theta}_j) + 2d_{ej}$$

This still leaves a problem when DIC differences are small. In an attempt to describe what differences in AIC or DIC count as significant, 'rules of thumb' options such as

$$\Delta DIC_{jk} = DIC_j - DIC_k > 4$$

(Spiegelhalter *et al.*, 2002) or $\Delta AIC_{jk} > 10$ (Anderson *et al.*, 2000), have been suggested.

If estimates of d_{ej} and d_{ek} are obtained from a preliminary run, then parallel sampling enables one to obtain the density of the ΔDIC_{jk} in a subsequent run, and hence an informed decision to be made about significant differences in DIC or AIC. We take the iteration-specific equivalent of the DIC to be $D(\theta_j^{(t)}) + d_{ej}$, and obtaining a density for ΔDIC_{jk} then involves repeated sampling of

$$\kappa_{jk}^{(t)} = D(\theta_j^{(t)}) - D(\theta_k^{(t)}) + (d_{ej} - d_{ek}) \qquad (2.25)$$

In fact, some analytic results are available to define the densities of intermodel fit statistics. Gelman *et al.* (2003) mention another estimate of effective parameters, based on the asymptotic distribution of the deviance difference between θ_j and the value θ_j^* providing the minimum deviance; thus

$$C_j = D(\theta_j) - D(\theta_j^*) \sim \chi^2(d_{ej}) \qquad (2.26)$$

Since the variance of a chi-square variable is twice the degrees of freedom, one may obtain an effective parameter estimate d_{ej} as half the variance of a stream of sampled deviances

$$d_{ej} = 0.5 \, \mathrm{Var}[D_j^{(t)}]$$

It follows that the difference $C_j - C_k$ between models is distributed as a difference of chi-square variables $\chi^2(d_{ej})$ and $\chi^2(d_{ek})$, so if $\delta_{jk}^{(t)} = D(\theta_j^{(t)}) - D(\theta_k^{(t)})$, then

$$0.5 \, \mathrm{Var}[\delta_{jk}^{(t)}] \approx d_{ej} + d_{ek}$$

So the variance of the deviance difference $\delta_{jk}^{(t)}$ measures the 'total complexity' of models being compared. Since the DIC is the posterior mean of the deviance plus d_{ej} (which is viewed as an unknown constant or perhaps a constant subject to small Monte Carlo variation), the difference in DICs is obtained as the posterior mean of $\kappa_{jk}^{(t)} = \delta_{jk}^{(t)} + (d_{ej} - d_{ek})$.

The variance of $\delta_{jk}^{(t)}$ and $\kappa_{jk}^{(t)}$ may not exactly equal $2(d_{ej} + d_{ek})$, since in repeated sampling of models in parallel it will be necessary to adjust for correlation r_{jk} between the series of sampled values $D_j^{(t)}$ and $D_k^{(t)}$, though this correlation tends to diminish as longer sampling chains are taken. Also, since it is a random correlation between two series and not an autocorrelation, it is generally small (e.g. between -0.05 and 0.05). The actual variance of $\kappa_{jk}^{(t)}$ will be

$$\text{Var}[\kappa_{jk}^{(t)}] = 2(d_{ej} + d_{ek} - w_{jk})$$

where w_{jk} is the covariance between series, so that the variance of the DIC difference is obtained as

$$\text{Var}[\Delta\text{DIC}_{jk}] = \text{Var}[\kappa_{jk}^{(t)}] - 2w_{jk} \qquad (2.27)$$

Consider also a stream of comparison measures on the log-likelihood scale such as the log-likelihood ratio $\lambda_{jk}^{(t)}$ or the log of evidence ratio

$$\varepsilon_{jk}^{(t)} = \log[E_{jk}^{(t)}] \qquad (2.28)$$

From above it follows that

$$2\text{Var}(\varepsilon_{jk}^{(t)}) = 2\text{Var}(\lambda_{jk}^{(t)}) \approx d_{ej} + d_{ek} - w_{jk} \qquad (2.29)$$

That is, the variance of the log evidence ratio, as obtained by repeated sampling involving comparison of two models, is also a measure of total complexity.

2.9   MODEL PROBABILITY ESTIMATES FROM PARALLEL SAMPLING

Suppose from parallel sampling, or less conveniently from pooling output from separate sequential sampling, that one obtains parameter samples for all J models. The collected output at iterations $t = 1, \ldots, T$ is $\{\theta^{(1)}, \ldots, \theta^{(T)}\}$ where $\theta^{(t)} = (\theta_1^{(t)}, \ldots, \theta_J^{(t)})$. Using such output, one may obtain estimates of $P(m = j|Y)$ of the posterior model probabilities, and hence estimated Bayes factors B_{jk} from the relation

$$P(m=j|Y)/P(m=k|Y) = [P(Y|m=j)/P(Y|m=k)][P(m=j)/P(m=k)]$$
$$= B_{jk}[P(m=j)/P(m=k)]$$

Averaging over the posterior $P(\theta|Y)$ of $\theta = (\theta_1, \theta_2, \ldots, \theta_J)$ one may obtain $P(m = j|Y)$ as

$$P(m=j|Y) = \int P(m=j, \theta|Y)\, d\theta = \int P(m=j|Y, \theta)P(\theta|Y)\, d\theta$$

A Monte Carlo estimate $\hat{P}(m = j|Y)$ of $P(m = j|Y)$ is therefore provided by

$$\hat{P}(m = j|Y) = \sum_{t=1}^{T} P(m = j|Y, \theta^{(t)})/T \qquad (2.30)$$

Consider the numerator in the following re-expression of $P(m = j|Y, \theta)$:

$$\begin{aligned} P(m = j|Y, \theta) &= P(Y, \theta, m = j)/P(Y, \theta) \\ &= P(Y, \theta, m = j)/\sum_{k} P(Y, \theta, m = k) \end{aligned} \qquad (2.31)$$

namely

$$\begin{aligned} P(Y, \theta, m = j) &= P(Y|\theta, m = j)P(\theta, m = j) = P(Y|\theta, m = j) \\ &\quad \times P(\theta|m = j)P(m = j) \end{aligned}$$

The first component is the likelihood when $m = j$, and reduces to

$$l_j = l(\theta_j|Y) = P(Y|\theta_j, m = j)$$

(cf. Carlin and Chib, 1995). Also assume

$$P(\theta_k|m = j) = 1 \quad (k \neq j) \qquad (2.32)$$

This is reasonable in terms of a subjective prior view since model $j \neq k$ is not meant to provide information about different values of θ_k.

The cross-model priors $P(\theta_k|m = j)$ are also termed pseudo-priors by Carlin and Chib (1995), but their role in this case is not strictly as priors but as linking densities in a product space search algorithm – see also Godsill (2001) on pseudo-priors in the metropolized version of the Carlin–Chib algorithm. However, Carlin and Chib argue that pseudo-priors can be set arbitrarily and so the choice (2.32) can be seen as consistent with this while also simplifying model probability calculations.

Thus with (2.32)

$$P(\theta|m = j) = \prod_{k=1}^{J} P(\theta_k|m = j) = P(\theta_j|m = j)$$

and (2.31) becomes

$$P(m = j|Y, \theta) = l_j P(\theta_j|m = j)P(m = j)/\left[\sum_{k=1}^{J} l_k P(\theta_k|m = k)P(m = k)\right]$$

$$= F_j/\sum_{k=1}^{J} F_k$$

where $F_j = P(Y, \theta_j, m = j)$. Note that the same simplification follows from assuming that the cross-model priors are equal without actually specifying them, i.e. that $P(\theta_{k_1}|m = j_1) = P(\theta_{k_2}|m = j_2)$ for all $k_1 \neq j_1, k_2 \neq j_2$.

Except possibly for small samples, $R_j = \log F_j$ is the quantity that is monitored in practice. Let m^* be the model with the maximum $R_j^{(t)}$ at a particular iteration, and define $\Delta R_j^{(t)} = R_j^{(t)} - R_{m^*}^{(t)}$. Then

$$w_j^{(t)} = \exp(-0.5\Delta R_j^{(t)})/\sum_k [\exp(-0.5\Delta R_k^{(t)})] \qquad (2.33)$$

is an estimator of $P(m = j|Y, \theta^{(t)})$ at iteration t. The averages

$$\bar{w}_j = \sum_t w_j^{(t)}/T \qquad (2.34)$$

may be used to approximate posterior model probabilities. The $\bar{w}_j$ are termed Bayesian model weights, and are Bayesian in the sense that they clearly incorporate prior information on θ_j into model choice and averaging, whereas for criteria such as the AIC this is not directly apparent (Kadane and Lazar, 2004). Another estimator for model choice is provided by considering the differences $\Delta \bar{R}_j$ between $\bar{R}_j$ and $\max(\bar{R}_j)$ and then calculating

$$\exp(-\Delta \bar{R}_j)/\sum_k [\exp(-\Delta \bar{R}_k)]$$

Model performance is also summarized by differences $\rho_{jk} = R_j - R_k$ in the posterior averages of the log total model likelihoods. For example, if the 95% interval of $\{\rho_{jk}^{(t)}, t = 1, \ldots, T\}$ is confined to positive values this would support model j.

Consider the iteration-specific average of a functional of parameters and data $\Delta(\theta_j^{(t)}, Y)$ obtained by using weights $w_j^{(t)}$, namely

$$\Delta_w(\theta^{(t)}, Y) = \sum_j w_j^{(t)} \Delta(\theta_j^{(t)}, Y)$$

Then the posterior mean $E[\Delta_w(\theta^{(t)}, Y)|Y]$, as estimated by $\sum_t \Delta_w (\theta^{(t)}, Y)/T$, provides a model averaged estimate of $\Delta(\theta, Y)$ that takes account of both model and parameter uncertainty. A particular application is illustrated by regression selection. Define inclusion weights $\pi_r^{(t)}$ for regressor x_r at iteration t based on totalling $w_k^{(t)}$ over those models k that include x_r. Then a density for π_r is obtained by averaging over iterations, while the posterior mean $E(\pi_r|y)$ may be used to form particular kinds of model. The regression model defined by those predictors with $E(\pi_r|y) > 0.5$ is sometimes known as the median probability model (Barbieri and Berger, 2004).

2.10   WORKED EXAMPLE

As a simple demonstration of the calculation in (2.34), consider the 2×2 factorial study (patient frailty and treatment) with a binomial mortality response, undertaken by Dellaportas *et al.* (2002) and Green (2003). All five possible models are considered, namely mean only, mean + frailty, mean + treatment, mean + frailty + treatment and mean + frailty + treatment + frailty * treatment. Priors are as in Dellaportas *et al.* The posterior model probabilities from the last 45 000 of a 50 000 iteration two-chain run are (0.008, 0.507, 0.017, 0.405, 0.063). These are close to those reported by existing studies (Green, 2003; Dellaportas *et al.*, 2002) using RJMCMC or the metropolized version of the Carlin–Chib algorithm.

The code in WINBUGS is

```
model { for (i in 1:N) { # J=5 models with differing predictors
logit (p[i,1]) <-b1[1]
logit (p[i,2]) <-b2[1] +b2[2] *equals(frai[i],1)
logit (p[i,3]) <-b3[1] +b3[2] *equals(trt[i],1)
logit (p[i,4]) <-b4[1] +b4[2] *equals(frai[i],1) +b4[3] *equals(trt[i],1)
logit (p[i,5]) <-b5[1] +b5[2] *equals(frai[i],1) +b5[3] *equals(trt[i],1)
               +b5[4] *equals(frai[i],1) *equals(trt[i],1)
for (j in 1:J) {Y[i,j] <- y[i]; Y[i,j] ~ dbin(p[i,j],n[i])
LL[i,j] <-logfact(n[i]) -logfact(n[i] -y[i]) -
logfact(y[i]) +y[i] * log(p[i,j]) +(n[i] -y[i]) * log(1-p[i,j])}}

# priors and prior ordinates
for (j in 1:d[1]) {b1[j] ~dnorm(B0[j],P[j]); Pr[j,1] <- 0.5* log(P[j] /6.28) -
0.5* P[j] *pow(b1[j] -B0[j],2)}
for (j in 1:d[2]) {b2[j] ~ dnorm(B0[j],P[j]); Pr[j,2] <- 0.5* log(P[j] /6.28) -
0.5* P[j] *pow(b2[j] -B0[j],2)}
for (j in 1:d[3]) {b3[j] ~dnorm(B0[j],P[j]); Pr[j,3] <-0.5* log(P[j] /6.28) -
0.5* P[j] *pow(b3[j] -B0[j],2)}
for (j in 1:d[4]) {b4[j] ~dnorm(B0[j],P[j]); Pr[j,4] <-0.5* log(P[j] /6.28) -
0.5* P[j] *pow(b4[j] -B0[j],2)}
for (j in 1:d[5]) {b5[j] ~dnorm(B0[j],P[j]); Pr[j,5] <-0.5* log(P[j] /6.28) -
0.5* P[j] *pow(b5[j] -B0[j],2)}

# scaled likelihoods and model probabilities at each iteration
for (j in 1:J) {L[j] <-sum(LL[,j]) +sum(Pr[1:d[j],j]) +log(PriorMod[j]
```

```
# underflow protection for disparate model likelihoods
      SL[j] <- max(L[j]-maxL,-700); expSL[j] <- exp(SL[j])
```

 # model probabilities at iteration t

```
      w[j] <- expSL[j]/sum(expSL[])}
```

 # maximum of model likelihoods

```
   maxL <- ranked(L[],J)}
```

The data are as follows:

list(PriorMod=c(0.2,0.2,0.2,0.2,0.2),d=c(1,2,2,3,4),N=4,J=5,P=c(0.125,

0.125,0.125,0.125,0.125),B0=c(0,0,0,0,0),y=c(15,22,5,7),n=c(21,26,20,

12),frai=c(1,1,0,0),trt=c(1,0,1,0))

Initial values for one chain are provided by the list

list(b1=c(0), b2=c(0,0), b3=c(0,0), b4=c(0,0,0), b5=c(0, 0, 0, 0))

with those for the other chain generated from the priors.

REFERENCES

Anderson, D., Burnham, K. and Thompson, W. (2000) Null hypothesis testing: problems, prevalence, and an alternative. *Journal of Wildlife Management*, **64**, 912–923.

Barbieri, M. and Berger, J. (2004) Optimal predictive model selection. *Annals of Statististics*, **32**, 870–897.

Bayarri, M. and Berger, J. (2000) P-values for composite null models. *Journal of the American Statistical Association*, **95**, 1127–1142.

Brooks, S. (2002) Discussion to Spiegelhalter et al *Journal of the Royal Statistical Society*, **64B**, 616–618.

Buck, C. and Sahu, S. (2000) Bayesian models for relative archaeological chronology building. *Applied Statistics*, **49**, 423–444.

Burnham, K. and Anderson, D. (2002) *Model Selection and Multimodel Inference: A Practical Information-theoretic Approach 2nd Edition*. Springs: New York.

Carlin, B. and Chib, S. (1995) Bayesian model choice via Markov chain Monte Carlo methods. *Journal of the Royal Statistical Society, Series B*, **57**, 473–484.

Carlin, B. and Louis, T. (1996) *Bayes and Empirical Bayes Methods for Data Analysis*. Chapman and Hall: London.

Carlin, B. and Louis, T. (2000) *Bayes and Empirical Bayes Methods for Data Analysis*, 2nd Edition. Chapman and Hall: London.

Chib, S. (1995) Marginal likelihood from the Gibbs output. *Journal of the American Statistical Association*, **90**, 1313–1321.

Congdon, P. (2005a) Bayesian predictive model comparison via parallel sampling. *Computational Statistics and Data Analysis*, **48**, 735–753.

Congdon, P. (2005b) Bayesian model choice based on Monte Carlo estimates of posterior model probabilities. *Computational Statistics and Data Analysis* (in press).

Congdon, P. (2005c) A model for nonparametric spatially varying regression effects. *Computational Statistics and Data Analysis* (in press).

Congdon, P. (2005d) Bayesian model comparison via parallel model output. *Journal of Statistical Computation and Simulation*, **75** (in press).

Dellaportas, P., Forster, J. and Ntzoufras, I. (2002) On Bayesian model and variable selection using MCMC. *Statistics and Computing*, **12**, 27–36.

Dempster, A. (1997) The direct use of likelihood for significance testing. *Statistics and Computing*, **7**, 247–252.

Draper, D. (1995) Assessment and propagation of model uncertainty. *Journal of the Royal Statistical Society, Series B* **57**, 45–97.

Fernandez, C., Ley, E. and Steel, M. (2001) Model uncertainty in cross-country growth regressions. *Journal of Applied Econometrics*, **16**, 563–576.

Geisser, S. and Eddy, W. (1979) A predictive approach to model selection. *Journal of the American Statistical Association*, **74**, 153–160.

Gelfand, A. (1996) Model determination using sampling based methods. Chapter 9 in *Markov Chain Monte Carlo in Practice*, Gilks, W., Richardson, S. and Spieglehalter, D. (eds). Chapman and Hall: London/CRC Press: Boca Raton, FL.

Gelfand, A. and Dey, D. (1994) Bayesian model choice: asymptotics and exact calculations. *Journal of the Royal Statistical Society, Series B*, **56**, 501–514.

Gelfand, A. and Ghosh, S. (1998) Model choice: a minimum posterior predictive loss approach. *Biometrika*, **85**, 1–11.

Gelfand, A., Dey, D. and Chang, H. (1992) Model determination using predictive distributions with implementation via sampling based methods (with discussion). In *Bayesian Statistics 4*, Bernardo J. et al (eds). Oxford University Press: Oxford, 147–167.

Gelman, A. and Meng, X.-L. (1998) Simulating normalizing constants: from importance sampling to bridge sampling to path sampling. *Statistical Science*, **13**, 163–185.

Gelman, A., Carlin, J., Stern, H. and Rubin, D. (1995) *Bayesian Data Analysis*, Chapman and Hall: London.

Gelman, A., Meng, X.-L. and Stern, H. (1996) Posterior predictive assessment of model fitness via realized discrepancies. *Statistica Sinica*, **6**, 733–807.

Gelman, A., Carlin, J., Stern, H. and Rubin, D. (2003) *Bayesian Data Analysis*, 2nd Edition. CRC Press: Boca Raton, FL.

Godsill, S. (2001) On the relationship between Markov chain Monte Carlo for model uncertainty. *Journal of Computational and Graphical Statistics*, **10**, 230–248.

Green, P. (2003) Trans-dimensional Markov chain Monte Carlo. In *Highly Structured Stochastic Systems*, Green, P., Hjort, N. and Richardson, S. (eds). Oxford University Press: Oxford, 179–198.

Hoeting, J., Madigan, D., Raftery, A. and Volinsky, C. (1999) Bayesian model averaging: a tutorial. *Statistical Science*, **14**, 382–401.

Hoeting, J., Raftery, A. and Madigan, D. (2002) A method for simultaneous variable and transformation selection in linear regression. *Journal of Computational and Graphical Statistics*, **11**, 485–507.

Ibrahim, J., Chen, M. and Sinha, D. (2001) Criterion-based methods for Bayesian model assessment. *Statistica Sinica*, **11**, 419–443.

Jeffreys, A. (1961) *The Theory of Probability*, 2nd Edition. Cambridge: Cambridge University Press.

Kadane, J. and Lazar, N. (2004) Methods and criteria for model selection. *Journal of the American Statistical Association*, **99**, 279–290.

Kass, R. and Raftery, A. (1995) Bayes factors. *Journal of the American Statistical Association*, **90**, 773–795.

Kuo, L. and Peng, F. (1999) A mixture model approach to the analysis of survival data. In *Generalized Linear Models: A Bayesian Perspective*, Dey, K., Ghosh, S. and Mallick, B. (eds). Marcel Dekker: New York.

Laud, P. and Ibrahim, J. (1995) Predictive model selection. *Journal of the Royal Statistical Society, Series B*, **57**, 247–262.

Lewis, S. and Raftery, A. (1997) Estimating Bayes factors via posterior simulation with the Laplace-Metropolis estimator. *Journal of the American Statistical Association*, **92**, 648–655.

Marshall, E. and Spiegelhalter, D. (2003) Approximate cross-validatory predictive checks in disease mapping models. *Statistics in Medicine*, **22**, 1649–1660.

McCullagh, P. and Nelder, J. (1989) *Generalized Linear Models*, 2nd Edition. Chapman and Hall: London.

Meyer, M. and Laud, P. (2002) Predictive variable selection in generalized linear models. *Journal of the American Statistical Association*, **97**, 859–871.

Newton, M. and Raftery, A. (1994) Approximate Bayesian inference by the weighted likelihood bootstrap. *Journal of the Royal Statistical Society, Series B*, **56**, 3–48.

Raftery, A., Madigan, D. and Hoeting, J. (1997) Bayesian model averaging for regression models. *Journal of the American Statistical Association*, **92**, 179–191.

Rousseeuw, P. (1985) Multivariate estimators with high breakdown point. In *Mathematical Statistics and its Applications* (vol. B), Grossmann, W., Pflug, G., Vincze, I. and Wertz, W. (eds). Dordrecht: Reidel, 283–297.

Schwartz, G. (1978) Estimating the dimension of a model. *Annals of Statistics*, **6**, 461–464.

Sinha, D., Patra, K. and Dey, D. (2003) Modelling accelerated life test data by using a Bayesian approach. *Applied Statistics*, **52**, 249–259.

Sinharay, S. and Stern, H. (2003) Posterior predictive model checking in hierarchical models. *Journal of Statistical Planning and Inference*, **11**, 209–221.

Smith, A. and Gelfand, A. (1992) Bayesian statistics without tears: a sampling-resampling perspective. *American Statistician*, **46**, 84–88.

Song, X. and Lee, S. (2002) A Bayesian model selection method with applications. *Computational Statistics and Data Analysis*, **40**, 539–557.

Spiegelhalter, D., Best, N., Carlin, B. and van der Linde, A. (2002) Bayesian measures of model complexity and fit. *Journal of the Royal Statistical Society, Series B*, **64**, 583–639.

Tierney, L. and Kadane, J. (1986) Accurate approximations for posterior moments and marginal densities. *Journal of the American Statistical Association*, **81**, 82–86.

Trevisani, M. and Gelfand, A. (2003) Inequalities between expected marginal log likelihoods with implications for likelihood-based model comparison. *Canadian Journal of Statistics*, **31**, 239–250.

Vernardinelli, I. and Wasserman, L. (1991) Bayesian analysis of outlier problems using the Gibbs sampler. *Statistics and Computing*, **1**, 105–117.

Wasserman, L. (2000) Bayesian model selection and model averaging. *Journal of Mathematical Psychology*, **44**, 92–107.

Weiss, R. (1994) Pediatric pain, predictive inference and sensitivity analysis. *Evaluation Review*, **18**, 651–678.

West, M. (1984) Outlier models and prior distributions in Bayesian linear regression. *Journal of the Royal Statistical Society, Series B*, **46**, 431–439.

Wintle, B., McCarthy, M., Volinsky, C. and Kavanagh, R. (2003) The use of Bayesian model averaging to better represent uncertainty in ecological models. *Conservation Biology*, **17**, 1579–1590.

CHAPTER 3

Regression for Metric Outcomes

3.1 INTRODUCTION: PRIORS FOR THE LINEAR REGRESSION MODEL

Regression models seek to quantify the relationship between two or more variables, with the simplest framework assuming a one-way dependence between a response or dependent variable y and one or more independent variables. The latter are taken to be observed without error. Then conditional on X of dimension p, the normal linear regression model assumes

$$y_i \sim N(\mu_i, \sigma^2)$$
$$\mu_i = \beta_0 + \beta_1 x_{i1} + \beta_2 x_{i2} + \cdots + \beta_p x_{ip}$$

or equivalently

$$y_i = \beta_0 + \beta_1 x_{i1} + \beta_2 x_{i2} + \cdots + \beta_p x_{ip} + \varepsilon_i$$
$$\varepsilon_i \sim N(0, \sigma^2) \tag{3.1}$$

The function predicting y is linear in X, and the conditional variance $V(y_i|X_i) = V(\varepsilon_i)$ is constant regardless of the values of y_i, $X_i = (x_{i1}, x_{i2}, \ldots, x_{ip}\}$ or other information. Analytic results may be obtained for the posterior densities of the parameters of the normal linear model without needing to involve MCMC sampling. These often involve reference priors, such as a uniform prior on $\{\beta, \log \sigma\}$ where $\beta =$

Bayesian Models for Categorical Data   P. Congdon
© 2005 John Wiley & Sons, Ltd

$(\beta_0, \beta_1, \ldots, \beta_p\}$. Equivalently (Lee, 1997; Gelman *et al.*, 2003) this prior may be expressed

$$p(\beta, \sigma^2) \propto \sigma^{-2}$$

under which the conditional posterior for β is

$$\beta|\sigma^2, y, X \sim N(\tilde{\beta}, \sigma^2(X'X)^{-1})$$

with $\tilde{\beta} = (X'X)^{-1}X'y$, the usual ordinary least squares estimate. The conditional posterior for $\tau = \sigma^{-2}$ is a gamma density[1]

$$\tau|\beta, y, X \sim Ga\left(0.5(n - p - 1), 0.5\sum e_i^2\right)$$

where e_i is the estimated error, $e_i = y_i - \beta X_i = y_i - \mu_i$.

Prior interdependence between β and σ^2 of the form $p(\beta, \sigma^2) = p(\beta|\sigma^2)p(\sigma^2)$ may be expressed in the conjugate multivariate normal prior of dimension $p + 1$ with prior mean m_β and covariance $\sigma^2 V$. Among options proposed for V are $V = g(X'X)^{-1}$ where g is taken large to guarantee that the prior contains relatively little information as compared with that in the data; for example, g between 10 and 100 (Smith and Kohn, 1996) or $g = \max([p + 1]^2, n)$ as in Fernandez *et al.* (2001a, 2001b). The prior on the precision $\tau = \sigma^{-2}$ is taken as $\tau \sim Ga(a, b)$. There is debate about the appropriate values for $\{a, b\}$ or whether the prior should be set on some transformation of τ, such as a uniform prior on $\log(\tau)$ or σ (Gelman *et al.*, 2003), in which case conjugacy no longer applies. Independent priors on β and σ^2 as in $p(\beta, \sigma^2) = p(\beta)p(\sigma^2)$, such as $\beta \sim N_{p+1}(B_0, T_0^{-1})$ and $\tau \sim Ga(a, b)$ lead to a posterior density $p(\beta|\sigma^2, y, X)$ with variance $\Sigma_\beta = (T_0 + \tau X'X)^{-1}$ and mean $\Sigma_\beta(T_0 B_0 + \tau X'y)$.

Another approach to specifying priors in the normal linear model (and other types of general linear model) is to set them in such a way that the Bayes factor will be fairly insensitive to changes in the prior. As noted in Chapter 2, the Bayes factor is the formal method of Bayesian model choice but may be sensitive to prior specification, especially to alternative diffuse priors. Raftery *et al.* (1996) propose proper priors for GLM regression models, based on the data, that are relatively flat over a range

---

[1] The WINBUGS code for this model when $p = 1$ (i.e. intercept and one predictor) is

```
model { for (i in 1:n) { y[i] ~ dnorm(mu[i] ,tau);     e[i] <- y[i] -mu[i];
                         mu[i] <- a + b* (x[i] -mean(x[] ))}
A1 <- 0.5* (n - p - 1);   A2 <- 0.5* inprod(e[] ,e[] )
a ~ dflat()               b ~ dflat()
tau ~ dgamma(A1,A2)       sig2 <- 1/tau}
```

of plausible range of values for β_j; see also Hoeting *et al.* (2002). Their approach is related to the g priors of Zellner (1986) that are used in Fernandez *et al.* (2001b). Thus the priors on each β_j are independent, with $\beta_0 \sim N(\tilde{\beta}_0, s_y^2)$ and $\beta_j \sim N(0, \phi^2/s_j^2)$ $(j = 1, \ldots, p)$, where s_y^2 is the observed variance of Y and s_j^2 the variance of X_j. The prior on τ has the form $\tau \sim Ga(0.5\nu, 0.5\nu\lambda)$ with expected precision $1/\lambda$. Raftery *et al.* establish default values $\lambda = 0.28$, $\nu = 2.58$ and $\phi = 2.85$.

The model in (3.1) is appropriate if the assumptions of linearity, normal errors and constant variance hold, at least approximately. It also assumes the effect of predictors is the same across all subjects. In practice departures such as outlier points, non-linear effects of predictors, non-constant error variances or heavier tailed errors than normal will suggest modified models. Another scenario is when regression effects differ between subjects or subsets of subjects (local regression effects), suggesting perhaps a discrete regression mixture. Even in the linear model there may be collinearity between predictors such that a parsimonious model is required that selects only a subset of the available predictors. Because of the wide variety of possible model features, as well as alternative priors, selection among models becomes necessary.

Example 3.1 Radiata pines Here we consider linear model selection using Bayes factors. Carlin and Chib (1995) and others have investigated prediction of Y, strength of radiata pines $(n = 42)$, using either x = density or w = adjusted density in a simple linear regression (3.1). So under M_1, $p = 1$ with x as predictor and under M_2, $p = 1$ with w as predictor, and the models are linear regressions

$$M_1 \qquad y_i = \alpha_1 + \beta_1 x_i + \varepsilon_{i1}$$
$$M_2 \qquad y_i = \alpha_2 + \beta_2 w_i + \varepsilon_{i2}$$

with $\varepsilon_{ik} \sim N(0, 1/\tau_k)$. Carlin and Chib obtain a Bayes factor of 4420 in favour of M_2 by allowing jumps between these models and using pseudo-priors on θ_2 when model 1 is selected (and vice versa) so that MCMC conditions are satisfied – see Congdon (2001, section 10.4).

Here the first code in Program 3.1 is used to obtain the pseudo marginal likelihood (2.16). $N(3000, 10^6)$ priors are assumed for $\{\alpha_1, \alpha_2\}$, $N(185, 10^4)$ priors for $\{\beta_1, \beta_2\}$, and $Ga(1, 0.0000111)$ priors for $\{\tau_1, \tau_2\}$ (essentially the same priors as used by Carlin and Chib and in several other analyses of these data). A two-chain run of 10 000 iterations is taken with default initial values in one run and values from a pilot run in the other. This provides (with convergence at under 500 iterations in both models) pseudo marginal likelihoods of -306.9 and -298.1 on models 1 and 2.

This leads to a log (to base e) of the pseudo Bayes factor of 8.84 (see section 3.3.1). Spreadsheet3_1.xls shows calculation of this likelihood for M_1 and a suggested exercise is to obtain it for M_2. One may note that the scaled CPO for observation 41 under M_1 is 0.01, as Y is overpredicted; under M_2 the scaled CPO for this observation is 0.26, and so is no longer an outlier.

The path sampling method described in Chapter 2 is applied with 41 equally spaced points between 0 and 1 ($s_0 = 0$, $s_1 = 0.025$, etc., to $s_{40} = 1$) and is then applied using code B in Program 3.1. Applying this requires one to set a prior on 41 values of α, β_1, β_2 and τ in the path model specified in (2.8). This gives $\log(\text{BF}) = 8.51$. So model 2 is conclusively preferred again.

A Monte Carlo parallel sampling approach is then applied using the approximate model probabilities in (2.30)–(2.31). Prior model probabilities $P(M_1) = 0.9995$ and $P(M_2) = 0.0005$ are assumed to compensate for the low support for model 1. Convergence in the separate models is reached early (at under 1000 iterations) in a two-chain run of 20 000 iterations for each model, and with divergent starting values on θ_1 and θ_2 where $\theta_k = (\alpha_k, \beta_k, \tau_k)$. With $T = 19\,000$, average weights $\bar{w}_1$ and $\bar{w}_2$ are obtained as 0.34 and 0.66. Taking account of the unequal prior model probabilities, the estimated Bayes factor is 3880 and $\log(\text{BF}) = 8.26$.

The WINBUGS code for this analysis is as follows:

```
model { # Centre Predictors
  for (i in 1:42) { cx[i] <- X[i] -mean(X[] ); cw[i] <- W[i] -mean(W[])
  # repeat data for two models
  for (j in 1:2) { Y[i,j] <- y[i]}}
  # sampling model
   for (i in 1:42) { for (j in 1:2) { Y[i,j] ~ dnorm(mu[i,j],tau[j] )
  # log LKD contribution
  LL[i,j] <- -0.5*tau[j]*pow(y[i] -mu[i,j],2) + 0.5*log(tau[j]/6.2832) }
  # means for M1 and M2
       mu[i,1] <- a[ 1] + b[1]*cx[i];     mu[i,2] <- a[2] + b[2]*cw[i]}
  # scaled likelihoods and model probabilities w[ k] at each iteration
  for (k in 1:2) { L[k] <- sum(LL[,k] )+sum(logPrior[1:3,k] )+log(Prior-
  Mod[k] )
       SL[k] <- max(L[k] -maxL,-700)
       expSL[k] <- exp(SL[k] );
  # model weights
  w[k] <- expSL[k] /sum(expSL[] )}
  # maximum of model likelihoods
       maxL <- ranked(L[] ,2)
  # priors
```

```
for (j in 1:2) { a[j] ~dnorm(M[1],T[1]); b[j] ~dnorm(M[2],T[2]); tau[j]
~ dgamma(r,m)
# prior ordinates
    logPrior[1,j] <- 0.5*log(T[1]/6.28)-0.5*T[1]*pow(a[j]-M[1],2)
    logPrior[2,j] <- 0.5*log(T[2]/6.28)-0.5*T[2]*pow(b[j]-M[2],2)
    logPrior[3,j] <- log(pow(tau[j],r-1)*exp(-m*tau[j])*
pow(m,r))-loggam(r)}}
```

Initial values are given in the list vectors:

list(a=c(0,0),b=c(0,0),tauc(1,1))

list(a=c(3000,3000),b=c(200,200),tau=c(0.001,0.001))

3.2 REGRESSION MODEL CHOICE AND AVERAGING BASED ON PREDICTOR SELECTION

While approaches based on the penalized fit (e.g. Burnham and Anderson, 2002), or on predictions (e.g. Laud and Ibrahim, 1995), or on marginal likelihood approximations have been applied to regression model selection, more directed methods have been proposed for particular model choice questions occurring in regression. Variable selection is indicated when predictors $X_1, X_2, \ldots, X_p$ are highly correlated (approximately collinear) and a subset of predictors is required in order to obtain a precisely identified model. With correlated predictors, coefficients on particular X_j in a model including all predictors may be reduced to 'insignificance' (a 95% interval straddling the null zero value), or take signs opposite to what is expected on substantive grounds.

The simplest choice schemes specify priors on binary indicators $\delta_j, j = 1, \ldots, p$ relating to the inclusion ($\delta_j = 1$) or exclusion ($\delta_j = 0$) of the jth predictor. Thus the linear regression model becomes

$$y_i = \beta_0 + \delta_1\beta_1 x_{i1} + \delta_2\beta_2 x_{i2} + \cdots + \delta_p\beta_p x_{ip} + \varepsilon_i \qquad (3.2)$$

The prior probability $\pi_j = P(\delta_j = 1)$ may be preset. Typically, the value $\pi_j = 0.5$ is assumed, ensuring equal probabilities for the 2^p possible models (assuming the intercept is included by default). Gibbs sampling is based on the conditionals

$$\beta|\delta, Y, \sigma^2, \quad \sigma^2|\beta, Y, \quad \delta_j|\delta_{[-j]}, \beta$$

where $\delta_{[-j]} = \{\delta_1, \ldots, \delta_{j-1}, \delta_{j+1}, \ldots, \delta_p\}$. From a long run the marginal posterior probabilities that $\delta_j = 1$ (i.e. that X_j be included) are obtained. Also obtained are posterior probabilities on each of the $J = 2^p$ regression models. This approach assumes a common prior for σ^2 and β_0 even though these parameters would in fact differ across the J models if each were fitted separately.

Specifically, a run of T samples will visit a particular model at iteration t as defined by that subset of $\{\delta_1^{(t)}, \dots, \delta_p^{(t)}\}$ which are ones. The probability that $\beta_j \neq 0$ is then obtained as $1/T$ times the number of iterations where $\delta_j = 1$ or equivalently as the number of models visited that contain X_j. If models $M_1, \dots, M_J$ are visited $T_1, \dots, T_J$ times, where $T = \sum_j T_j$, then posterior model probabilities are estimated as

$$P(M_j|Y) = T_j/T$$

Fernandez et al. (2001b) and Koop (2003, chapter 11) compare the posterior model inferences under variable selection with formal Bayes factor methodology.

An equivalent procedure selects the model from a multinomial probability vector with equal prior probabilities $1/2^p$, where model 1 might be the intercept only, model 2 a model with X_1 only, model 3 a model with X_2 only and so on till the final model M_J including all predictors. A particular form of model selection used when p is large is the MC^3 algorithm of Madigan and York (1995). Under this a Metropolis step is used under which the chain moves from the current model M_c to new model M_n with probability

$$p = \min\{1, [P(Y|M_n)P(M_n)]/[P(Y|M_c)P(M_c)]\}$$

where $P(Y|M_n)$ is the marginal likelihood of model n. The chain remains at M_c with probability $1 - p$. To reduce the range of new models M_n may be confined to models with one fewer or one more predictor than M_c.

If the object is to choose the best model then a second run might be confined to those predictors X_j with posterior probabilities π_j greater than 0.5. If the object is rather to allow for model uncertainty (i.e. to estimate parameters under a model averaging perspective) then the focus is on the products $\theta_j = \delta_j\beta_j$ in (3.2). MCMC sampling will average over iterations when $\delta_j = 0$ as well as iterations where a model including X_j is chosen. For example, if the posterior density estimates of θ_j have 95% intervals straddling zero, this would indicate that X_j is not important in predicting Y.

Selection may be extended to multiparameter coefficients where the number of selection indicators δ_k is less than the number of predictors. For example, in log-linear regression for cross-classified counts y_{ijk} $(i = 1, \dots, I;\ j = 1, \dots, J;\ k = 1, \dots, K)$, there would typically be a single inclusion indicator for the totality of $(I - 1)(J - 1)$ parameters $u_{12(ij)}$ describing second-order interactions. One might also apply this idea to standard linear regression for metric y with a single γ applying to the inclusion of a subset of predictors (e.g. for $p = 5$, $\gamma_1 = 1$ if $\{X_1, X_3\}$ are to be included and $\gamma_2 = 1$ if $\{X_2, X_4, X_5\}$ are to be included). In this case prior probabilities π_j would be $(0.4, 0.6)$ because the subsets are of different size. One might also extend the selection approach to sets of random effects (e.g. $\delta = 1$ if random three-way effects $u_{123(ijk)}$ are to be included in a log-linear model for counts y_{ijk}).

George and McCullough (1993, 1997) develop a stochastic search variable selection (SSVS) method based on a mixture prior

$$P(\beta_j|\delta_j) = \delta_j P(\beta_j|\delta_j = 1) + (1 - \delta_j)P(\beta_j|\delta_j = 0)$$

whereby β_j has a vague prior centred at zero when $\delta_j = 1$. However, when $\delta_j = 0$ is selected, the prior is centred at zero (or some other value close to zero) with high precision. For instance, taking a univariate normal prior for $\beta_j|\delta_j = 1$

$$\beta_j|\delta_j = 1 \sim N(0, V_{1j})$$

one might assume that V_{1j} is large (leading to a diffuse prior) but multiply V_{1j} by a small (near zero) constant c_j, $V_{0j} = c_j V_{1j}$ such that the prior

$$\beta_j|\delta_j = 0 \sim N(0, V_{0j})$$

is closely constrained around zero. c_j should be chosen so that the range of β_j under $P(\beta_j|\delta_j = 0)$ is confined to substantively insignificant values. Ntzoufras (1999) recommends applying an 'automatic scaling' for all j (e.g. $c_j = c = 0.01$), possibly with calibration of c to assist the chain to move better between models.

In the approaches of Kuo and Mallick (1998) and Smith and Kohn (1996) the selection indicators δ_j and coefficients β_j are independent rather than being governed by mixture priors as in the SVSS approach, such that $\delta_j = 0$ if $\beta_j = 0$, and $\delta_j = 1$ if $\beta_j \neq 0$. Kuo and Mallick suggest independent normal priors on the β_j, though not overly diffuse, e.g. $\beta_j \sim N(0, \phi_j)$ where $0.4 \leq \phi_j \leq 16$. The prior on β may also be specified as a form of g prior as $N_{p+1}(0, g\sigma^2(XX)^{-1})$ where g is a known constant. In the linear regression model (3.2) let X_δ be the predictor matrix corresponding to those δ_j selected as 1, and set

$$S(\delta) = Y'Y - [g/(1 + g)]Y'X_\delta(X_\delta X_\delta)^{-1}X_\delta Y \qquad (3.3)$$

The conditional probability for δ_j given all other indicators $\delta_{[-j]}$ under this prior is then

$$P(\delta_j|\delta_{[-j]}, Y) \propto \pi_j^{\delta_j}(1 - \pi_j)^{1-\delta_j}(1 + g)^{-p/2}S(\delta) - 0.5n \qquad (3.4)$$

and the conditional probabilities $P(\delta_j = 1 | \delta_{[-j]}, Y)$ are obtained by setting $\delta_j = 1$ in (3.4) and normalizing.

Finally the parallel sampling approach (see section 2.7) may be used in forward selection. Consider a normal linear regression with p potential predictors apart from the intercept. An algorithm based on parallel sampling of models (see Example 3.3) may be stated as follows:

1. To obtain the first predictor to be entered into the model, select from $J = p + 1$ parallel models defined by

$$y_i = \alpha_j + \beta_j x_{ij} + e_{ij} \qquad e_{ij} \sim N(0, \phi_j), \quad j = 1, \ldots, p$$
$$y_i = \alpha_{p+1} + e_{i,p+1} \qquad e_{i,p+1} \sim N(0, \phi_{p+1})$$

Obtain the weights (2.19)–(2.21) and (2.33), the comparison probabilities (2.22) and the best model indicator B_j as in (2.23) or (2.24). If model $j_1 \leq p$ with predictor x_{ij_1} is preferred according to these criteria go to step 2, while if model $M = p + 1$ is preferred then stop. Thus if $Q(p+1, j) > 0.5$ for all $j < p + 1$, stopping is justified because a lower dimension model has a probability over 0.5 of giving better predictions than all competing higher dimension models.

2. Select from $J = p$ parallel models, with the first $p - 1$ models being

$$y_i = \alpha_j + \gamma_j x_{ij_1} + \beta_j x_{ik} + e_{ij} \qquad e_{ij} \sim N(0, \phi_j)$$

where $j = 1, \ldots, p - 1$ as k varies between 1 and p excluding $k = j_1$. The lower dimension model is

$$y_i = \alpha_p + \gamma_p x_{ij_1} + e_{ip} \qquad e_{ip} \sim N(0, \phi_p)$$

If model $j (j \leq p - 1)$ is selected via predictive criteria with predictor x_{ij_2} go to 3, or if model p is selected then stop.

3. Select from $J = p - 1$ parallel models, with $p - 2$ new models

$$y_i = \alpha_j + \gamma_{1j} x_{ij_1} + \gamma_{2j} x_{ij_2} + \beta_j x_{ik} + e_{ij} \qquad e_{ij} \sim N(0, \phi_j)$$

where $k \neq \{j_1, j_2\}$. Model $p - 1$ is

$$y_i = \alpha_{p-1} + \gamma_{1,p-1} x_{ij_1} + \gamma_{2,p-1} x_{ij_2} + e_{i,p-1} \qquad e_{i,p-1} \sim N(0, \phi_p)$$

Then select model $j (j \leq p - 2)$ and extra predictor j_3 via the criteria used in step 1, or if model $p - 1$ is selected then stop. This process continues until the majority of weights are higher for model J, and $Q(J, k)$ exceeds 0.5 for $k < J$.

There may be a conflict between choice based on likelihood weights (which are not penalized by a parameter count) and the other weights which are so penalized. The AIC weights may also tend to select more heavily parameterized models than BIC weights and model probability weights (2.33). The latter may tend (like the formal Bayes factor) to select relatively simple models.

Example 3.2   Electricity usage   Chatterjee *et al.* (1995, p 292) present data on electricity usage ($y = \log$ kilowatt-hours) by month for a house in Westchester, New York. The predictor variables are powers of average monthly Fahrenheit temperature. The choice is between combinations of a linear term in temperature (X_1), a quadratic term (X_2) and a cubic term (X_3). These three variables are standardized. Note that X_1 and X_2 are highly correlated (0.993) and X_1 is correlated slightly less with X_3 (0.975).

Fit is assessed by the predictive criteria (2.10) and (2.12), by posterior model probabilities from parallel model sampling, and by indicator/ model selection as in section 3.2. The eight models being compared are $\{1\}$, $\{1 + X_1\}$, $\{1 + X_2\}$, $\{1 + X_3\}$, $\{1 + X_1 + X_2\}$, $\{1 + X_2 + X_3\}$, $\{1 + X_1 + X_3\}$ and $\{1 + X_1 + X_2 + X_3\}$. The precisions in the normal priors for the coefficients on X_1 to X_3 are 10 000, while the prior for the intercept is set at N(43, 10 000) where 43 is the OLS intercept in $\{X_0, X_1, X_2, X_3\}$. The criterion (2.10), which is equivalent to (2.12) when k is very large, prefers model 5, namely $\{1 + X_1 + X_2\}$. The three lowest values of the criterion, respectively 10 098, 10 213 and 10 224, are for model 5, model 7 and model 8. Taking $k = 1$ in (2.12) still leaves model 5 rated as the best model.

The approximate posterior model probabilities from parallel model sampling (section 2.8) are for models 5, 7 and 8, namely 0.45, 0.43 and 0.002, so the model with $p = 3$ receives much less weight. This is based on two chains of 10 000 iterations and a 1000 burn-in. Note that whereas the coefficients on powers of temperature in both 5 and 7 are well identified, the coefficients β_2 and β_3 in model 8 are less well identified. There is also a probability of 0.10 on the two-predictor model 6, namely $1 + X_2 + X_3$.

With binary selection indicators on the β_j as in (3.2) the respective probabilities on models 5 to 8 are 0.488, 0.056, 0.278 and 0.178. The models corresponding to various combinations of $\delta_j = 1$ are obtained by defining a binary model indicator m_j (i.e. $m_j = 1$ when model j is selected)

$$m_j = I(\delta_1 = A_{j1})I(\delta_2 = A_{j2})I(\delta_3 = A_{j3})$$

where A_{jk} is one when model j includes variable k. The direct model selection approach, selecting j as a categorical variable from a multinomial vector with prior probabilities 1/16, yields similar probabilities on models 5 to 8 of 0.477, 0.071, 0.268 and 0.183.

Example 3.3 Malnutrition in cystic fibrosis Altman (1991) and Everitt and Rabe-Hesketh (2001) consider $n = 25$ observations on continuous measures of malnutrition in cystic fibrosis patients. There are nine possible predictors apart from $x_{i0} = 1$, $i = 1, \ldots, n$ (see Table 3.1).

Table 3.1 Predicting malnutrition in cystic fibrosis patients

y	x_1 Age	x_2 Sex	x_3 Height	x_4 Weight	x_5 BMP	x_6 FEV	x_7 RV	x_8 FRC	x_9 TLC
95	7	0	109	13.1	68	32	258	183	137
85	7	1	112	12.9	65	19	449	245	134
100	8	0	124	14.1	64	22	441	268	147
85	8	1	125	16.2	67	41	234	146	124
95	8	0	127	21.5	93	52	202	131	104
80	9	0	130	17.5	68	44	308	155	118
65	11	1	139	30.7	89	28	305	179	119
110	12	1	150	28.4	69	18	369	198	103
70	12	0	146	25.1	67	24	312	194	128
95	13	1	155	31.5	68	23	413	225	136
110	13	0	156	39.9	89	39	206	142	95
90	14	1	153	42.1	90	26	253	191	121
100	14	0	160	45.6	93	45	174	139	108
80	15	1	158	51.2	93	45	158	124	90
134	16	1	160	35.9	66	31	302	133	101
134	17	1	153	34.8	70	29	204	118	120
165	17	0	174	44.7	70	49	187	104	103
120	17	1	176	60.1	92	29	188	129	130
130	17	0	171	43.6	69	38	172	130	103
85	19	1	156	37.2	72	21	216	119	81
85	19	0	174	54.6	86	37	184	118	101
160	20	0	178	64	86	34	225	148	135
165	23	0	180	73.8	97	57	171	108	98
95	23	0	175	51.1	71	33	224	131	112
195	23	0	179	71.5	95	52	225	127	101

Sex: = 0 for males
BMP Basic Metabolic Panel
FEV Forced Expiratory Volume
RV Residual Volume
FRC Functional Residual Capacity
TLC Total Lung Capacity

Non-informative $N(0, 10^4)$ priors are adopted on the regression coefficients in a normal errors linear regression, with all continuous predictors being centred. Selection is based on likelihood, AIC, BIC and Bayesian model weights (2.19), (2.21) and (2.33), the predictive comparison probability $Q(J, k)$, $k < J$, and the best model indicator B_j in (2.24). These are compared with the posterior average $C_m^{0.5}$ in (2.10), as used by Laud and Ibrahim (1995). The model with weight, BMP, FEV and RV is selected by classical backward and forward selection methods (Everitt and Rabe-Hesketh, 2001, p 194).

At the first selection stage $(J = 10)$, nine models containing just one predictor and an intercept (e.g. model 9 is $1 + X_9$) are compared with each other and to a tenth model containing just an intercept. A two-chain run of 10 000 iterations shows that all criteria prefer the model $1 + X_4$.

At the second stage $(J = 9)$ there are eight new models all containing an intercept and X_4, with the extra variable being just one of the eight remaining predictors. So models 1 to 8 are $1 + X_4 + X_j$ $(j = 1, 2, 3, 5, 6, 7, 8, 9)$ and model 9 is $1 + X_4$. The predictive probability estimates $\{Q(9, k), k = 1, 8\}$ obtained at this stage (Table 3.2) show that five higher dimension models appear to provide no gain over $1 + X_4$ and are overparameterized. The likelihood and AIC prefer $1 + X_4 + X_5$, i.e. BMP combined with weight, while the BIC weights and Bayes model probability estimates prefer the lower dimension model $1 + X_4$ and would not proceed further.

If another stage with $J = 8$ is pursued, there are seven new models including new predictors in addition to X_4 and X_5. Only one model $1 + X_4 + X_5 + X_6$ improves on $1 + X_4 + X_5$ according to the probabilities $Q(J, k)$, where $J = 8$. Hence the selected model involves $x_6 = $ FEV, $x_5 = $ BMP and $x_4 = $ weight. In fact the more complex model $1 + X_4 + X_5 + X_6$ is preferred by the BIC weight though (2.31) still prefers the lower dimension model.

As an illustration of the approximation (2.29) we monitor ε_{8k} and find the variances to range between 4.66 and 5.02, or combined complexity 9.3 to 10. This is slightly higher than the nominal total of 9, though the approximation may be affected by the small sample.

At the fourth stage $(J = 7)$ the lower dimension seventh model $1 + X_4 + X_5 + X_6$ is preferred by the BIC weight and model probability estimate. The AIC weight is indecisive between $1 + X_4 + X_5 + X_6$ and $1 + X_4 + X_5 + X_6 + X_7$, while the likelihood weight and predictive criteria prefer the larger model. This is consistent with the choice via classical forward selection. The fifth stage shows $Q(J, k) > 0.5$ where model J is $1 + X_4 + X_5 + X_6 + X_7$, so selection stops by all criteria.

Table 3.2   Cystic fibrosis data, forward selection diagnostics

Stage	Model	$C^{1/2}$	B	Likel'd	AIC	BIC	BMW	$Q(J, k)$
				Model weights				
1	1. $1 + X_1$	188.5	0.258	0.287	0.286	0.285	0.245	0.146
	2. $1 + X_2$	227.9	0	0.003	0.003	0.003	0.001	0.460
	3. $1 + X_3$	190.7	0.143	0.223	0.222	0.221	0.196	0.160
	4. $1 + X_4$	184.5	0.599	0.452	0.451	0.449	0.372	0.125
	5. $1 + X_5$	231.5	0	0.002	0.002	0.002	0.002	0.485
	6. $1 + X_6$	213.9	0	0.017	0.017	0.017	0.013	0.322
	7. $1 + X_7$	226.9	0	0.004	0.004	0.004	0.001	0.444
	8. $1 + X_8$	217.7	0	0.011	0.010	0.010	0.003	0.357
	9. $1 + X_9$	235.9	0	0.001	0.001	0.001	0.001	0.522
	10. 1	234.0	0	0.001	0.004	0.007	0.165	
2	1. $1 + X_4 + X_1$	186.9	0.020	0.076	0.068	0.059	0.009	0.529
	2. $1 + X_4 + X_2$	183.9	0.073	0.120	0.102	0.089	0.012	0.497
	3. $1 + X_4 + X_3$	187.0	0.013	0.069	0.065	0.057	0.009	0.529
	4. $1 + X_4 + X_5$	177.2	0.413	0.270	0.233	0.206	0.021	0.413
	5. $1 + X_4 + X_6$	183.4	0.105	0.140	0.118	0.103	0.017	0.484
	6. $1 + X_4 + X_7$	185.9	0.024	0.080	0.075	0.065	0.010	0.513
	7. $1 + X_4 + X_8$	188.3	0.012	0.069	0.060	0.052	0.007	0.529
	8. $1 + X_4 + X_9$	185.6	0.025	0.082	0.073	0.064	0.010	0.520
	9. $1 + X_4$	184.5	0.315	0.094	0.206	0.305	0.906	
3	1. $1 + X_4 + X_5 + X_1$	178.9	0.029	0.087	0.076	0.067	0.008	0.525
	2. $1 + X_4 + X_5 + X_2$	176.9	0.077	0.117	0.103	0.091	0.015	0.509
	3. $1 + X_4 + X_5 + X_3$	181	0.044	0.078	0.069	0.061	0.008	0.539
	4. $1 + X_4 + X_5 + X_6$	164.1	0.611	0.438	0.400	0.365	0.076	0.382
	5. $1 + X_4 + X_5 + X_7$	181.6	0.010	0.061	0.053	0.046	0.009	0.538
	6. $1 + X_4 + X_5 + X_8$	180.2	0.020	0.064	0.056	0.049	0.008	0.538
	7. $1 + X_4 + X_5 + X_9$	179.1	0.010	0.068	0.060	0.053	0.012	0.535
	8. $1 + X_4 + X_5$	177.2	0.200	0.088	0.184	0.268	0.864	0.520
4	1. $1 + X_4 + X_5 + X_6 + X_1$	164.8	0.084	0.124	0.108	0.095	0.012	0.506
	2. $1 + X_4 + X_5 + X_6 + X_2$	168.6	0.027	0.078	0.067	0.059	0.010	0.545
	3. $1 + X_4 + X_5 + X_6 + X_3$	166.5	0.058	0.100	0.087	0.076	0.010	0.526
	4. $1 + X_4 + X_5 + X_6 + X_7$	158.4	0.272	0.250	0.221	0.196	0.057	0.432
	5. $1 + X_4 + X_5 + X_6 + X_8$	161.3	0.145	0.171	0.149	0.132	0.034	0.467
	6. $1 + X_4 + X_5 + X_6 + X_9$	162.2	0.134	0.162	0.141	0.125	0.033	0.475
	7. $1 + X_4 + X_5 + X_6$	164.1	0.280	0.115	0.227	0.318	0.845	
5	1. $1 + X_1 + X_4 + X_5 + X_6 + X_7$	160.8	0.119	0.165	0.136	0.115	0.018	0.525
	2. $1 + X_2 + X_4 + X_5 + X_6 + X_7$	162.3	0.083	0.139	0.114	0.097	0.018	0.540
	3. $1 + X_3 + X_4 + X_5 + X_6 + X_7$	159.6	0.122	0.170	0.140	0.119	0.018	0.512
	4. $1 + X_4 + X_5 + X_6 + X_7 + X_8$	162.1	0.093	0.143	0.118	0.100	0.019	0.535
	5. $1 + X_4 + X_5 + X_6 + X_7 + X_9$	160.2	0.134	0.179	0.147	0.126	0.024	0.516
	6. $1 + X_4 + X_5 + X_6 + X_7$	158.4	0.449	0.204	0.345	0.443	0.902	

3.3   ROBUST REGRESSION METHODS: MODELS FOR OUTLIERS

As discussed in Chapter 1, many data sets present complex features that militate against the application of a single parametric density, such as a normal or any other density for continuous data (e.g. lognormal, gamma, exponential) serving to summarize adequately all the data. Observations may show multimodality, or contain outlier points, either isolated or clustered (Justel and Pena, 2001). In a regression model where variability in y_i is related to predictors x_i, multimodality will often have a substantive equivalent, e.g. heterogeneity over subpopulations which have different regression relationships. Thus different models may hold in subsets of the data, so increasing the uncertainty in predicting new observations on the basis of the sampled observations.

Bayesian methods for robust regression fall into two broad sets of procedures. Outlier diagnostic methods seek a central model and control for the potentially distorting effect on estimates of that model of outlying or influential observations with a low probability of being generated by it. Model extension methods seek to model more directly the source of aberrant observations, for instance by adopting a discrete mixture or by replacing a single normal density by a scale mixture (i.e. an error distribution with heavier tails). Preliminary detection of aberrant or influential points may be undertaken with criteria (e.g. CPOs) considered in section 2.4. Thus for normal linear regression with n observations and p predictors, and under flat priors, Pettit (1990) shows that the conditional predictive ordinate is proportional to the standardized residual test for detecting outliers. Thus

$$P(y_i|y_{[i]}) \propto \sqrt{1 - h_i}\left(1 + \frac{1 + r_i^2}{n - p - 1}\right)/s_{[i]}$$

where $s_{[i]}^2$ is the residual variance when observation i is excluded, h_i is the leverage of observation i, namely the ith diagonal element of $H = X(X'X)^{-1}X'$, and r_i is the studentized residual

$$r_i = (y_i - X_i\beta)/[\sqrt{1 - h_i}/s_{[i]}]$$

Observations with large studentized residuals will have a small CPO and be detected as outliers, while observations with high leverage will also have small conditional predictive ordinates, regardless of whether they are outliers, since then h_i tends to one.

The diagnostic approach is illustrated by a contaminated normal regression model, in which each observation is a potential outlier with a low probability ω and outliers have shifted location and or variances (Verdinelli and Wasserman, 1991). A shifted location model means the density of y is then

$$f(y_i|\mu, \sigma^2, \omega, A) = (1 - \omega)\phi(y_i|\mu, \sigma^2) + \omega\phi(y_i|\mu + A_i, \sigma^2)$$

where ϕ is the normal density, and $A = (A_1, \ldots, A_n)$ is the shift in location for an outlier. An inflated variance model is

$$f(y_i|\mu, \sigma^2, \omega, \kappa) = (1 - \omega)\phi(y_i|\mu, \sigma^2) + \omega\phi(y_i|\mu, \kappa\sigma^2)$$

where $\kappa > 0$. Often κ is preset, say at 5 or 10, since ω and κ may be difficult to identify if both are unknowns. Computation and inferences are aided by adding latent variables $\delta_i \sim \text{Bern}(\omega)$ so that the location shift model becomes

$$y_i|\mu, \sigma^2, \omega, A \sim \text{N}(\mu + \delta_i A_i, \sigma^2)$$

If regressors are included then the shifted location model is

$$y_i|\sigma^2, \beta, \omega, A \sim (1 - \omega)\text{N}(y_i|X_i\beta, \sigma^2) + \omega\text{N}(y_i|X_i\beta + A_i, \sigma^2).$$

Model extension methods to mitigate against the impact of outliers or influential points on the central model include: (a) adopting heavier tail densities than the normal for the error distribution by scale mixing; and (b) discrete regression mixtures (section 3.7).

   The scale mixture of Andrews and Mallows (1974) augments the data in such a way as to produce an analysis equivalent to assuming Student t errors. Thus

$$y_i \sim \text{N}(\mu_i, \sigma^2/\lambda_i) \tag{3.5}$$

where $\lambda_i \sim \text{Ga}(0.5\nu, 0.5\nu)$. This is equivalent to assuming regression errors in (3.1) following the Student t density with ν degrees of freedom. Other mixture densities may be used (West, 1987). If $\lambda_i \sim \text{Exp}(0.5)$, i.e. an exponential density with mean 2, then the regression errors in (3.1) have a double-exponential density, another heavy-tailed alternative to the normal. Low values of λ_i for particular points (in relation to the average for all points) suggest observations that are only included in the model by inflating the observation-specific variance $V_i = \sigma^2/\lambda_i$ (West, 1984) and so suggest these points are outliers or not adequately represented by the model.

   While ν in the Student t option may be preset – for example, $\nu = 4$ is a robust option considered by Gelman et al. (1995, p 352) – a prior may also be adopted. Geweke (1993) points out problems with adopting diffuse priors for ν. Possibilities are discrete priors (Knorr-Held, 1999) with maximum possible ν equivalent to a de facto normal density (e.g.

$\nu = 100$) or an exponential density such as $\nu \sim E(0.1)$ as proposed by Fernandez and Steel (1998a) or Sahu *et al.* (2003). Example 3.4 considers the efficacy of such priors in recovering Student *t* parameters when data are generated under known parameters.

Example 3.4 Simulated regression with Student *t* errors To illustrate the operation of possible priors for the Student *t* scale mixture model 100 observations are generated from the model

$$y_i \sim N(\mu_i, \sigma^2/\lambda_i)$$
$$\mu_i = -2 + 0.5x_i$$
$$\lambda_i \sim Ga(5,5)$$

where the x_i are N(0,1) variates and $\sigma^2 = 0.2$; this is equivalent to a linear regression with Student *t* errors and $\nu = 10$ degrees of freedom. The goal is to re-estimate the model parameters $(\beta_0, \beta_1, \sigma^2, \nu)$ with the sampled y_i and x_i. The λ_i generated as above range from 0.16 to 2.3 with variance 0.17.

We consider three possible priors for ν. The first is a discrete prior over 81 possible values $\{2^k, k = -1, -0.9, -0.8, \ldots, 6.9, 7\}$ (Knorr-Held, 1999). The second is an exponential prior for $\nu/2$ with parameter κ, which is in turn obtained as the inverse of a parameter ω, uniformly distributed between 0 and 2. The third is an exponential prior with κ preset at 0.2 (equivalent to a mean on $\nu/2$ of 5).

Under the discrete prior, a two-chain run of 2500 iterations (convergent from 1500) shows ν overestimated at 49 and the variance of the λ_i understated at 0.08, whereas σ^2 is overstated at 0.26. Under the exponential prior with underlying uniform prior on $\omega = 1/\kappa$, the ν and σ^2 parameters are more accurately reproduced, with respective means in a two-chain run of 2500 iterations of 10.6 and 0.23. The variance of the λ_i stands at 0.20, closer to the actual 0.17, and the posterior means for $\{\beta_0, \beta_1\}$ are -1.92 (s.d. $= 0.05$) and 0.42 (0.05). Under the exponential prior with κ preset at 0.1, the variance of the λ_i is understated at 0.07 and the degrees of freedom overestimated at 38.

Example 3.5 Puromycin experiment To illustrate the use of scale mixing within a normal errors model as in (3.5), consider the non-linear regression model used by Carlin and Polson (1991) for data from an experiment in biochemical kinetics. The response is an enzymatic reaction (in counts/min^2) predicted as a function

$$\mu_i = \alpha + \beta x_i/(\gamma + x_i)$$

of a substrate concentration (in ppm). The enzyme has been treated with puromycin. The mixtures considered are a Student t with 2 degrees of freedom, equivalent to normal scale mixing with $\lambda_i \sim Ga(1, 1)$, and a double exponential (DE), namely scale mixing with $\lambda_i \sim Exp(0.5)$. One goal of the analysis is prediction at an unobserved x value, namely $x = 0.5$. To summarize model fit, the DIC and mean absolute deviation (MAD) between y_i and μ_i are used.

The double exponential has a DIC of 90 and an MAD of 92.5, with a possible discrepancy for observation 2 with $Y = 47$ but $\mu = 62$. This point has the lowest λ_i, namely 1.16, against an average of 2. The prediction at $x = 0.5$ has mean 188 with a 95% interval (169,205). The Student t mixture again points up observation 2, with $\lambda_i = 0.7$ (against an average of 1) and with $\mu_2 = 61$. This model has a worse DIC and MAD (respectively 93.7 and 95) than the double exponential.

Both the DIC and MAD may understate the impact on fit of the imprecision of parameter estimates under the double exponential. For example, Carlin and Polson (1991) adopt a different model choice method: they include a model node in their analysis and obtain a higher posterior probability $P(m = 1|y) = 0.73$ for the Student t as against $P(m = 2|y) = 0.27$ for the double exponential.

Here the second half of a two-chain run of 100 000 iterations using the parallel sampling approach with diverse starting values for both models gives posterior probabilities of 0.68 on the Student t mixture and 0.32 on the DE mixture – far from conclusive. Following Carlin and Polson (1991), one might be interested in parameter averaging (or averaging some other quantity, not itself a parameter) over the alternative models.

Thus consider the average of β over the Student t and DE models using iteration-by-iteration weights (2.31a). The densities of the two parameters and their average are summarized in Table 3.3.

It can be seen that the averaged parameter has greater precision (in terms of a narrower 95% interval) than the parameters from either model, with mean closer to the estimate from the Student t model. In many cases, though, model averaging may lead to greater uncertainty (less precision)

Table 3.3 Averages of β

	Mean	2.5%	Median	97.5%
β_1	193.9	160.2	192.0	237.0
β_2	195.4	157.9	192.7	249.7
$\bar{\beta}$ (average over models)	194.5	163.7	193.0	233.3

as compared with estimates from a single model (Hoeting *et al.*, 2002, p 499).

3.4 ROBUST REGRESSION METHODS: MODELS FOR SKEWNESS AND HETEROSCEDASTICITY

Robust regression may also consider questions such as non-constant variance and non-normality due to skewness as well as, or instead of, heavier tailed error densities. Often these problems are interrelated, and solutions to them may be combined with other procedures: for example, predictor selection as in Hoeting *et al.* (2002) and Smith and Kohn (1996), or with model selection as in Laud and Ibrahim (1995).

A well-known method for skew responses or predictors involves a skew-minimising transformation such as the Box–Cox transform

$$y(\lambda) = (y^\lambda - 1)/\lambda$$

for $\lambda \neq 0$, and $y(0) = \log(y)$ for $\lambda = 0$ (e.g. Pericchi, 1981; Hinkley and Runger, 1984; Weiss, 1994). If the analysis includes transforming the response then a discrete grid of λ values may be assumed (e.g. $\lambda = -1$, $-0.5, 0.5$ and 1), as in Hoeting *et al.* (2002) and Smith and Kohn (1996). Hoeting *et al.* (2002) also consider a change point transformation on predictors, under which there is no change in the expectation of y above or below a certain level of X. Skewness may also be tackled by adopting flexible asymmetric densities, such as the gamma density. For example, one may assume

$$Y_i \sim \text{Ga}(\eta, \eta/\mu_i)$$

where $\log(\mu_i) = \beta X_i$ (McCullagh and Nelder, 1989). Modifications of the normal and Student t densities have also been suggested. Sahu *et al.* (2003) introduce an extra random effect δ_i (with known scale as for a latent trait in factor analysis) into the regression mean. This extra effect is constrained to be positive. Thus a skewed normal model is

$$y_i = \beta X_i + \lambda \delta_i + \varepsilon_i \qquad i = 1, \ldots, N \qquad (3.6)$$

where $\varepsilon_i \sim \text{N}(0, \sigma^2)$,

$$\delta_i \sim \text{N}(0, 1)I(0)$$

and λ is a loading that is positive when there is right skew in the data and negative when there is left skew. Equivalently

$$Y_i \sim \text{N}(\mu_i, \sigma^2)$$
$$\mu_i = \beta X_i + \lambda \delta_i \qquad\qquad (3.7)$$

A positive density such as the gamma might also be used for δ_i, e.g.

$$\delta_i = \delta_i^* - 1 \qquad (3.8)$$

where $\delta_i^* \sim \text{Ga}(1, 1)$ as this accommodates additional non-normality. Introducing such effects may induce excess parameterization and an option is to assume a Dirichlet process prior (DPP) (section 3.5.3) with a maximum number of say $N/10$ or $N/20$ clusters for the skew effects.

Fernandez and Steel (1998b) consider methods to address both skewness and fat tails together. Thus the baseline variance σ^2 (or precision $\tau = \sigma^{-2}$) is scaled differentially according to whether the residual $\varepsilon_i = y_i - \mu_i = y_i - \beta X_i$ is negative or positive. For positive residuals the variance is scaled by a positive factor $1/\gamma^2$, while for negative residuals the scaling is by γ^2. So when there is positive skewness in the errors ε, values of $\gamma > 1$ are selected. By contrast, $\gamma = 1$ corresponds to a symmetric density, and values of γ under 1 to negative skewness. This model for skewness can be combined with a scale mixture form of the Student t density to allow for both skewness and heavy tails.

A frequently occurring form of heteroscedasticity is when the variance of the residuals increases with the size of the fitted values. As for skewness, one solution to this involves transformation of the response, such as square root or log. Direct modelling of changes in the variance $V(y|X)$ between observations is also possible (Boscardin and Gelman, 1996). Among possible schemes for such heteroscedasticity, consider

$$y_i = \beta X_i + \varepsilon_i = \mu_i + w_i \varepsilon_i$$

where $\varepsilon_i \sim N(0, 1)$. Then if the variance of the residuals increases with the fitted values one might set

$$w_i = \exp(\mu_i)$$
$$w_i = |\mu_i|^\gamma$$

or

$$w_i = (\alpha_0 + \alpha_2 \mu_i^2)^{0.5}$$

Alternatively, if heteroscedasticity is related to predictors (Aitkin, 1987; Cepeda and Gamerman, 2000) consider

$$y_i = \mu_i + \varepsilon_i$$

where $\varepsilon_i \sim N(0, \eta_i)$ and let $\log(\eta_i) = \gamma Z_i$ where Z_i are additional predictors (or include some of the X_i). If the heteroscedasticity is related to one predictor (e.g. X_{i1}) then one could set $\log(\eta_i) = \gamma_0 + \gamma_1 X_{i1}$. Homoscedasticity would be shown by values of γ_1 not clearly differing from zero.

Dynamic linear (state-space) priors of various types for volatility in regression errors have been proposed, especially for time series but with relevance to other types of regression. For example, if the data are arranged in ascending order of y then with $\kappa_i = \log(\sigma_i)$, an RW(1) prior would take $\kappa_i \sim N(\kappa_{i-1}, \tau_\kappa)$. A penalized spline method for heteroscedasticity is proposed by Ruppert *et al.* (2003) and illustrated below.

Example 3.6   Cherry Wood   Aitkin (1987) considers $n = 31$ observations on the volume V of usable wood in cherry trees, in terms of the tree heights H and diameters D. The actual model transforms V to reduce skewness and heteroscedasticity, i.e.

$$V_i^{1/3} = \beta_0 + \beta_1 H_i + \beta_2 D_i + \varepsilon_i$$

where $\varepsilon_i \sim N(0, \sigma^2)$. However, examination of the residuals from this model (model 1) suggests that their variation increases with height. Therefore an alternative model (model 2) is heteroscedastic $\varepsilon_i \sim N(0, \exp(\gamma_0 + \gamma_1 H_i + \gamma_2 D_i))$. The criterion (2.12) shows that this model has worse fit, with $PrL_2(1) = 0.40$ and $PrL_2(1000) = 0.49$, whereas $PrL_1(1) = 0.33$ and $PrL_1(1000) = 0.42$. (Results are based on the second half of a two-chain run of 100 000 iterations.) The coefficients γ_1 and γ_2 are not well identified, i.e. 0.104 (with s.d. $= 0.076$) and 0.036 (0.133) from the second half of a two-chain run of 100 000 iterations.

The residual plots from both models 1 and 2 show the residuals most widely scattered at middle values of $\mu_i = \beta_0 + \beta_1 H_i + \beta_2 D_i$ with their dispersion tapering off for larger and smaller μ. Therefore one might try, as model 3, taking $\log(w_i) = (\alpha_0 + \alpha_1 \mu_i + \alpha_2 \mu_i^2)$. This yields $PrL_3(1) = 0.35$ and $PrL_2(1000) = 0.45$, with posterior means (s.d.) for α_1 and α_2 of 14.7 (5.5) and -2.1 (0.9).

Example 3.7   Student interview data   To illustrate alternative models for skew data, this example follows Sahu *et al.* (2003) in considering scores of 584 students on a non-academic skills rating scale. The data suggest some negative skew (Figure 3.1). Relevant predictors are three dummy indicators for gender–ethnicity. Thus $X_1 = 1$ for white male (0 otherwise), $X_2 = 1$ for non-white male (0 otherwise), $X_3 = 1$ for non-white female (0 otherwise), so the reference category with average score given by the intercept is white females. The fit of the different models is assessed by sampling new data and obtaining total predictive errors squared, as in (2.10), as well as by the DIC, defined as minus twice the log likelihood. Sahu *et al.* (2003) cite a DIC of 2751 for the conventional normal model.

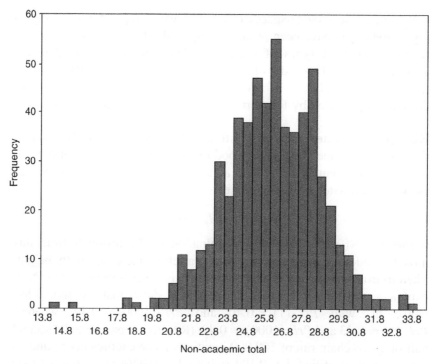

Figure 3.1   Observed student ratings

For the skew normal a two-chain run of 10 000 iterations (convergent at under 1000) shows a predictive criterion of 2125 and reduced DIC of 2697, obtained as $2\bar{D} - D(\bar{\mu}, \bar{\tau})$ where $\tau = 1/\sigma^2$. The probability that λ in (3.6) is negative is obtained as 0.88; the parameter itself has a positively skewed density. The β_j are all negative (with means -0.63, -0.88 and -1.05, and 95% intervals confined to negative values) indicating that the highest non-academic scores are for white female student interviewees.

The alternative gamma prior for δ_i in (3.8) gives $\bar{D} = 2536$, $D(\bar{\mu}, \bar{\tau}) = 2379$ and a DIC of 2688; the predictive criterion is 1965. The low first observation $Y_1 = 14$ is predicted better under this option.

The method of Fernandez and Steel (1998b) gives $\bar{D} = 2739$, $D(\bar{\mu}, \bar{\tau}) = 2733.3$ and a DIC of 2745, slightly better than the usual normal regression. However, the predictive criterion (2.10) worsens to 3760. The scaling parameter γ is estimated as 0.91 with 95% interval (0.85,0.98).

These data provide an illustration of a DPP prior (see section 3.7) as there is repetition in both the predictor variables and the responses. So instead of assuming $n = 584$ random effects, one may assume clustering of students into a smaller total of random effects, such that many students with the same or similar response/predictors have the same effect. Accordingly replace the individual level prior for the skew normal, i.e. $\delta_i^* \sim \mathrm{Ga}(1,1)$, with $\delta_i = \delta_i^* - 1$ $(i = 1, \ldots, n)$ by assuming

$$G_i \sim \mathrm{Cat}(p)$$

where the probability vector $p = (p_1, \ldots, p_K)$ governs allocation to one of $K = 30$ possible clusters. The K cluster effects follow

$$\delta_k = \delta_k^* - 1; \quad \delta_k^* \sim \mathrm{Ga}(1,1)$$

and the sequence of p_k follows (3.10). Then at a particular iteration, having sampled δ_k and G_i, one sets $\delta_i = \delta_{G_i}$. Instead of $n = 584$ random effects there are a maximum of K cluster effects (not all the clusters need to be chosen). A two-chain run of 5000 iterations (convergent from 1000) shows an average of 21 clusters (with mean Dirichlet concentration parameter 5.2), and a more clearly negative λ in (3.21) with a 95% interval $(-2.9, -0.3)$. This model produces $\bar{D} = 2590$, $D(\bar{\mu}, \bar{\tau}) = 2502$ and a DIC of 2678, so there appears to be a benefit from reducing the parameterization of the δ_i.

3.5   ROBUSTNESS VIA DISCRETE MIXTURE MODELS

Finite mixture models provide a way to model complex observed structures, approximating the sampling density by a mixture of K latent subpopulations

$$f(y_i) = \sum_{j=1}^{K} \pi_j f_j(y_i | \theta_j), \quad i = 1, \ldots, n \tag{3.9}$$

where $f_j(\cdot | \theta_j)$ is a given parametric density such as the normal. At least one parameter component of the density will vary over subpopulations: for example, a mixture of normals might have a common variance $\sigma_j^2 = \sigma^2$, but would then necessarily have varying means. In the absence of covariates there would be K different means μ_j to estimate, and several Bayesian model selection analyses focus on the optimal number of subgroups in such situations (Richardson and Green, 1997). Applications without predictors are usually seen as providing a smoothing of the mean of each data value towards a value consistent with the full set of observations.

Using the Gibbs sampler in discrete mixture estimation involves augmenting the data with allocation indicators, and this means one can express the likelihood of the model in terms of 'complete data'. It may be necessary to use parameter constraints for identifiability (e.g. the prior on the means of components is ordered) and even without constrained priors, some information will be needed in the priors – improper priors lead to undefined posterior densities (Robert, 1996).

The most flexible procedures for finite mixture models allow switches in the number of subgroups K at each iteration, as in the reversible jump algorithm (e.g. Hurn $et\ al.$, 2003). This will produce a posterior density of the number of subgroups K, such that one can derive a Bayes factor on (say) a three-group mixture as against a two group mixture. A very similar consequence follows the application of a truncated Dirichlet process (see below), in that the density of non-empty clusters will indicate the appropriate number of subpopulations.

In regression applications with predictors $X_i = (x_{i1}, x_{i2}, \ldots, x_{ip})$ together with an intercept, means may differ over subjects according to their membership probabilities of groups with differing regression parameters $\beta_j = \{\beta_{j1}, \ldots, \beta_{jp}\}$, $j = 1, \ldots, K$. Regression model selection may be applied to assess whether a particular regression parameter (e.g. for x_{ik}) is equal over all subgroups $\beta_{jk} = \beta_k$, $j = 1, \ldots, K$, or within subsets of the full set of groups (e.g. $\beta_{1k} = \beta_{2k}$, but β_{jk} unequal for $j > 2$). As illustrated by Dayton and Macready (1988), the unknown allocation indicators $G_i \in (1, \ldots, K)$ may also be modelled as functions of covariates $w_i = (w_{i1}, \ldots, w_{iq})$ such that the multinomial probabilities are defined by parameters

$$\lambda_{ij} = \exp(\phi_j w_i) / \sum_k \exp(\phi_k w_i)$$

3.5.1 Complete data representation

As mentioned by Diebolt and Robert (1994) and Dempster $et\ al.$ (1977) a finite mixture can be expressed in terms of the original data $\{y_i, X_i\}$ and augmented data consisting of the unknown categories G_i of group membership. G_i may be equivalently stated in terms of binary indicators: $d_{ij} = 1$ when subject i is in group j and $d_{ij} = 0$ otherwise. The likelihood of the complete (observed and augmented) data $\{y_i, X_i, G_i\}$ is

$$\prod_{j=1}^{J} \prod_{i=1}^{n} \pi_j^{d_{ij}} [f_j(y_i | x_i \theta_j)]^{d_{ij}}$$

The allocation of subjects to groups at each iteration as well as the sampled values of the group parameters determine the complete data likelihood, which may be monitored as one index of model fit.

In estimation via repeated sampling the indicators d_{ij} may switch between subgroups ($d_{ij}^{(t)} = 1$ if subject i is allocated to component j at iteration t). This can reflect both uncertainty about which group subject i belongs to and an identification problem known as 'label switching' (see section 3.5.2). The complete data likelihood is unaffected by label switching (as are some functions of the parameters such as means for each observation).

By contrast to random effects models, the number of parameters will be known in finite mixture models so that AIC measures may be obtained. For DPP models the effective parameters may be estimated by comparing the deviance at the mean $D(\bar{\mu}|y)$ with the mean deviance or by counting the number of non-empty clusters.

The complete data representation means that a mixture model can be represented hierarchically, and such representations facilitate estimation via repeated sampling. The highest level specifies priors on the mixture density parameters θ_j, which in turn drive the densities for the latent class indicators

$$G_i \sim g(G_i|\theta_j)$$

and the likelihood is

$$y_i \sim f(y_i|\theta_j, d_{ij})$$

The posterior probabilities of group membership for each subject, i.e. $\lambda_{ij} = \Pr(d_{ij} = 1)$, have the form

$$\lambda_{ij} = \pi_j f_j(y_i|\theta_j)/\sum_k \pi_k f_k(y_i|\theta_k)$$

Gibbs sampling in a finite mixture model includes the augmented data d_{ij} as parameters and updates them according to

$$G_i \sim \text{Categoric}(\lambda_{i1}, \lambda_{i2}, \ldots, \lambda_{iK})$$

If $T_j = \sum_i d_{ij}$ subjects are allocated to group j at iteration t and a prior mass on group j is set at t_j then the subpopulation proportions are updated according to a Dirichlet

$$\pi \sim \text{D}(T_1 + t_1, T_2 + t_2, \ldots, T_K + t_K)$$

3.5.2 Identification issues

Repeated sampling of latent class mixture models where labelling is relevant to inference (e.g. finite mixture models) may be problematic in terms of obtaining well-identified solutions. Setting K too large will mean the solution is only identifiable with an informative prior that supplies an adequate prior sample size for all subgroups. There is also a possibility that 'label switching' will occur in MCMC estimation of finite mixture models. If two or more chains are run with an unconstrained prior, there is a strong chance that one chain will adopt a different labelling to the other(s), even though within chains there is no label switching. This will show in trace plots where (say) μ_1 in chain 1 is a high mean group, but in chain 2 μ_1 is a low mean group and they will never converge.

If an unconstrained prior with K groups is used, then the parameter space has $K!$ subspaces corresponding to possible ways of relabelling the states. To avoid this one may 'pin down' the analysis by selected parameter restrictions: for example, specifying that one mixture probability is always greater than another. Alternatively one or more density or regression parameters might be subject to a constraint (e.g. for a normal mixture one might specify a priori that $\mu_1 > \mu_2 > \ldots > \mu_K$, or alternatively that variances are ordered). A particular form of constraint may be more or less appropriate to a particular data set. Exploratory procedures have been proposed (Frühwirth-Schnatter, 2001) to assess whether a particular form of constraint will distort the solution more than another. If in fact the subgroup means (say) are well separated then a prior constraint to that effect will have little distorting effect on final parameter estimates. A constraint such as $\mu_1 > \mu_2 > \ldots > \mu_K$ may be needed for the effective operation of flexible mixture schemes that allow for K to be stochastic, as in applications of the reversible jump method (Richardson *et al.*, 2002).

Another option (Celeux *et al.*, 2000) applies clustering or other procedures to the MCMC output from an unconstrained prior. This involves first selecting a relatively short run of MCMC iterations (say $T = 100$ iterations) where there is no label switching. The means $\xi_{jk} = \sum_t \xi_{jkt}/T$ on parameters of type k in group j are then obtained from this sample. For a normal mixture there will be three types of parameters $\xi_j = \{\pi_j, \mu_j, \sigma_j^2\}$ and for K groups there will be $3K$ parameters. The initial run of sampled parameter values provides a reference labelling (any one arbitrarily selected labelling among the $K!$ possible). It also enables definition of means of $\{\pi, \mu, \sigma^2\}$ under alternative (non-reference) labelling schemes. In a subsequent run of R iterations where label switching might occur, iteration r is assigned to that scheme that is closest to it in distance terms and a relabelling is applied if there has been a switch away from the reference scheme. The means under the schemes are recalculated at each iteration $T + r$ (see Celeux *et al.*, 2000, p 965).

3.5.3 Dirichlet process mixture models

A flexible mixture modelling structure is based on various forms of Dirichlet prior process, which in applications may involve truncation at a large number of potential groups or clusters K^* to which individual observations i may belong. Ishwaran and James (2002) consider the use of the truncated Dirichlet process to approximate a formal Dirichlet process where the number of clusters is in theory infinite.[2] Unlike finite mixture analysis, which requires that all component subpopulations contain at least one sample member to be identified, the DPP model allows empty clusters. DPP mixture models define a baseline prior H_0 from which potential values for θ_j are drawn.

While more than n clusters are possible, in practice a smaller number $K^* \leq n$ of potential clusters may be envisaged in any application (e.g. $K^* = 10$ or $K^* = 20$). So $j = 1, \ldots, K^*$ values of θ_j will be drawn from H_0 and the most appropriate value for case i among the candidate values $\{\theta_1, \theta_2, \ldots, \theta_{K^*}\}$ is selected using a categorical prior for the subpopulation indicator G_i. In a normal mixture application without predictors the DPP might be applied to selecting the means μ_j most appropriate to each subject, and one might take H_0 to be a normal density N(0, V) with preset large variance (e.g. $V = 1000$) or V an additional unknown.

Then $G_i = j$ if group j is appropriate for subject i and

$$G_i \sim \text{Categorical}(p)$$

where $p = (p_1, \ldots, p_{K^*})$. The prior for the category probabilities p is defined by a 'stick-breaking' procedure. Let $r_1, r_2, \ldots, r_{K^*-1}$ be a sequence of Beta(1,κ) random variables (with $r_{K^*} = 1$)

$$r_1, r_2, \ldots, r_{K^*-1} \sim \text{Beta}(1, \kappa)$$

and set

$$
\begin{aligned}
p_1 &= r_1 \\
p_2 &= r_2(1 - r_1) \\
p_3 &= r_3(1 - r_2)(1 - r_1) \\
&\vdots \\
p_M &= r_{K^*}(1 - r_{K^*-1})(1 - r_{K^*-2}) \cdots (1 - r_1)
\end{aligned}
\tag{3.10}
$$

---

[2] If the sample size is n and truncation at K^* clusters is assumed then the marginal density for a normal mixture obtained under the truncated Dirichlet process can be compared with that under the infinite random measure of Ferguson (1973). The discrepancy in terms of an L_1 error bound is approximately $4n \exp[-(K^* - 1)/\alpha]$.

The number K of non-empty clusters (between 1 and K^*) depends in practice on the so-called concentration parameter κ, which may be preset or itself assigned a prior. Other names for κ include mass parameter and precision parameter.

Typical values of κ (if it is preset) are between 1 and 5. Smaller values of κ lead to solutions (when the data have no predictors) resembling a finite mixture model with a few subgroups, while large κ leads to a smoother appearance. Ishwaran and James (2002) suggest a gamma $Ga(\eta_1, \eta_2)$ prior for κ, with its updating taking place according to the conditional

$$f(\kappa|) \sim Ga\left(K^* + \eta_1 - 1, \eta_2 - \sum_{k=1}^{K^*-1} \log(1 - r_k) \right)$$

Let T_k be the number of subjects for whom $G_i = k$. Then the conditional for r_k (which defines the cluster probabilities p_k) is

$$r_k \sim Beta\left(1 + T_k, \kappa + \sum_{j=k+1}^{K^*} T_k \right)$$

A predictive density for a future observation y_{new} can be obtained by sampling (at each iteration) as follows: $G_{new} \sim \text{Categorical}(p)$ and then $y_{new} \sim f(y|\theta_{G_{new}})$. This provides a way of plotting the implied predictive density.

Example 3.8 Galaxy data   These data relate to $N = 82$ velocities of galaxies in the Corona Borealis region. Different analyses have suggested different numbers of components K in a discrete normal mixture for such data. With K given the mixture density is denoted

$$f(Y|\mu_{jK}, \tau_{jK}, \pi_{jK}) = \sum_{j=1}^{K} \pi_{jK} \phi(Y_i|\mu_{jK}, \tau_{jK})$$

Thus for $K = 2$ the parameters are $\{\mu_{12}, \tau_{12}, \pi_{12}, \mu_{22}, \tau_{22}, \pi_{22}\}$ while for $K = 3$ the parameters are $\{\mu_{13}, \tau_{13}, \pi_{13}, \mu_{23}, \tau_{23}, \pi_{23}, \mu_{33}, \tau_{33}, \pi_{33}\}$. Four models with $K = 2, 3, 4$ and 5 respectively are compared with priors $\mu_{jK} \sim N(20, 25)$ on the component means, while Dirichlet priors with K-vector components of 1 are used for π_{jK}. The precisions $1/\tau_{jK}$ are taken to be $Ga(1, 0.001)$.

The method of parallel sampling (section 2.8) is used to assess the posterior probabilities of the different mixture densities. The prior adopted on μ_{jK} means that model weights are unaffected by relabelling of parameters at a particular iteration. So it is not necessary to apply parameter constraints in order to obtain the optimal K. As well as the total model fit criteria, the estimated means for individual cases are unaffected by relabelling. These are obtained at iteration t as

$$\mu_{G_{iK}^{(t)}}$$

where $G_{iK}^{(t)}$ is a latent observation-specific index variable with values between 1 and K. Hence one might average the predicted means of selected cases (e.g. the minimum and maximum y) in order to assess how well the different mixture models are representing these extreme points and also to obtain a model averaged mean for these observations at a particular iteration. For example, assuming ranked data $y_n \geq y_{n-1} \geq \ldots \geq y_1$, the maximum value y_n is predicted at iteration t as

$$\mu_n^{(t)} = w_1^{(t)} \mu_{G_{n2}^{(t)}} + w_2^{(t)} \mu_{G_{n3}^{(t)}} + w_3^{(t)} \mu_{G_{n4}^{(t)}} + w_4^{(t)} \mu_{G_{n5}^{(t)}}$$

There are $J = 4$ models comparing $K = 2, 3, 4$ and 5 groups. A two-chain run of 5000 iterations (with a 1000 burn-in) and prior model probabilities $P(m = j) = 1/4$ gives $w_1 = 0.0005$, $w_2 = 0.907$, $w_3 = 0.084$ and $w_4 = 0.008$, so the $K = 3$ model is preferred, though not overwhelmingly, and different priors might affect the choice. Carlin and Chib (1995) compare $K = 3$ and $K = 4$ and obtain $P(K = 3|y)$ as 0.64; Moreno and Lisco (2003) compare $K = 1, 2, 3, 4, 5$ and obtain $P(K = 3|Y) = 0.84$, $P(K = 4|Y) = 0.1599$ with negligible weights on the other values of K. The maximum observed velocity (in thousands) is $y_{82} = 34.28$ and under models 1 to 4 this observation has means 21.9, 31.75, 31.8 and 31.9 respectively. So the model with $K = 3$ adequately represents the maximum observation. (Of course other model checks may be carried out in addition.) The model averaged μ_{82} (with 95% credible interval) is 31.77 (29.3, 34.1).

A DPP model is then applied with a cluster upper limit $K^* = 20$ and the values $\{\eta_1, \eta_2\} = (2, 4)$ for the prior on κ, as assumed by Escobar and West (1995) in their analysis of these data. Note that more diffuse priors, such as $\{\eta_1, \eta_2\} = (0.1, 0.1)$, will lead to larger numbers of non-empty clusters being identified. Let $K|\kappa, y$ be the number of distinct y values (number of non-empty clusters). Then a two-chain run of 5000 iterations shows nine as the modal number of clusters (see Table 3.4), considerably higher than implied by the standard non-parametric mixture approach.

The mean for μ_{82} is now 33.1, a more accurate estimate but at the cost of heavier parameterization. The predictive density for a new observation shows the greater heterogeneity implied by this model (Figure 3.2).

Table 3.4 Posterior probabilities for number of distinct y values (non-empty clusters)

k	6	7	8	9	10	11	12	>12
$\Pr(K = k)$	0.033	0.108	0.186	0.219	0.180	0.134	0.079	0.061

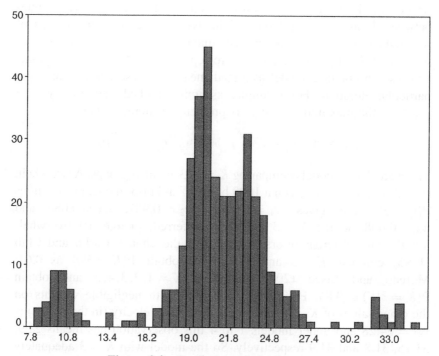

Figure 3.2 Predictive density for new y

3.6 NON-LINEAR REGRESSION EFFECTS VIA SPLINES AND OTHER BASIS FUNCTIONS

The linear model of (3.1), namely

$$y_i = \beta_0 + \beta_1 x_{i1} + \beta_2 x_{i2} + \cdots + \beta_p x_{ip} + \varepsilon_i$$

assumes a known form (an additive linear impact of the predictors) with the main issue being the identification of significant predictors given the a priori assumption of linearity. If one or more of the predictors, say x_{ik}, has a non-linear impact on y then known regression forms (e.g. polynomial regression or the Box–Cox transform in section 3.4) are also available. However, while sometimes non-linear functions may be based on subject matter knowledge, there is often little knowledge concerning an appropriate non-linear function.

One may instead simply assume that the regression surface for x_{ik} is smooth but try to estimate its unknown non-linear form. This approach, known as non-parametric regression, estimates a function that adapts to the underlying true form. Methods for modelling y_i as a general non-linear function of one or more predictors mostly assume linear combinations of basis functions $B(x_{ik})$ of predictors. Examples of basis functions are truncated polynomial or spline functions (Friedman and Silverman, 1989) or more recent models discussed by Denison *et al.* (2002), such as multivariate linear splines, wavelets and multivariate adaptive regression splines.

Assume a single predictor with values ordered

$$x_1 < x_2 < \cdots < x_n$$

and let the mean of y be represented as a function of x, with a random error representing residual effects

$$y_i = \gamma_0 + B(x_i) + \varepsilon_i$$

Assume also one or more predictors $w_{i1}, w_{i2}, \ldots, w_{im}$ have a conventional linear effect; then a semi-parametric model is obtained. For example, a single w predictor and an adaptive regression in a single x give

$$y_i = \gamma_0 + \beta_1 w_i + B(x_i) + \varepsilon_i$$

Spline functions to approximate $B_i = B(x_i)$ include quadratic and cubic splines, i.e. piecewise quadratic or cubic polynomials that interpolate $\mu(x)$ at K selected knot points $t_1, t_2, \ldots, t_K$ within the range of the variable x, such that $\min(x_i) < t_1 < t_2 < \ldots < t_K < \max(x_i)$. A cubic regression spline for metric y and with homoscedastic normal errors has typical form

$$y_i \sim N(\mu_i, \sigma^2)$$
$$\mu_i = \gamma_0 + B(x_i)$$

$$B(x_i) = \gamma_1 x_i + \gamma_2 x_i^2 + \gamma_3 x_i^3 + \sum_{k=1}^{K} 1(x_i - t_k)\beta_k(x_i - t_k)^3$$

where $1(x_i - t_k)$ is one if x_i exceeds the kth knot t_k, and zero otherwise. An alternative notation with the same meaning is

$$B(x_i) = \gamma_1 x_i + \gamma_2 x_i^2 + \gamma_3 x_i^3 + \sum_{k=1}^{K} \beta_k(x_i - t_k)_+^3 \qquad (3.11)$$

where $(x_i - t_k)_+ = \max(0, x_i - t_k)$. Sometimes the terms $\gamma_1 x_i, \gamma_2 x_i^2, \gamma_3 x_i^3$ are omitted, so that $\mu_i = \gamma_0 + B(x_i)$ with

$$B(x_i) = \sum_{k=1}^{K} \beta_k (x_i - t_k)_+^3 \qquad (3.12)$$

Denison *et al.* (2002, p 74) discuss issues around prior specification in (3.11) which are resolved in a two-sided cubic spline model

$$B(x_i) = \sum_{k=1}^{K} \beta_k (x_i - t_k)_+^3 + \sum_{k=K+1}^{2K} \beta_k (t_k - x_i)_+^3$$

There is no certainty on how many knot points to include or where to locate them. More knots are needed in regions where $B(x)$ is changing rapidly (Eubank, 1988). Sometimes subject knowledge may be relevant in placing knots where a change in the shape of the curve is expected: human mortality shows a 'bathtub' shape with minimum mortality around age 10. Using too few knots or poorly sited knots means the approximation to the true curve $B(x)$ will be degraded. By contrast, a spline using too many knots will be imprecise.

Knots may be based on selecting among the existing x values (e.g. Friedman and Silverman, 1989), might be equally spaced within the range $[\min(x), \max(x)]$ as in Biller (2000), or be taken as unknowns. Another approach is based on 'smoothing splines' whereby there is a knot, or potential knot, at each distinct value of x_i so that the number of knots may equal the sample size. For n large, selection is required so that $K \ll n$. For example, Ruppert *et al.* (2003) suggest a maximum of $K = 35$ or 40.

Starting with a relatively large number of candidate knot locations, regression selection may be used to select significant knot points (Smith and Kohn, 1996; Smith *et al.*, 2001). Smith and Kohn (1996) suggest indicator variables δ_{1k} (for $\gamma_1, \ldots, \gamma_3$ in the polynomial part of the cubic spline (3.11) and δ_{2k} (for the $k = 1, \ldots, K$ spline coefficients β_k) such that if, at a particular iteration, the indicator variables are zero (one) then the corresponding predictor is excluded (included).

Alternatively, under a free knot approach, the knots can take any values within the range $[\min(x), \max(x)]$ and are not tied to selecting from the observations (e.g. Denison *et al.*, 1998a). Note that for a known number of knots, a prior on knots with free location specifies their ordering and might take the form

$$t_1 \sim N(0, V)I(\min(x), t_2)$$
$$t_2 \sim N(0, V)I(t_1, t_3)$$
$$\vdots$$
$$t_K \sim N(0, V)I(t_{K-1}, \max(x))$$

$$(3.13)$$

where V is known (e.g. $V = 1000$) or an extra parameter. Biller (2000) and Denison *et al.* (1998a) use reversible jump MCMC to allow switching between models with different numbers and sitings of free knots. Denison *et al.* (1998a) make the simplification of calculating β and γ coefficients by standard least squares formulae rather than the full Bayesian prior/posterior updating procedure.

3.6.1 Penalized random effects for spline coefficients

Berry *et al.* (2002) and Ruppert *et al.* (2003) avoid regression selection among fixed effects β_k by using penalizing random effects priors on the β coefficients. These may link the variance σ_β^2 of the β_k to $\text{Var}(\varepsilon) = \sigma^2$ and so induce varying degrees of constraint on the β_k. Under this approach a linear spline is often appropriate except for highly non-linear regression effects, though it may involve increasing K till a satisfactory fit is obtained. Let n_u be the number of distinct x values. Ruppert *et al.* (2003, 126) recommend $K = \min(35, n_u/4)$, though values such as $K = 80$ may occasionally be needed, for n sufficiently large. The knots might correspond to percentiles in the observed x (Berry *et al.*, 2002): for example, t_k sited at the percentile $k/(K + 1)$. For example, if $K = 19$ then the knots are located at every fifth percentile of x. Then a spline of degree q is

$$y_i = \gamma_0 + \gamma_1 x_i + \cdots + \gamma_p x_i^q + \sum_{k=1}^{K} \beta_k (x_i - t_k)_{++}^q + \varepsilon_i \qquad (3.14)$$

where $\beta_k \sim N(0, \sigma_\beta^2)$ and $q = 1$ gives a linear spline. The full conditionals on $1/\sigma_\beta^2$ and $1/\sigma^2$ are

$$1/\sigma_\beta^2 \sim \text{Ga}\left(a_1 + 0.5K, b_1 + 0.5\sum_{k=1}^{K} \beta_k^2\right)$$

$$1/\sigma^2 \sim \text{Ga}\left(a_2 + 0.5n, b_2 + 0.5\sum_{i=1}^{n} \varepsilon_i^2\right)$$

where $1/\sigma_\beta^2 \sim Ga(a_1, b_1), 1/\sigma^2 \sim Ga(a_2, b_2)$ are the priors on the precisions. This approach may be extended to modelling heteroscedasticity and so provide a 'spatially adaptive' non-linear smooth. Thus let $\varepsilon_i \sim N(0, \sigma_i^2)$; then a non-constant variance would involve an additional spline model

$$\log(\sigma_i^2) = \varphi_0 + \varphi_1 x_i + \cdots + \varphi_q x_i^q + \cdots + \sum_{k=1}^{M} \phi_k (x_i - t_k)_+^q \qquad (3.15)$$

Usually $M < K$ (see Ruppert and Carroll, 2000).

3.6.2 Basis function regression

As noted above, one may replace the conceptual framework of the linear regression model

$$y_i = \beta_0 + \beta_1 x_{i1} + \beta_2 x_{i2} + \cdots + \beta_p x_{ip} + \varepsilon_i$$

by a more general class of basis function models (Denison *et al.*, 2002) with

$$y_i = \beta_0 + \sum_{m}^{M} \sum_{k}^{K_m} \beta_{mk} B_{mk}(X_i) + \varepsilon_i$$

where each $B_{mk}(X_i)$ is a function of one or more $x_{ij}, j = 1, \ldots, p$. In matrix terms

$$Y = 1\beta_0 + B\beta + \varepsilon$$

where Y, 1 and ε are $n \times 1$, β is $L \times 1$ where $L = \sum_{m}^{M} K_m$, and B is $n \times L$ with

$$B = \begin{bmatrix} B_{11}(X_1) & \cdots & B_{MK_M}(X_1) \\ \vdots & \ddots & \vdots \\ B_{11}(X_n) & \cdots & B_{MK_M}(X_n) \end{bmatrix}$$

For example, if $p = 1$ with $X_i = x_i$ then the truncated power splines can be written, following the simplified (3.12), as

$$y_i = \beta_0 + \sum_{k}^{K} \beta_k (x_i - t_k)_+^q + \varepsilon_i$$

Setting $q = 1$ gives a truncated linear spline and $q = 3$ the truncated cubic spline. A similar model with more than one predictor might allow differing numbers of knots and degrees between predictors, but still limited to univariate splines in each. Thus

$$y_i = \beta_0 + B_1(x_{i1}) + B_2(x_{i2}) + \cdots + B_p(x_{ip}) + \varepsilon_i$$

$$= \beta_0 + \sum_k^{K_1} \beta_{k1}(x_{i1} - t_{k1})_+^{q_1} + \sum_k^{K_2} \beta_{k2}(x_{i2} - t_{k2})_+^{q_2}$$

$$\qquad\qquad (3.16)$$

$$+ \cdots + \sum_k^{K_2} \beta_{kp}(x_{ip} - t_{k2})_+^{q_2} + \varepsilon_i$$

with $M = p$ and $L = K_1 + K_2 + \cdots + K_p$.

3.6.3 Special spline functions

Special spline forms have been proposed to deal with possible interactions between variables or reduce ill-conditioning in regression problems. To allow for interactions between predictors in such spline functions, one may consider products such as

$$B_k(x_{i1}, x_{i2}) = [x_{i1} - t_{k1}]_+^{q_1}[x_{i2} - t_{k2}]_+^{q_2}$$

for knots $k_1 = 1, \ldots, K_1$ and $k_2 = 1, \ldots, K_2$. With a linear form $(q_1 = q_2 = 1)$ such products figure in the Bayesian MARS approach (Denison *et al.*, 1998b):

$$B_k(x_{i1}, x_{i2}) = [x_{i1} - t_{k1}]_+[x_{i2} - t_{k2}]_+ \qquad\qquad (3.17)$$

A two-sided spline form allows one or both terms in the above example to be of the form $[t - x]_+$ rather than $[x - t]_+$, for instance

$$B_k(x_{i1}, x_{i2}) = [x_{i1} - t_{k1}]_+^{q_1}[t_{k2} - x_{i2}]_+^{q_2}$$

Alternatively one may write a typical term $(x - t_0)_+$ in a truncated univariate linear spline as a dot product $[(1, x)\,(-t_0, 1)]_+$. This suggests the multivariate generalization (Holmes and Mallick, 2001)

$$B(x_{i1}, x_{i2}, \ldots, x_{ip}) = [(1, x_1, x_2, x_3, \ldots,)(-t_0, t_1, t_2, t_3, \ldots)]_+$$

Without loss of generality, the minus sign may be omitted from t_0, giving

$$B(x_{i1}, x_{i2}, \ldots, x_{ip}) = \left(t_0 + \sum_j^p t_j x_{ij} \right)_+ \qquad\qquad (3.18)$$

So, in addition to 'main effect splines' as in (3.16) one might have $\binom{p}{2}$ interaction splines of order 2, as defined by (3.18). For example, if $p = 3$, then there are three second-order interaction splines

$$B_1(x_{i1}, x_{i2}) = (t_{10} + t_{11}x_{i1} + t_{12}x_{i2})_+$$
$$B_2(x_{i1}, x_{i3}) = (t_{20} + t_{21}X_{i1} + t_{23}x_{i3})_+$$
$$B_3(x_{i2}, x_{i3}) = (t_{30} + t_{31}x_{i2} + t_{33}x_{i3})_+$$

Then $M = 6$ since there are three main effect splines in this example and three second-order splines. Holmes and Mallick (2001) and Denison *et al.* (2002) consider a framework allowing jumps between interactions of order 1, 2 and higher, so that main effect splines are nested within interaction splines. A full model for $p = 3$ might take the form

$$
\begin{aligned}
y_i = \ &\beta_0 + \beta_1 B_1(x_{i1}, x_{i2}) + \beta_2 B_2(x_{i1}, x_{i3}) + \beta_3 B_3(x_{i2}, x_{i3}) + \beta_4 B_4(x_{1i}) \\
&+ \beta_5 B_5(x_{2i}) + \beta_6 B_6(x_{3i}) + \varepsilon_i
\end{aligned}
$$

raising identification issues[3] involving the β and t coefficients.

One possible problem with truncated power splines such as (3.11) is that they may be ill-conditioned in terms of the regression model, with the normal equations difficult to solve (Eubank, 1988). Specialized versions of cubic splines such as natural cubic splines have been suggested; natural splines reduce to straight lines when x is below the smallest knot or above the largest, namely t_K. An alternative basis less prone to ill-conditioning is provided by B splines. Consider a B spline of degree D, with K interior knots and define $m = D + 1$, and boundary

---

[3] Holmes and Mallick (2001) suggest sampling t_{kj} (for $j > 0$) from a $G(1,1)$ density and then normalizing by dividing each t_{kj} by the sum of squared t_{kj} involved in the spline. For example, in B_1 one obtains

$$t_{11}^* = t_{11}/[t_{11}^2 + t_{12}^2]$$
$$t_{12}^* = t_{12}/[t_{11}^2 + t_{12}^2]$$

The t_{k0} coefficients are obtained by sampling from a uniform (discrete) density over all n subjects, Let J_k be the number of $t_{kj}(j > 0)$ coefficients in spline k, e.g. in B_1, B_2 and B_3; J_k is 2. Also let W_{kj} denote the predictors included in the kth spline, e.g. in $B_1(X_1)$, $W_{11} = 1$ and $W_{12} = 2$. With probability $1/n$, t_{k0} takes the one among $\{1, \ldots, n\}$ of the subject-specific sums

$$-\sum_j^{J_k} X_{iW_{kj}} t_{kj}$$

This device ensures that the plane for basis k passes through at least one observed data point.

points $a \leq \min(x)$, $b \geq \max(x)$. Then $2m$ additional knots are required with

$$t_{-D} = t_{-D-1} = \cdots = t_0 = a$$
$$t_{K+1} = \cdots = t_{K+m} = b$$

The B spline has to be evaluated at the observed points between a and b, and may be defined by the recursion (see De Boor, 1978; MacNab and Dean, 2001)

$$B_{ji}(x) = (x - t_j)B_{j,i-1}(x)/(t_{j+i-1} - t_j) + (t_{j+i} - x)B_{j+1,i}(x)/(t_{j+i} - t_{j+1}),$$
$$j = -D, \ldots, k; i = 1, \ldots, m \qquad (3.19)$$

In this recursion any term of the two on the right involving division by $t_{j+i-1} - t_j$ when $t_{j+i-1} = t_j$ (or by $t_{j+i} - t_{j+1}$ when $t_{j+i} = t_{j+1}$) is defined as zero. The initial terms $B_{j1}(x)$ in the recursion are simply indicators defining a partition of the x values.

For example, for a cubic ($D = 3$) B spline with $K = 2$ knots there are three non-zero initial indicator functions

$$B_{01}(x) = 1 \quad \text{if } x \in (a, t_1), \quad 0 \text{ otherwise}$$
$$B_{11}(x) = 1 \quad \text{if } x \in (t_1, t_2), \quad 0 \text{ otherwise}$$
$$B_{21}(x) = 1 \quad \text{if } x \in (t_2, b), \quad 0 \text{ otherwise}$$

All other $B_{j1}(x)$ functions are zero, namely $B_{-3,1}(x)$, $B_{-2,1}(x)$ and $B_{-1,1}(x)$. There are four non-zero functions $B_{j,2}(x)$ defined by the recursion (3.19) with $j = -1, \ldots, 2$. For example,

$$B_{-1,2}(x) = (t_1 - x)B_{01}(x)/(t_1 - a) = (t_1 - x)/(t_1 - a)$$

There are five non-zero functions $B_{j3}(x)$, $j = -2, \ldots, 2$ and $m + K = 6$ non-zero polynomial functions $B_{j4}(x)$ (with $j = -3, -2, \ldots, 2$) that define the basis. The smooth at an observed value x is defined as

$$S(x_i) = \sum_{j=1}^{m+K} \gamma_j B_{j-m,m}(x_i)$$

where γ_j are parameters to be determined.

Example 3.9   Titanium and heat   Eubank (1988, p 395) presents data on a property of titanium as a function of heat. An optimal five-knot solution is known to be at the heat values (836, 876.4, 898.1, 916.3, 973.9) with these points representing the hump in y for values of x between 800 and 1000. Among ways to fit a smooth to these data might be truncated polynomial or B splines with regression selection among

Table 3.5 Posterior knot densities, $K = 5$

Mean	S.d.	2.5%	97.5%
835.2	2.5	830.4	840.0
878.4	2.7	874.2	884.1
895.4	4.3	887.1	902.2
917.8	3.1	911.6	923.4
973.3	4.9	964.2	982.7

knots at equally spaces in the range of x; for example, ten equally spaced points over 785 to 1055.

Here a knot total $K = 5$ is initially assumed (model A), and unknown knot locations with prior as in (3.13), with parameters $\{\beta, \gamma\}$ derived by conventional least squares formulae (cf. Denison *et al.*, 1998a). With early convergence in a two-chain run the estimates in Table 3.5 are obtained by pooling over iterations 1000–1500, and a close fit is obtained.

The total error sum of squares (between Y_i and the posterior means $\hat{S}_i$ of the S_i) is 0.0075.

A complete Bayesian analysis with both unknown knot locations and with parameters $\{\beta, \gamma\}$ subject to posterior updating is more computationally intensive and may pose identifiability issues. McCullagh and Nelder (1989, p 380) note such problems in non-linear models involving sums of exponentials. Here a fully Bayes implementation of this option includes predictor selection as in Smith and Kohn (1996), since an analysis without such selection suggested $\beta_1 - \beta_3$ to be effectively zero. Priors on the knots were based on the first model, though with downweighted precision (prior variance of 20). The prior for the intercept is based on that of Raftery *et al.* (1996). The last 50 000 iterations of a two-chain run of 250 000 iterations show an error sum of squares 0.045, inflated by the uncertainty introduced by making predictor effects additional unknowns (see Figure 3.3). The posterior means (and s.d.) of the knots are 829.5 (1.4), 867.8 (2.2), 906.1 (1), 917.4 (0.6) and 986.6 (4.5).

Finally two models taking the knots known at the optimal values (of model A) are applied. These are truncated cubic spline and a B spline (see Example3_9.xls for derivation of the spline coefficients). Predictor selection is not used. The B spline gives an error sum of squares of 0.035 and the truncated cubic spline one of 0.01.

Depending on the focus of interest one might for a simplified analysis apply the model A approach, with either truncated polynomial or B

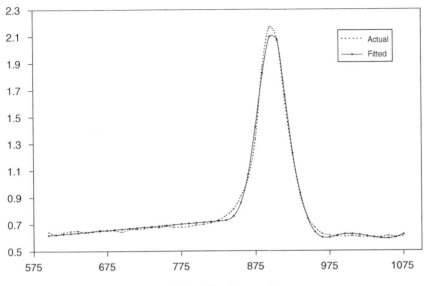

Figure 3.3   Titanium and heat

splines, for a few alternative numbers of knots (e.g. $K = 4$, $K = 5$, $K = 6$). Then model C can be applied with the knots assumed known (at their posterior means from model A).

Example 3.10   Vitamin in turnip green   Draper and Smith (1966) present data (Table 3.6) from a study concerning vitamin B_2 content in turnip green in relation to three predictors. These are X_1, radiation in relative gram calories per minute in the preceding half-day (divided by 100), X_2, the average soil moisture tension (divided by 100), and X_3, the Fahrenheit temperature (divided by 100). They refer to an earlier model chosen by Anderson and Bancroft (1959) of the form $y_i = \beta_0 + \beta_1 x_{i1} + \beta_2 x_{i2} + \beta_3 x_{i3} + \beta_4 x_{i1} x_{i2} + \varepsilon_i$.

Here we first fit (models A and B in Program 3.10) a normal linear regression with binary selection (see section 3.2) and with predictors $\{X_1, X_2, X_3, X_1 X_2, X_1 X_3, X_2 X_3\}$ calculated after centring $X_1 - X_3$. The results suggest that only X_2 merits inclusion and all other main effects and all interactions are not needed, though the interaction between X_1 and X_2 has the highest posterior probability of inclusion. The coefficient for X_2 has mean -78 and 95% interval $(-96, -60)$, and the deviance of a model with only X_2 as predictor is 48.5, with predictive error sum of squares 71. This is obtained (see section 2.4) by sampling new data Z_i and comparing it with the observations.

Table 3.6 Turnip green data

Obs.	X_1	X_2	X_3	Y	$X_1 * X_2$
1	1.76	0.07	7.8	110.4	0.123
2	1.55	0.07	8.9	102.8	0.109
3	2.73	0.07	8.9	101	0.191
4	2.73	0.07	7.2	108.4	0.191
5	2.56	0.07	8.4	100.7	0.179
6	2.8	0.07	8.7	100.3	0.196
7	2.8	0.07	7.4	102	0.196
8	1.84	0.07	8.7	93.7	0.129
9	2.16	0.07	8.8	98.9	0.151
10	1.98	0.02	7.6	96.6	0.040
11	0.59	0.02	6.5	99.4	0.012
12	0.8	0.02	6.7	96.2	0.016
13	0.8	0.02	6.2	99	0.016
14	1.05	0.02	7	88.4	0.021
15	1.8	0.02	7.3	75.3	0.036
16	1.8	0.02	6.5	92	0.036
17	1.77	0.02	7.6	82.4	0.035
18	2.3	0.02	8.2	77.1	0.046
19	2.03	0.474	7.6	74	0.962
20	1.91	0.474	8.3	65.7	0.905
21	1.91	0.474	8.2	56.8	0.905
22	1.91	0.474	6.9	62.1	0.905
23	0.76	0.474	7.4	61	0.360
24	2.13	0.474	7.6	53.2	1.010
25	2.13	0.474	6.9	59.4	1.010
26	1.51	0.474	7.5	58.7	0.716
27	2.05	0.474	7.6	58	0.972

A multivariate linear spline (MLS) analysis with binary selection among the six basis functions for $\{X_1, X_2, X_3, X_1X_2, X_1X_3, X_2X_3\}$ is then applied (model C in Program 3.10). The results of this analysis replicate the findings based on the standard linear model, since only the basis for X_2 has a posterior probability of inclusion (namely, 0.97) exceeding the prior probability of 0.5. However, the interactions X_1X_2 and X_2X_3 have the second and third highest posterior probabilities, around 0.52 and 0.42. These probabilities are considerably higher than under the linear model.

It is possible to rerun the model only with the significant effects (as in model D in Program 3.10). From the viewpoint of obtaining predictions of y that reflect model uncertainty, a form of model averaging results from allowing all possible basis functions to be relevant, as in model C (Denison *et al.*, 2002, p 56; Smith and Kohn, 1996). A two-chain run of 10 000 iterations (convergent by 1000) of model C shows 2.75 as the average number of basis functions retained, with mean deviance as 46.9 and predictive error sum of squares of 59.3. The latter criterion is more favourable to the MLS model than a deviance comparison.

Finally model E applies a MARS model involving single knot terms in X_1, X_2 and X_3 and spline products for $\{X_1, X_2\}$ and $\{X_2, X_3\}$ as in (3.17):

$$y_i = \beta_0 + \beta_1[x_{i1} - t_1]_+ + \beta_2[x_{i2} - t_2]_+ + \beta_3[x_{i3} - t_3]_+$$
$$+ \beta_4[x_{i1} - t_4]_+[x_{i2} - t_5]_+ + \beta_5[x_{i2} - t_6]_+[x_{i3} - t_7]_+ + \varepsilon_i$$

This gives a mean deviance and predictive error sum of squares similar to that for model C, namely 47.4 and 62.6. The coefficient β_2 is clearly significant as we would expect from previous analysis, while the knot estimate $t_2 = 0.1$ essentially separates observations 1–18 from observations 19–27.

Example 3.11   Penalized spline for light detection and ranging Ruppert *et al.* (2003) consider 221 observations on reflection of light from lasers in order to measure mercury in the atmosphere (Figure 3.4). The independent variable is the distance travelled before the light is reflected back and the response is the log of the ratio of received light

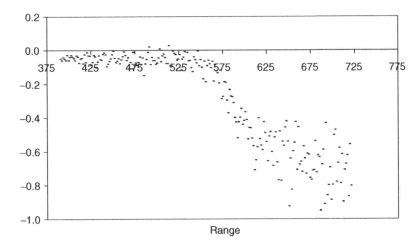

Figure 3.4   LIDAR data

from two lasers. One has a resonance frequency equal to that of mercury, the other has an alternate frequency.

The data show clear non-linearity and an increased scatter in the y–x relationship at higher x. A penalized spline model is applied assuming a constant conditional variance $\mathrm{Var}(y|x)$ (model 1) and then a heteroscedastic model (model 2). $K = 23$ knots are used, located at the 8th, 12th, 16th, ..., 96th percentiles. Thus in the first model a linear spline for the regression means is

$$y_i = \gamma_0 + \gamma_1 x_i + \sum_{k=1}^{K} \beta_k (x_i - t_k)_+ + \varepsilon_i$$

where $\beta_k \sim N(0, \sigma_\beta^2)$ and $\varepsilon_i \sim N(0, \sigma^2)$. $\mathrm{Ga}(0.1, 0.1)$ priors are assumed on both precision parameters. In the second model the regression mean is the same but the variance is non-constant according to

$$\log(1/\sigma_i^2) = \varphi_0 + \varphi_1 x_i + \sum_{k=1}^{K} \phi_k (x_i - t_k)_+$$

The fit for both models is assessed using predictions of new data and the criterion in (2.12) with $k = 1$ and $k = 1000$.

For model 1, the second half of a two-chain run of 20 000 iterations shows $PrL_1(1) = 2.51$ and $PrL_1(1000) = 3.28$. By contrast, model 2 shows $PrL_1(1) = 2.36$, $PrL_1(1000) = 3.05$ and a close fit to the data (Figure 3.5).

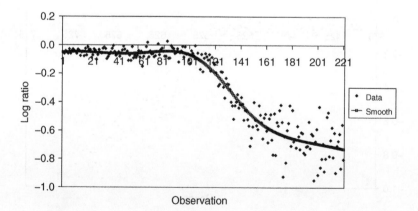

Figure 3.5 LIDAR data smooth via heteroscedastic model

3.7 DYNAMIC LINEAR MODELS AND THEIR APPLICATION IN NON-PARAMETRIC REGRESSION

For univariate or multivariate outcomes a dynamic linear model describes the evolution of observations y_t in terms of unobserved continuous states θ_t. Such models may apply equally to time or over the space of predictors arranged in ascending order. For simplicity a time frame is initially adopted. Thus a dynamic linear model (DLM) consists of an observation equation specifying the distribution of y_t, conditional on the states θ_t, and a state equation specifying how the states change dynamically (Berliner, 1996; Meyer, 2000). A first-order Markov dependence in θ_t leads to a model such as

$$y_t|\theta_t = f_1(\theta_t, \alpha) + \varepsilon_t$$
$$\theta_t = f_2(\theta_{t-1}, \beta) + \omega_t$$

where f_1 and f_2 may be linear or non-linear functions and typically the ε_t are normal with variance σ^2. Linear forms for these two equations are often used and typically involve a design matrix F_t specifying which latent states and covariates impact on the outcome, while the state equation involves a transition matrix G_t for describing how latent state values at successive times are related. Thus

$$y_t = F_t\theta_t + \varepsilon_t \tag{3.20a}$$
$$\theta_t = G_t\theta_{t-1} + \omega_t \tag{3.20b}$$

Suppose y_t is multivariate of dimension m and θ_t of dimension d, so that F_t is $m \times d$ and G_t is $d \times d$. Even though y_t might be univariate $(m = 1)$, θ_t may be of dimension d greater than one. In this case some of the design matrix elements will be zero.

Unless the analysis conditions on some early observations, initializing prior assumptions are needed for the initial latent state values. In a first-order Markov scheme for θ_t these would consist of a single parameter θ_0 which is usually assigned a diffuse prior. The errors ε_t and ω_t in (3.20) are generally taken to be mutually uncorrelated and not correlated with the initial latent state values.

Different MCMC sampling schemes have been proposed for DLMs according to the form of outcome. Carlin *et al.* (1992) suggest a Gibbs sampling scheme where the states are updated individually based on the conditional densities of the components $p(\theta_{jt}|\theta_{[-j]t}, \phi, Y)$, where ϕ specifies the observation (co)variances and state (co)variances. More efficient M–H updating schemes (Carter and Kohn, 1994; Frühwirth-

Schattner, 1994a; 1994b) involve block updating for the state vector based on the densities $p(\theta_t|\phi, Y)$ but may reduce modelling flexibility.

Other computational aspects are relevant to the identifiability of models involving state-space priors. For example, priors such as (3.20) do not usually specify a level for the θ_t series, and so devices such as recentring at each MCMC iteration, to ensure the effects sum to zero, assist in convergence. Other options might be to set one or more initial parameters to zero (Koop, 2003).

3.7.1 Some common forms of DLM

The general form (3.20) may be illustrated by commonly used models for a univariate outcome. An additive model involving an underlying trend α_t ($\equiv \theta_{1t}$), a periodic component $\gamma_t (\equiv \theta_{2t})$ and an uncorrelated error ε_t is

$$y_t = \alpha_t + \gamma_t + \varepsilon_t$$

One possible trend scheme involves a linear trend through time and an additional unknown series κ_t

$$\alpha_t = \alpha_{t-1} + \kappa_{t-1} + \omega_{1t}$$
$$\kappa_t = \kappa_{t-1} + \omega_{2t}$$

where ω_{1t} and ω_{2t} are uncorrelated over time and independent of each other. For trend or regression coefficients other commonly used schemes are the first- and second-order random walks, typically taken to be normal (Fahrmeier and Knorr-Held, 2000). The first-order random walk, sometimes denoted RW(1), has the form

$$\alpha_t = \alpha_{t-1} + \omega_t$$

where $\omega_t \sim N(0, \tau_\alpha)$ or equivalently $\alpha_t \sim N(\alpha_{t-1}, \tau_\alpha)$. The first-order random walk penalizes large discontinuities between successive values, especially if the prior on τ_α favours relatively small variances. The second-order random walk, or RW(2) model, has the form

$$\alpha_t = 2\alpha_{t-1} - \alpha_{t-2} + \omega_t$$

or equivalently $\alpha_t \sim N(2\alpha_{t-1} - \alpha_{t-2}, \tau_\alpha)$. This prior penalizes large deviations from the linear trend $2\alpha_{t-1} - \alpha_{t-2}$. As an example of how F_t and G_t in (3.20) are specified, consider a model with $y_t = \alpha_t + \varepsilon_t$, and α_t following an RW(2) prior $\alpha_t \sim N(2\alpha_{t-1} - \alpha_{t-2}, \tau_\alpha)$. Then

$$\theta_t = \begin{bmatrix} \alpha_t \\ \alpha_{t-1} \end{bmatrix} = \begin{bmatrix} 2 & -1 \\ 1 & 0 \end{bmatrix} \begin{bmatrix} \alpha_{t-1} \\ \alpha_{t-2} \end{bmatrix} + \begin{bmatrix} \omega_t \\ 0 \end{bmatrix}$$

and $y_t = (1, 0)\theta_t + \varepsilon_t$.

In the RW(1) and RW(2) models there are respectively one and two initial values to consider, namely $\theta_0 = \{\alpha_0\}$ and $\theta_0 = \{\alpha_0, \alpha_{-1}\}$. These are typically assigned diffuse priors – though see Carlin *et al.* (1992) and Berzuini and Larizza (1996) for examples of more informative initial priors. In the RW(1) model one might take $\alpha_0 \sim N(0, V_0)$ with V_0 large and known (say $V_0 = 1000$).

While apparently asymmetric these priors may be written in undirected form. For example, assume normal errors ω_t and ε_t

$$y_t = \alpha_t + \gamma_t + \varepsilon_t$$
$$\alpha_t = \alpha_{t-1} + \omega_t$$

with precisions $\psi = 1/\sigma^2$ and $\psi_\alpha = 1/\tau_\alpha$. Then the full conditionals for α_t $(t = 2, \ldots, T-1)$ are normal with means

$$(\psi_\alpha(\alpha_{t+1} + \alpha_{t-1}) + \psi y_t)(2\psi_\alpha + \psi)^{-1}$$

and variances $1/(2\psi_\alpha + \psi)$, making clear a pooling of strength both forward and backward in time. The conditional for α_1 has mean $(\psi_\alpha \alpha_2 + \psi y_t)(\psi_\alpha + \psi)^{-1}$, and that for α_T has mean $(\psi_\alpha \alpha_{T-1} + \psi y_t)$ $(\psi_\alpha + \psi)^{-1}$.

3.7.2   Robust errors

The assumption of normal errors in the dynamic linear model may not be robust against sudden shifts in the series or outlying observations. One alternative is a Student t density based on scale mixing (Carter and Kohn, 1994; Knorr-Held, 1999). This may be used in the observation equation, some or all components of the state equation, or both. For example, under a scale mixture prior on the trend component of the state equation, an RW(1) prior for α_t would become

$$\alpha_t = \alpha_{t-1} + \omega_t$$
$$\omega_t \sim N(0, \tau_\alpha/\lambda_t)$$
$$\lambda_t \sim G(0.5\nu, 0.5\nu)$$

and ν is the degrees of freedom parameter of the Student t density. More flexibility in the shape of the error density may be achieved by a discrete mixture (see section 3.5), possibly a two-group mixture with known mixture probabilities on components (West, 1997). Thus one might have an observation equation with errors

$$\varepsilon_t \sim (1 - \pi)N(0, \sigma^2) + \pi N(0, \varphi\sigma^2) \tag{3.21}$$

where $\pi = 0.05$ or 0.01 and φ is large (say between 10 and 100) to accommodate outliers. The probability $\Pr(G_t = 2)$ of belonging to the minority group will be below the prior probability (i.e. below 0.05 if $\pi = 0.05$) for most observations.

3.7.3 General additive models

DLMs have considerable utility in regression applications that modify the usual assumption of linear predictor effects. Using these priors, adaptive smoothing functions can explore the form of underlying non-linear relationships between predictors X_j and outcome y. Predictor effects are represented by univariate smooth functions in single predictors $s(X_j)$, though two-dimensional smoothing functions (e.g. in latitude and longitude) are possible (Bowman and Azzalini, 1997, chapter 8). Generalized additive models assume that smooth functions combine additively, and like the non-parametric models considered in section 3.6 they avoid the usual assumption of a global linear effect applicable to all n observations and also avoid modelling non-linearity by again assuming non-linear effects that apply globally. Additive models for metric outcomes may be extended to generalized additive models (GAMs) for discrete outcomes, such as binary or count-dependent variables (Chapter 4).

For a metric outcome $y_1, \ldots, y_n$ assume corresponding values of a single predictor $x_1, \ldots, x_n$ ordered such that

$$x_1 < x_2 < \cdots < x_n$$

Note that the observations $\{y_t, x_t\}$ need not be a time series. The GAM seeks to represent the mean $s(x)$ as a locally smooth function of x as x varies through its range. The plot of $s(x)$ against x is the non-parametric analogue of the usual linear regression plot.

The observation model error term is typically assumed to be parametric (e.g. normal, Student t), though it can also be modelled non-parametrically, e.g. via a Dirichlet process. The GAM with this single predictor is then

$$y_t = \beta_0 + s(x_t) + \varepsilon_t$$

where typically $\varepsilon_t \sim N(0, \sigma^2)$. Let $s_t = s(x_t)$ be the smooth function representing the locally changing impact of x on y as it varies over its range. A convenient prior might then be provided by normal or Student random walks in the first, second or higher differences of s_t (Fahrmeier and Lang, 2001; Koop and Poirier, 2001). Note that centring of the s_t or some other constraint (e.g. $s_1 = 0$) may be needed for identifiability since otherwise the level of the s_t is confounded with β_0. Because there is only local smoothing, inferences may also be sensitive to priors assumed for evolution variance τ^2 and other aspects of the model.

If there is equal spacing then the first- and second-order random walk priors are just

$$s_t \sim N(s_{t-1}, \tau^2) \tag{3.22}$$

$$s_t \sim N(2s_{t-1} - s_{t-2}, \tau^2) \tag{3.23}$$

For metric responses the parameterization $\tau^2 = \sigma^2 \lambda$ is often used; smaller values of λ result in a smoother curve. Koop and Poirier (2001) outline a method for eliciting λ based on the cross-validation function

$$\sum_{t=1}^{n} [y_t - \beta_0 - s(x_t)]^2$$

In general regression applications the x_t are usually unequally spaced, and the random walk prior is then modified to weight each preceding point differently. The precision of s_t is reduced the larger the gap between x_t and its predecessor (remembering that the x_t are ordered). Suppose gaps between points are $\delta_1 = x_2 - x_1, \delta_2 = x_3 - x_2, \ldots, \delta_{n-1} = x_n - x_{n-1}$. A first-order normal random walk would then be

$$s_t \sim N(s_{t-1}, \delta_t \tau^2) \tag{3.24}$$

and a second-order one would be

$$s_t \sim N(\nu_t, \delta_t \tau^2) \tag{3.25a}$$

where

$$\nu_t = s_{t-1}(1 + \delta_t/\delta_{t-1}) - s_{t-2}(\delta_t/\delta_{t-1}) \tag{3.25b}$$

(Fahrmeir and Lang, 2001). Separate, usually fixed effect priors are assumed for the initial values (e.g. s_1 in a first-order random walk).

If there are ties in the x variables with only n_u distinct x values, denoted $\{x_t^*, t = 1, \ldots, n_u\}$ then the above priors would be on the differences $\delta_t^* = x_t^* - x_{t-1}^*$ in the unique x and it would be necessary to specify a grouping index O_i (ranging between 1 and n_u) for each observation $i = 1, \ldots, n$.

If there is more than one predictor then a semi-parametric model might be adopted with smooth functions $s_j(x_j)$ on a subset $j = 1, \ldots, q$ of p predictors, with the remainder modelled by assuming global linearity. So

$$y_t = \beta_0 + s_1(x_{1t}) + s_2(x_{2t}) + \cdots + s_q(x_{qt}) + \beta_1 x_{q+1,t} + \cdots + \beta_{p-q} x_{p,t} + \varepsilon_t$$

If non-parametric functions are estimated by state space methods for several regressors $x_{1t}, x_{2t}, x_{3t} \ldots$, then an ordering index O_{1t}, O_{2t},

$O_{3t}, \ldots, O_{pt}$ for each of p regressors is necessary. These indices range between 1 and $n_{u1}, n_{u2}, \ldots, n_{up}$ (rather than between 1 and n) if there are tied predictor values.

3.7.4 Alternative smoothness priors

Other smoothness priors have been proposed. A scheme analogous to (3.24) but allowing a choice between RW(1) and RW(2) dependence for unequally spaced x is proposed by Berzuini and Larizza (1996), i.e.

$$s_t \sim N(\Gamma_t, V_t)$$

where

$$\Gamma_t = s_{t-1}[1 + (\delta_t/\delta_{t-1})\exp(-\alpha\delta_t)] - s_{t-2}[(\delta_t/\delta_{t-1})\exp(-\alpha\delta_t)]$$
$$V_t = \delta_t^2 \tau^2 [1 - \exp(-2\alpha\delta_t)]$$

and $\alpha > 0$. Larger values of α, such that $\exp(-\alpha\delta_t)$ tends to zero, imply an approximate RW(1) prior and less smoothness.

In a Fourier series prior (Lenk, 1999) with x_t defined on the interval $[a, b]$, $s(x_t)$ may be defined by a truncated series

$$s(x) = \sum_{k=1}^{K} \omega_k \varphi_k(x)$$

where

$$\varphi_k(x) = \left(\frac{2}{b-a}\right)^{0.5} \cos\left[\pi k\left(\frac{x-a}{b-a}\right)\right]$$

In line with local smoothing, the prior for ω_k assumes decay as k increases; for instance, in the geometric smoother prior

$$\omega_k \sim N(0, \tau^2 \exp(-\psi k))$$

where $\psi > 0$ determines the rate of decay of θ_k.

Another approach (Wood and Kohn, 1998; Wahba, 1983) is the state-space version of spline smoothing. For a spline of order $2m - 1$, $s_t = s(x_t)$ is generated by a differential equation

$$\frac{d^m s_t}{dt^m} = \tau \frac{dW_t}{dt} \tag{3.26}$$

with W_t a Weiner process and τ^2 the evolution variance. The state vector

$$V_t = \left(s_t, \frac{ds_t}{dt}, \frac{d^2 s_t}{dt^2}, \ldots, \frac{d^{(m-1)} s_t}{dt^{(m-1)}}\right)$$

is then of order m, evolving stochastically according to

$$V_t = F_t V_{t-1} + u_t \qquad (3.27a)$$

where F_t is an $m \times m$ transition matrix and u_t is a multivariate error. Consider $m = 2$ (corresponding to a cubic spline), then $V_t = (s_t, ds_t/dt)$ is bivariate and the transition matrix is

$$F_t = \begin{bmatrix} 1 & \delta_t \\ 0 & 1 \end{bmatrix} \qquad (3.27b)$$

where $\delta_t = x_{t+1} - x_t$. When $m = 2$, the u_t are bivariate (e.g. MVN) with zero mean and covariance $\tau^2 U_t$, where

$$U_t = \begin{bmatrix} \delta_t^3/3 & \delta_t^2/2 \\ \delta_t^2/2 & \delta_t \end{bmatrix} \qquad (3.27c)$$

As usual there may be grouping in the x values and the prior (3.27) would be on the number of distinct values $n_u \le n$. Each observation would have a grouping index O_t with values between 1 and n_u. The full regression in x for $m = 2$ can then be written

$$y = \beta_0 + \beta_1 x + s(O_t) + \varepsilon_t$$

Example 3.12   Static expiratory pressure in cystic fibrosis patients
Consider data from Everitt and Rabe-Hesketh (2001) on the maximum static expiratory pressure (y) for $n = 25$ cystic fibrosis patients in relation to four predictors: weight, BMP, FEV and RV (Table 3.7). A GAM is assumed with the impacts of all these regressors modelled non-parametrically via RW(2) priors. The gaps between successive ordered values of the predictors are unequal so the prior is as in (3.27).

Only one predictor, namely weight, has 25 distinct values, while BMP has only 16 distinct values so O_{2t} has only 16 values (Table 3.7). The prior in s_{2t} will have the form

$$s_{2t} \sim N(\nu_{2t}, \delta_{2t}\tau_2^2) \quad t = 2, 16$$

where $\nu_{2t} = s_{2,t-1}(1 + \delta_{2t}/\delta_{2,t-1}) - s_{2,t-2}(\delta_{2t}/\delta_{2,t-1})$ and $\delta_{22} = 1$ (i.e. 65 minus 64), $\delta_{23} = 1, \ldots, \delta_{2,16} = 2$ (i.e. 97 minus 95). The priors on $\tau_j^2 (j = 1, 4)$ and the measurement error variance σ^2 are linked via

$$\tau_j^2 = \sigma^2 \lambda_j$$

where $G(0.1,1)$ priors on λ_j are adopted favouring lower evolution variances as compared with σ^2.

A standard multiple linear regression has a DIC of 234 (see code A). By contrast a GAM in all four predictors (code B in Program 3.12) has a

Table 3.7 Original values and observation ranks, cystic fibrosis data

Subject	Weight	BMP	FEV	RV	Order (Weight)	Order (BMP)	Order (FEV)	Order (RV)	y
1	13.1	68	32	258	2	5	11	16	95
2	12.9	65	19	449	1	2	2	24	85
3	14.1	64	22	441	3	1	4	23	100
4	16.2	67	41	234	4	4	17	14	85
5	21.5	93	52	202	6	14	21	8	95
6	17.5	68	44	308	5	5	18	19	80
7	30.7	89	28	305	9	11	8	18	65
8	28.4	69	18	369	8	6	1	21	110
9	25.1	67	24	312	7	4	6	20	70
10	31.5	68	23	413	10	5	5	22	95
11	39.9	89	39	206	14	11	16	10	110
12	42.1	90	26	253	15	12	7	15	90
13	45.6	93	45	174	18	14	19	4	100
14	51.2	93	45	158	20	14	19	1	80
15	35.9	66	31	302	12	3	10	17	134
16	34.8	70	29	204	11	7	9	9	134
17	44.7	70	49	187	17	7	20	6	165
18	60.1	92	29	188	22	13	9	7	120
19	43.6	69	38	172	16	6	15	3	130
20	37.2	72	21	216	13	9	3	11	85
21	54.6	86	37	184	21	10	14	5	85
22	64	86	34	225	23	10	13	13	160
23	73.8	97	57	171	25	16	22	2	165
24	51.1	71	33	224	19	8	12	12	95
25	71.5	95	52	225	24	15	21	13	195

DIC of 219 with 19 effective parameters. This is on the basis of a two-chain run of 20 000 iterations, with convergence after 5000. Plots of the smooths are given in Figures 3.6 to 3.9.

The same model was also fitted using truncated cubic splines in the four predictors. The predictors are taken in standard form and five knot points at $\{-2, -1, 0, 1, 2\}$ are assumed. Without selection on the regressors, there are apparently many non-significant coefficients. The smooth plots, however, resemble those from the GAM. Code D uses the regression selection method of Smith and Kohn (1996) in a cubic spline and also produces non-linear smooths with similar characteristics to the GAM.

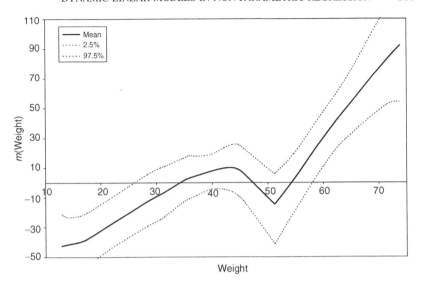

Figure 3.6   Smooth in weight

Example 3.13   Electricity use   To illustrate the use of alternative smoothness priors, specifically the state-space version of spline regression in (3.27), consider again the data on electricity use by temperature. The full gamma conditional for $1/\tau^2$ is used to update this parameter, with prior values $\eta = (1, 1)$. There are 37 distinct temperature values and it is

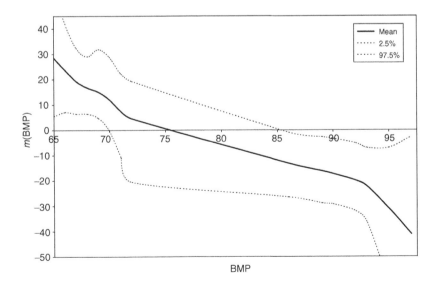

Figure 3.7   Smooth for BMP

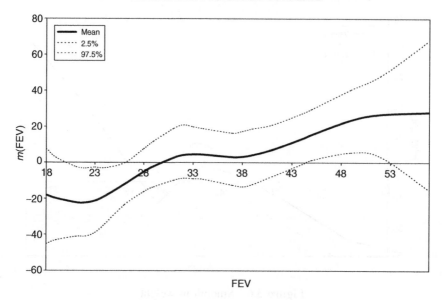

Figure 3.8   Smooth for FEV

Example 3.13 (continued) ... To illustrate the use of smoothing splines, which are polynomials, the discrepancy smooth of splines, in ... ... 3.9 ... smooth spline for ... smoothing ... a ... ... on ... ... the full posterior cumulated and ... ... in this posterior ... prior values preset ... ... There are ... ... at ... covariate values, and ...

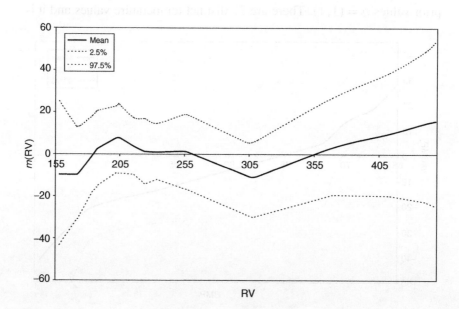

Figure 3.9   Smooth for RV

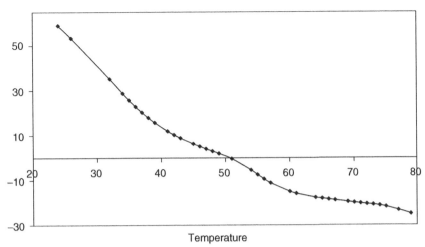

Figure 3.10 Electricity use and temperature

necessary to supply a group index O_i to each of the 55 cases. Since an intercept is included it is necessary to centre the 37 smooth effects. Two chains are run for 10 000 iterations to produce a smooth (Figure 3.10) showing a sharp fall in use at lower temperatures.

EXERCISES

1. Using model A in Program 3.1 obtain the pseudo marginal likelihood of M_1 (with predictor W = adjusted density) and so derive the pseudo Bayes factor assuming $P(M_0) = P(M_1) = 0.5$. (This involves monitoring g[] in Program 3.1.) Also obtain the DIC for models M_1 and M_0 by monitoring the deviance (Dv in Program 3.1) as well as the model means and precisions (mu[i] and tau); mu[] can be monitored via inference/summary in BUGS rather than the full monitoring option.

2. In Example 3.1 try the model $M_3 : y_i = \alpha_3 + \beta_3 x_i + \gamma_3 w_i + \varepsilon_{i3}$ and assess its fit against M1 and M2 using the pseudo marginal likelihood, the path sampling method for estimating the Bayes factor, and via the posterior model probability based on parallel sampling.

3. In Example 3.5 try fitting the puromycin data with a gamma scale mixture $\lambda_i \sim Ga(\nu/2, \nu/2)$ where the degrees of freedom are unknown. Does this improve over the Ga(1,1) option?

4. In Example 3.6 try a normal scale mixture model rather than a heteroscedastic model and detect which observations are most 'outlying' with regard to the model

$$y_i = V_i^{1/3} = \beta_0 + \beta_1 H_i + \beta_2 D_i + \varepsilon_i$$

Also try an outlier model with variance inflation, namely

$$f(y_i|\mu, \sigma^2, \omega, \kappa) = (1 - \omega)\phi(y_i|\mu, \sigma^2) + \omega\phi(y_i|\mu, (1 + \kappa)\sigma^2)$$

where $\kappa = 4$ and $\delta_i \sim \text{Bern}(\omega)$ where $\omega = 0.05$. Which observation has the highest posterior average for δ_i.

5. In Example 3.8 try a DPP mixture with $\{\eta_1, \eta_2\} = (1, 1)$ and find the modal number of clusters.

6. In Example 3.9 try a truncated cubic and B spline models with ten equally spaced knots at 800, 825, 850, 875, 900, 925, 950, 975, 1000 and 1025, and with predictor selection applied both to β and γ coefficients. How does the fit obtained compare with that obtained by using the optimal knots with locations assumed known but with predictor selection?

7. In Example 3.10 obtain the DIC of the MLS model C with basis function selection. This requires monitoring the means (m[i] in the program) and precision τ in order to estimate the deviance at the posterior mean, where the deviance is defined as

$$\tau \sum (Y_i - \mu_i)^2 - n \log(\tau)/2\pi$$

8. In Example 3.13 try fitting the same model but without using an intercept. This means it is no longer necessary to centre the smooth effects

REFERENCES

Aitkin, M. (1987) Modelling variance heterogeneity in normal regression using GLIM. *Applied Statistics*, **36**, 332–339.

Altman, D. (1991) *Practical Statistics for Medical Research*. Chapman and Hall: London.

Anderson, R. and Bancroft, T. (1959) *Statistical Theory in Research*. McGraw-Hill: New York.

Andrews, D. and Mallows, C. (1974) Scale mixtures of normal distributions. *Journal of the Royal Statistical Society, Series B*, **36**, 99–102.

Berliner, L. (1996) Hierarchical Bayesian time series models. In *Maximum Entropy and Bayesian Methods*, Hanson, K. and Silver, R. (eds). Kluwer Academic: Dordrecht.

Berry, S., Carroll, R. and Ruppert, D. (2002) Bayesian smoothing and regression splines for measurement error problems, *Journal of the American Statistical Association*. **97**, 160–169.

Berzuini, C. and Larizza, C. (1996) A unified approach for modelling longitudinal and failure time data, with application in medical monitoring. *IEEE Transactions on Pattern Analysis and Machine Intelligence*, **18**, 109–123.

Biller, C. (2000) Adaptive Bayesian regression splines in semiparametric generalized linear models. *Journal of Computational and Graphical Statistics*, **9**, 122–140.

Boscardin, W. and Gelman, A. (1996) Bayesian computation for parametric models of heteroscedasticity in the linear model. In *Advances in Econometrics, Volume 11A, Bayesian Computational Methods and Applications*, Hill, R (ed.). JAI Press: Greenwich, CT.

Bowman, A. and Azzalini, A. (1997) *Applied Smoothing Techniques for Data Analysis*. Oxford University Press: New York.

Burnham, K. and Anderson, D. (2002) *Model Selection and Multimodel Inference: A Practical Information-theoretic Approach*, 2nd Edition. New York: Springer.

Carlin, B. and Chib, S (1995) Bayesian model choice via the Markov chain Monte Carlo methods, *Journal of the Royal Statistical Society, Series B*, **57**, 473–484.

Carlin, B. and Polson, N (1991) Inference for nonconjugate Bayesian models using the Gibbs sampler. *Canadian Journal of Statistics*, **19**, 399–405.

Carlin, B., Polson, N. and Stoffer, D. (1992) A Monte Carlo approach to nonnormal and nonlinear state-space modeling. *Journal of the American Statistical Association*, **87**, 493–500.

Carter, C. and Kohn, R. (1994) On Gibbs sampling for state space models. *Biometrika*, **81**, 541–553.

Celeux, G. Hurn, M. and Robert, C. (2000) Computational and inferential difficulties with mixture posterior distributions. *Journal of the American Statistical Association*, **95**, 957–970.

Cepeda, E. and Gamerman, D. (2000) Bayesian modeling of joint regressions for the mean and covariance matrix. *Technical Report* 135, Statistical Laboratory LES-UFRJ.

Chatterjee, S., Handcock, M. and Simonoff, J. (1995) *A Casebook for a First Course in Statistics and Data Analysis*. John Wiley & Sons: New York.

Congdon, P. (2001) *Bayesian Statistical Modelling*. John Wiley & Sons: Chichester.

Dayton, C. and Macready, G. (1988) Concomitant-variable latent class models. *Journal of the American Statistical Association*, **83**, 173–178.

De Boor, C. (1978) *A Practical Guide to Splines*. Springer: Berlin.

Dempster, A., Laird, N. and Rubin, D. (1977) Maximum likelihood from incomplete data via the EM algorithm. *Journal of the Royal Statistical Society, Series B*, **34**, 1–38.

Denison, D., Mallick, B. and Smith, A. (1998a) Automatic Bayesian curve fitting. *Journal of Royal Statistical Society, Series B*, **60**, 333–359.

Denison, D., Mallick, B. and Smith, A. (1998b) Bayesian MARS. *Statistics and Computing*, **8**, 337–346.

Denison, D., Holmes, C., Mallick, B. and Smith, A. (2002) *Bayesian Methods for Nonlinear Classification and Regression*. John Wiley & Sons: New York.

Diebolt, J. and Robert, C. (1994) Estimation of finite mixture distributions through Bayesian sampling. *Journal of the Royal Statistical Society, Series B*, **56**, 363–375.

Draper, N. and Smith, H. (1966) *Applied Regression Analysis*. John Wiley & Sons: New York.

Escobar, M. and West, M. (1995) Bayesian density estimation and inference using mixtures. *Journal of the American Statistical Association*, **90**, 577–588.

Eubank, R. (1988) *Spline Smoothing and Nonparametric Regression*. Marcel Dekker: New York.

Everitt, B. and Rabe-Hesketh, S. (2001) *Analyzing Medical Data Using S-Plus*. Springer: New York.

Fahrmeier, L. and Knorr-Held, L. (2000) Dynamic and semiparametric models. In *Smoothing and Regression: Approaches, Computation and Application*, Schimek, M (ed.). John Wiley & Sons: New York, 513–544.

Fahrmeier, L. and Lang, S. (2001) Bayesian inference for generalized additive mixed models based on Markov random field priors. *Journal of the Royal Statistical Society, Series C*, **50**, 201–220.

Ferguson, T. (1973) A Bayesian analysis of some nonparametric problems. *Annals of Statistics*, **1**, 209–230.

Fernandez, C. and Steel, M. (1998a) On the dangers of modelling through continuous distributions: a Bayesian perspective. In *Bayesian Statistics 6*, Bernardo, J., Berger, J., Dawid, A. and Smith, A (eds). Oxford University Press: Oxford, 1–16.

Fernandez, C. and Steel, M. (1998b) On Bayesian modeling of fat tails and skewness. *Journal of the American Statistical Association*, **93**, 359–371.

Fernandez, C., Ley, E. and Steel, M. (2001a) Benchmark priors for Bayesian model averaging. *Journal of Econometrics*, **100**, 381–427.

Fernandez, C., Ley, E. and Steel, M. (2001b) Model uncertainty in cross-country growth regressions. *Journal of Applied Econometrics*, **16**, 563–576.

Friedman, J. and Silverman, B. (1989) Flexible parsimonious smoothing and additive modeling. *Technometrics*, **31**, 1–39.

Frühwirth-Schnatter, S. (1994a) Applied state space modelling of non-Gaussian time series using integration-based Kalman filtering. *Statistics and Computing*, **4**, 259–269.

Frühwirth-Schnatter, S. (1994b) Data augmentation and dynamic linear models. *Journal of Time Series Analysis*, **15**, 183–202.

Frühwirth-Schnatter, S. (2001) MCMC estimation of classical and dynamic switching and mixture models. *Journal of the American Statistical Association*, **96**, 194–209.

Gelman, A., Carlin, J., Stern, H. and Rubin, D. (1995) *Bayesian Data Analysis*. Chapman and Hall: London.

Gelman, A., Carlin, J., Stern, H. and Rubin, D. (2003) *Bayesian Data Analysis, 2nd Edition. CRC Press: Boca Raton, FL.*

George, E. and McCulloch, R. (1993) Variable selection via Gibbs sampling. *Journal of the American Statistical Association*, **88**, 881–889.

George, E. and McCulloch, R. (1997) Approaches for Bayesian variable selection. *Statistica Sinica*, **7**, 339–373.

Geweke, J. (1993) Bayesian treatment of the independent Student-t linear model. *Journal of Applied Econometrics*, **8S**, 19–40.

Hinkley, D. and Runger, G. (1984) The analysis of transformed data. *Journal of the American Statistical Association*, **79**, 302–319.

Hoeting, J., Raftery, A. and Madigan, D. (2002) Bayesian variable and transformation selection in linear regression. *Journal of Computational and Graphical Statistics*, **11**, 485–507.

Holmes, C. and Mallick, B. (2001) Bayesian regression with multivariate linear splines. *Journal of the Royal Statistical Society, Series B*, **63**, 3–17.

Hurn, M., Justel, A. and Robert, C. (2003) Estimating mixtures of regressions. *Journal of Computational and Graphical Statistics*, **12**, 55–79.

Ishwaran, H. and James, L. (2002) Approximate Dirichlet process computing in finite normal mixtures: smoothing and prior information. *Journal of Computational and Graphical Statistics*, **11**, 508–532.

Justel, A. and Pena, D. (2001) Heterogeneity and model uncertainty in Bayesian regression models. Dept. of Mathematics, Universidad Autonoma de Madrid.

Knorr-Held, L. (1999) Conditional prior proposals in dynamic models. *Scandinavian Journal of Statistics*, **26**, 129–144.

Koop, G. (2003) *Bayesian Econometrics*, John Wiley & Sons: Chichester.

Koop, G. and Poirier, D. (2001) Bayesian variants of some classical semiparametric regression techniques. *Working Papers*, University of California Irvine – School of Social Sciences.

Kuo, L. and Mallick, B. (1998) Variable selection for regression models. *Sankhya* **60B**, 65–81.

Laud, P. and Ibrahim, J. (1995) Predictive model selection. *Journal of the Royal Statistical Society, Series B*, **57**, 247–262.

Lee, P. (1997) *Bayesian Statistics: An Introduction*, 2nd Edition. Arnold: London.

Lenk, P. (1999) Bayesian inference for semiparametric regression using a Fourier representation. *Journal of the Royal Statistical Society, Series B*, **61**, 863–879.

MacNab, Y. and Dean, C. (2001) Autoregressive spatial smoothing and temporal spline smoothing for mapping rates. *Biometrics*, **57**, 949–956.

McCullagh, P. and Nelder, J. (1989) *Generalized Linear Models*, 2nd Edition. Chapman and Hall: London.

Madigan, D. and York, J. (1995) Bayesian graphical models for discrete data. *International Statistical Review*, **63**, 215–232.

Meyer, R. (2000) Applied Bayesian data analysis using state-space models. In *Data Analysis: Scientific Modeling and Practical Applications*, Gaul, W., Opitz, O. and Schader, M (eds). Springer: New York, 259–271.

Moreno, E. and Lisco, B. (2003) A default Bayesian test for the number of components in a mixture. *Journal of Statistical Planning and Inference*, **111**, 129–142.

Ntzoufras, I. (1999) Aspects of Bayesian model and variable selection using MCMC. Dept. of Statistics, Athens University of Economics and Business.

Pericchi, L. (1981) A Bayesian approach to transformations to Normality. *Biometrika*, **68**, 35–43.

Pettit, L. (1990) The conditional predictive ordinate for the Normal distribution. *Journal of the Royal Statistical Society, Series B*, **52**, 175–184.

Raftery, A., Madigan, D. and Hoeting, J. (1996) Bayesian model averaging for linear regression models. *Journal of the American Statistical Association*, **92**, 179–191.

Richardson, S. and Green, P. (1997) On the Bayesian analysis of mixtures with an unknown number of components. *Journal of the Royal Statistical Society, Series B*, **59**, 731–792.

Richardson, S., Viallefont, V. and Green, P. (2002) Bayesian analysis of Poisson mixtures. *Journal of Nonparametric Statistics*, **14**, 181–202.

Robert, C. (1996) Mixtures of distributions: inference and estimation. In *Markov Chain Monte Carlo in Practice*, Gilks, W., Richardson, S. and Spiegelhalter, D. (eds). Chapman and Hall: London, 441–464.

Ruppert, D. and Carroll, R. (2000) Spatially-adaptive penalties for spline fitting. *Australian & New Zealand Journal of Statistics*, **42**, 205–223.

Ruppert, D., Wand, M. and Carroll, R. (2003) *Semiparametric Regression*. Cambridge University Press: Cambridge.

Sahu, S., Dey, D. and Branco, M. (2003) A new class of multivariate skew distributions with applications to Bayesian regression models. *Canadian Journal of Statistics*, **31**, 129–150.

Smith, M. and Kohn, R. (1996) Nonparametric regression using Bayesian variable selection. *Journal of Econometrics*, **75**, 317–334.

Smith, M., Kohn, R. and Yau, P. (2001) Nonparametric bivariate surface estimation. In *Smoothing and Regression: Approaches, Computation and Application*, Schimek, M. (ed.). John Wiley & Sons: New York, 545–580.

Verdinelli, I. and Wasserman, L. (1991) Bayesian analysis of outlier problems using the Gibbs sampler. *Statistics and Computing*, **1**, 105–117.

Wahba, G. (1983) Bayesian confidence intervals for the cross-validated smoothing spline. *Journal of the Royal Statistical Society, Series B*, **45**, 133–150.

Weiss, R. (1994) Pediatric pain, predictive inference, and sensitivity analysis. *Evaluation Review*, **18**, 651–677.

West, M. (1984) Outlier models and prior distributions in Bayesian linear regression. *Journal of the Royal Statistical Society, Series B*, **46**, 431–439.

West, M. (1987) On scale mixtures of normal distributions. *Biometrika*, **74**, 646–648.

West, M. (1997) Bayesian time series: models and computations for the analysis of time series in the physical sciences. In *Maximum Entropy and Bayesian Methods 15*, (Hanson K. and Silver R. (eds). Kluwer: Dordrecht.

Wood, S. and Kohn, R. (1998) A Bayesian approach to robust nonparametric binary regression. *Journal of the American Statistical Association*, **93**, 203–213.

Zellner, A. (1986) On assessing prior distributions and Bayesian regression analysis with g-prior distributions. In *Bayesian Inference and Decision Techniques*, Goel, P. and Zellner, A. (eds). North-Holland: Amsterdam, 233–243.

CHAPTER 4

Models for Binary and Count Outcomes

4.1 INTRODUCTION: DISCRETE MODEL LIKELIHOODS VS. DATA AUGMENTATION

A major motivation for the general linear model has been for the analysis of counts and binary data. Aggregations of binary data provide binomial data – see Table 4.1 of McCullagh and Nelder (1989) on ways of presenting binary data. In the analysis of count, binary and binomial data, classical methods rely especially on maximum likelihood (ML) or empirical Bayes (EB) techniques applied to densities such as the Poisson, negative binomial, Bernoulli, binomial and beta–binomial.

A Bayesian approach to binary, binomial and count data may have advantages when regression coefficients are not necessarily symmetrically (e.g. normally) distributed, and the actual 'exact' shape of posterior distributions of such coefficients can be assessed from repeated sampling from the posterior. The asymptotic approximations of classical likelihood methods may be particularly suspect in small samples or in samples with binary data (Zellner and Rossi, 1984). As usual the Bayes sampling approach also permits assessment of exact densities of derived statistics (e.g. functions of odds ratios derived from logit regression coefficients) (Fahrmeier and Knorr-Held, 2000).

A fully Bayes analysis is facilitated by MCMC methods, though simple Gibbs sampling is confined to conjugate analysis without regressors. For binary and Poisson regressions the full conditionals of the regression parameters are not in closed form, but the likelihood is log-concave and adaptive rejection sampling (ARS) can be used (Gilks and Wild, 1992).

Bayesian Models for Categorical Data   P. Congdon
© 2005 John Wiley & Sons, Ltd

An alternative is to introduce augmented (e.g. normal) data that underlie the discrete data, especially in binary regression, so that analysis reduces to the metric regression in terms of residual analysis and other techniques discussed in Chapter 3 (Albert and Chib, 1993). Other avenues explored in recent Bayesian research include modifications to standard link functions, and methods for correlated multivariate binary and count responses.

4.1.1   Count data

To illustrate the fully Bayesian approach in terms of direct likelihood updating of the prior (which gives similar results to ML or EB methods if diffuse priors are used), suppose observed counts $y = \{y_1, y_2, y_3, \ldots, y_n\}$ are a sample from a Poisson density with common mean μ. Suppose also that the data are not over- or underdispersed (variance greater than the mean and less than the mean respectively), though moderate departures from the variance = mean relationship may be allowed. Since μ is positive, the prior for μ might be taken as a gamma with parameters α and β, and so mean α/β. In a simple Poisson analysis with μ constant over subjects, α and β often take preset values (e.g. $\alpha = \beta = 0.001$ provides a diffuse prior). The Poisson likelihood for y given μ is proportional to

$$\left[\prod_{i=1}^{n} \exp(-\mu)\mu^{y_i} \right]$$

and the posterior in the three parameters $\{\mu, \alpha, \beta\}$, with $\{\alpha, \beta\}$ as unknowns, is proportional to

$$\left[\prod_{i=1}^{n} \exp(-\mu)\mu^{y_i} \right] [\alpha^{\beta} \mu^{\alpha-1} \exp(-\beta\mu)/\Gamma(\alpha)]P(\alpha, \beta)$$

where $P(\alpha, \beta)$ is the prior for $\{\alpha, \beta\}$. As a function of μ alone, the posterior is proportional to

$$\left[\prod_{i=1}^{n} \exp(-\mu)\mu^{y_i} \right] \mu^{\alpha-1} \exp(-\beta\mu) = \mu^{\alpha + \sum_i y_i - 1} \exp[-\mu(\beta + n)]$$

so the posterior for μ is a gamma, namely $Ga(\alpha + \sum_i y_i, \beta + n)$. Since prior and posterior densities have the same form, this provides a conjugate analysis.

It might not be sensible to assume a common rate. One option assumes that the varying means μ_i are separate fixed effects, typically with diffuse gamma priors such as $\mu_i \sim \text{Ga}(a, a)$ where a is small (e.g. $a = 0.001$) and taken as known. An alternative is to seek pooling of information over the units in order to stabilize estimates of unit-level Poisson rates μ_i especially if event totals in each unit i are small. If a gamma prior for μ_i is assumed as the pooling density, then α and β are taken as unknown hyperparameters governing the degree of pooling. The posterior density is then proportional to

$$\left[\prod_{i=1}^{n} \exp(-\mu_i)\mu_i^{y_i}\right] \alpha^\beta \mu_i^{\alpha-1} \exp(-\beta\mu_i)\Gamma(\alpha)$$

and the posterior density for each μ_i is $\text{Ga}(\alpha + y_i, \beta + 1)$. Such mixture models are considered more extensively in Chapter 5.

Poisson data may be in the form of observed counts in relation to expected counts E_i, as in disease mapping or hospital mortality applications (Albert, 1999) or as counts observed for certain exposure times t_i (McCullagh and Nelder, 1989, pp 193–208). So in the disease mapping case

$$y_i \sim \text{Poi}(\mu_i)$$
$$\mu_i = E_i \nu_i$$

where ν_i are unknowns. Where relevant, E_i may be taken as random (Wakefield and Best, 1999) but are usually taken as known, with $\nu_i = \mu_i/E_i$ amounting to relative risks of disease. With a Gamma prior $\nu_i \sim \text{Ga}(\alpha, \beta)$, and E_i as known constants, the posterior density is proportional to

$$\left[\prod_{i=1}^{n} \exp(-E_i\nu_i)\nu_i^{y_i}\right] \alpha^\beta \nu_i^{\alpha-1} \exp(-\beta\nu_i)/\Gamma(\alpha)$$

and the posterior density for each ν_i is $\text{Ga}(\alpha + y_i, \beta + E_i)$. The posterior mean of each underlying rate, namely

$$E(\nu_i|\alpha, \beta) = (\alpha + y_i)/(\beta + E_i)$$

may be written

$$(1 - B_i)y_i/E_i + B_i(\alpha/\beta)$$

that is, a weighted average of the crude fixed effects estimate and the prior mean α/β of the pooling density. The ratio $B_i = \alpha\beta/(\alpha\beta + E_i)$ amounts to a shrinkage factor which leads to greater pooling of smoothed rates $E(\nu_i)$ to the prior mean for smaller expected counts E_i.

If covariates $x_i = (x_{i1}, \ldots, x_{ip})'$ are introduced to explain variations between Poisson rates then a conjugate analysis is no longer possible. As outlined in Chapter 1, binomial, binary and Poisson regression usually involves introducing a link to ensure that the Poisson means or binary/binomial probabilities are appropriately constrained. Thus for count data assumed to be Poisson with mean μ_i, a link $g()$ is needed to convert the linear predictor $\eta_i = \beta_0 + \beta x_i$ onto a positive scale for μ_i. The link most commonly used is the $\log_e$ transform, so that

$$g(\mu_i) = \log_e(\mu_i) = \beta_0 + \beta x_i$$

since the inverse link $g^{-1} = \exp$ is analytically simple. For data with exposures or expected counts E_i then

$$\mu_i = E_i \nu_i$$
$$g(\nu_i) = \beta_0 + \beta x_i$$

or equivalently

$$g(\mu_i) = g(E_i \nu_i)$$

so that, for example,

$$\log(\mu_i) = \log(E_i) + \beta_0 + \beta x_i \tag{4.1}$$

E_i is then called an offset (see Example 4.4). A regression model may be combined with pooling of information via random effects (see Chapter 5). For example, one may assume $y_i \sim \text{Po}(\nu_i E_i)$, and a conjugate prior $\nu_i \sim \text{Ga}(\alpha, \alpha/\lambda_i)$ where $\lambda_i = \exp(\gamma x_i)$. Conditional on α and γ, the posterior mean of ν_i is $(y_i + \alpha)/(E_i + \alpha/\lambda_i)$. Albert (1999) presents approximate marginal likelihoods for comparing such a hierarchical model against the fixed effects alternative defined as $\alpha \to 0$.

While direct specification of priors on coefficients $\beta_j(j = 1, \ldots, p)$ in count regression is widely used, several methods have been suggested to include more directly historic or elicited information. Bedrick *et al.* (1996) and others have suggested data augmentation (or conditional means) priors. For count data this involves eliciting a mean value μ_{ijr} at $r = 1, \ldots, R$ values of the jth predictor x_{ij} and then including this information as implicit 'prior data' in the form of a gamma density. For a large number of covariates the mean values might just be elicited for a given number (e.g. p) of predictor combinations. Suppose $p = 1$ in (4.1) and that x_i is standardized, so that $\beta_0 \approx 0$ when $\sum_i y_i = \sum_i E_i$. Taking $R = 2$, the relative risk ν might be elicited as 1.5 for $x = 1$ but as 0.75 when $x = -1$. If one were willing to assign five prior observations to each of these elicitations then the prior is $\text{Ga}(7.5, 5)$ for $x = 1$ and $\text{Ga}(3.75, 5)$

when $x = -1$. These priors induce priors on β. If there were two predictors, both standardized and both factors that increase relative risk, then one might consider just two combinations $(x_1, x_2) = (-1, -1)$ and $(1, 1)$ of the predictors at which to obtain elicitations. In related work, Meyer and Laud (2002) propose conjugate priors for β in Poisson regression of the form

$$g(\beta_1, \ldots, \beta_p) \propto \exp\left\{\sum_i \gamma_0 [\mu_{i0} \beta x_i - \exp(\beta x_i)]\right\}$$

where μ_{i0} is an elicited mean and γ_0 measures strength of belief in the elicitation.

4.1.2 Binomial and binary data

For binomial data the occurrences of a certain event are counted among totals at risk n. The stylized version of the binomial refers to numbers of 'successes' y in relation to total trials n. Let π denote the average chance of occurrence of an event or the average success rate. Under the binomial model the likelihood is

$$p(y|\pi) = \binom{n}{y} \pi^y (1 - \pi)^{n-y}$$

or in terms of the unknown parameter

$$p(y|\pi) \propto \pi^y (1 - \pi)^{n-y}$$

The conjugate prior density for a common binomial probability is the beta density with parameters a and b (both positive), $\pi \sim \text{Be}(a, b)$, such that

$$p(\pi) \propto \pi^{a-1}(1 - \pi)^{b-1}$$

Note that a symmetric prior on π is obtained under the simplification $a = b$ and $a = b = 1$ reduces to a uniform prior. The posterior density of π is then also a beta with parameters $a + y$ and $b + n - y$: specifically

$$p(\pi|y, n) \propto \pi^{a+y-1}(1 - \pi)^{b+n-y-1}$$

with mean $\bar{\pi} = [a + y]/[a + b + n]$.

Alternatively, suppose several binomial observations are taken, $\{y_1, n_1\}, \{y_2, n_2\}, \ldots, \{y_G, n_G\}$. For example, in Kahn and Raftery (1996) the n_i are total stroke discharges from a set of hospitals $i = 1, \ldots, G$ and y_i are discharges to one type of post-hospital care, namely skilled nursing facilities. Assume there is a single underlying occurrence rate π across all units. Then with a $\mathrm{Be}(a, b)$ prior for π, with a and b known, the posterior density is

$$\pi \sim \mathrm{Be}\left(a + \sum_{i=1}^{G} y_i, b + \sum_{i=1}^{G} n_i - \sum_{i=1}^{G} y_i\right)$$

so that the posterior mean is

$$\bar{\pi} = \left[a + \sum_{i=1}^{G} y_i\right] \Big/ \left[a + b + \sum_{i=1}^{G} n_i\right] \tag{4.2}$$

A more sensible model for such data might not assume there is a single underlying rate but varying rates. A fixed effects analysis might take $\pi_i \sim \mathrm{Be}(a, b)$ with a and b known, such as the uniform prior $\mathrm{Be}(1, 1)$. Alternatively a population of similar or related units (e.g. hospitals) provides a basis (via exchangeability) for smoothing the unit specific rates. In smoothing applications the parameters in the $\mathrm{Be}(a, b)$ prior would usually be taken as unknowns and the full conditionals for the unit-specific proportions would be

$$\mathrm{Be}(a + y_i, b + n_i - y_i)$$

For both binomial and count data, fit is generally measured in terms of the likelihood or deviance. Bayesian measures of fit (e.g. the DIC) compare the posterior mean of these measures to their values at the posterior means of the parameters or at the posterior means of μ_i and π_i. For binomial data, the log likelihood for the data given π_i is

$$L = \sum_i \left[\log\binom{n_i}{y_i} + y_i \log(\pi_i) + (n_i - y_i) \log(1 - \pi_i)\right]$$

Letting $p_i = y_i/n_i$ then the saturated log likelihood is

$$L_s = \sum_i \left[\log\binom{n_i}{y_i} + y_i \log(p_i) + (n_i - y_i) \log(1 - p_i)\right]$$

and the GLM deviance (McCullagh and Nelder, 1989) is obtained as $D = -2(L - L_s)$, namely

$$D = \sum_i \left[y_i \log(p_i/\pi_i) + (n_i - y_i) \log\left(\frac{1 - p_i}{1 - \pi_i}\right)\right]$$

The deviance is sometimes also defined as minus twice the likelihood.

Underlying binomial totals r_i and n_i in aggregated binomial unit data are binary responses for individuals within clusters (e.g. for the jth patient in the ith hospital the responses are $y_{ij} = 1$ or $y_{ij} = 0$), and so equivalently an estimate for a common occurrence rate is

$$\bar{\pi} = \left[a + \sum_{i=1}^{G} y_{ij} \right] / \left[a + b + \sum_{i=1}^{G} n_i \right]$$

A disaggregation to the underlying binary outcomes is relevant when the latent scale approach to binary regression is adopted (section 4.2). For binary data both definitions of the deviance, namely as $-2L$ or $-2(L - L_s)$, are problematic as they reduce to functions of the fitted probabilities π_i and do not involve the observations y_i (Collett, 2003). Therefore other methods of assessing fit may be needed; for example, sensitivity and specificity rates based on the proportions of successes and failures correctly classified when new data are sampled from the model. The DIC method may still be used to assess effective model dimensions.

When predictors are introduced for binomial or binary data, possible links must ensure a transformation from $\eta_i = \beta x_i$ on the real line to the $\pi \in [0, 1]$ probability scale. The most common links meeting this requirement are the logit, probit and complementary log–log links. This raises questions about which link is most appropriate or whether one might average over different links (see section 4.3). The probit and logit links are symmetric about $\pi = 0.5$, and the relationship

$$g(\pi) = -g(1 - \pi)$$

holds for both. By contrast, the complementary log–log link allows asymmetry by specifying

$$\log[-\log(1 - \pi_i)] = \beta x_i$$

or equivalently

$$\pi_i = 1 - \exp[-\exp(\beta x_i)]$$

This transform tends to one faster than it tends to zero. One may compare different links and use the standard model assessment criteria to choose the best link. Alternatively, flexible and non-parametric link modelling for binary data has also been considered from a Bayesian perspective (Chapter 5).

As for count regression, default priors for regression parameters might be univariate normal $\beta_j \sim N(0, V_j)$ where V_j are known, or a multivariate normal prior on $(\beta_1, \dots, \beta_p)$. However, one may also elicit prior probabilities associated with chosen predictor values. Consider again the case $p = 1$, and a standardized predictor. Suppose a prior guess for π

when $x = 1$ was $\pi_1 = 0.75$ but $\pi_2 = 0.25$ when $x = -1$. Assume a prior sample size (strength of belief in the guess) of $m = 5$ observations for both elicitations. Then a conditional means prior (CMP), following Bedrick *et al.* (1996), might consist of the two beta densities $Be(\pi_i m, [1 - \pi_i]m)$, i.e. $Be(3.75, 1.25)$ when $x = 1$ and $Be(1.25, 3.75)$ when $x = -1$. The Meyer and Laud prior for logistic regression has the form

$$g(\beta_1, \ldots, \beta_p) \propto \exp\left\{ \sum_i \gamma_0 [\pi_{i0} \beta x_i - \log(1 + \exp(\beta x_i))] \right\}$$

where μ_{i0} is an elicited probability for π based on the predictor vector x_i and γ_0 measures the strength of belief in the elicitation.

Special designs may require adaptations of the standard logistic framework. For example, Holford *et al.* (1978) consider a conditional logistic model for matched case–control studies in epidemiology. Each case $k = 1, \ldots, K$ is matched to M_k controls, so the data can be considered as K strata, each containing $1 + M_k$ individuals. Define $y_{ik} = 1$ for the case in stratum k and $y_{ik} = 0$ for the M_k controls. The appropriate analysis takes case–control status y_{ik} as Poisson with mean $(\mu_{ik}) = \exp(\alpha_k + \beta x_{ik})$, where x_{ik} is a vector of risk factors. The α_k are usually taken as fixed effects representing the level of the outcome in stratum k after conditioning on the risk factors. This method may be extended to situations where the kth stratum contains N_k cases and M_k controls by conditioning on the observed predictors in each matched set.

Example 4.1 Binomial clustering, bovine trypanosomiasis Böhning (1999) considers prevalence rates of bovine trypanosomiasis in $G = 50$ Ugandan farms, and the impact of clustering (higher infection rates may be related to particular herds) on the appropriateness of a pooled rate, as in (4.2). The total $\sum_{i=1}^{G} y_i$ is 87 and $\sum_{i=1}^{G} n_i$ is 487, so with a $Be(1, 1)$ prior on the pooled mean rate π the posterior mean will be $88/489 = 0.18$ with standard deviation 0.01735 under the variance formula for the beta density. If a pooled mean rate π is not appropriate then predictions (replicate data) $z_i(i = 1, \ldots, G)$ from this model will have a smaller variance V_z than actually observed, namely $V_y = 7.58$.

A standard binomial is applied under a 50 000 two-chain run and new infection totals $z_i(i = 1, \ldots, 50)$ sampled at each iteration. It is seen that the binomial underpredicts the variance: the 95% interval for V_z is $\{1.43, 4.34\}$ with a probability of zero that $V_z > V_y$. Pooling strength models for heterogeneity (e.g. discrete mixture models or beta–binomial models) will therefore provide a better fit and a better check against the actual dispersion (see Chapter 5).

4.2   ESTIMATION BY DATA AUGMENTATION: THE ALBERT–CHIB METHOD

Bayesian simulation facilitates the introduction of latent data ('data augmentation') on a different scale to the actual observed data. For example, for binary data y the latent data w is defined on a metric scale which yields $y = 1$ when $w \geq 0$ and $y = 0$ when $w < 0$ (Albert and Chib, 1993). The introduction of augmented data may assist in residual analysis and in multivariate analysis combining data of different types (e.g. a mixture of binary and ordinal responses). An underlying scale may be introduced in terms of utilities U_{i1} and U_{i0} of options 1 and 0 with

$$U_{ij} = V_{ij} + \varepsilon_{ij} = \beta_j^* x_i + \varepsilon_{ij}$$
$$w_i = U_{i1} - U_{i0}$$

The probability that option 1 is selected is then

$$\Pr(y_i = 1) = \Pr(w_i \geq 0) = \Pr(\varepsilon_{i0} - \varepsilon_{i1} \leq V_{i1} - V_{i0})$$

Assume ε_{ij} is normal with mean 0 and variance σ^2 and define $\beta = \beta_1^* - \beta_0^*$. Then the comparison of utilities leads to a probit link with

$$\Pr(y_i = 1) = \Phi(\beta x_i / \sigma)$$

It is apparent that β and σ cannot be separately identified and the usual approach is to set $\sigma^2 = 1$. It is then possible to sample the latent differences w_i.

The augmented data approach to estimating parameters for binary responses is particularly relevant in regression problems where the values of predictors in combination with the response provide information on the latent scale. If y_i is one, then a probit link (ε normal) means that the latent response w_i is constrained to be positive and sampled from a normal with mean βx_i and variance $\sigma^2 = 1$. If $y_i = 0$, w_i is sampled from the same normal density but constrained to be negative (with 0 as a ceiling value). So

$$\begin{aligned} w_i &\sim N(\beta x_i, 1) I(0, \infty) & y_i &= 1 \\ w_i &\sim N(\beta x_i, 1) I(-\infty, 0) & y_i &= 0 \end{aligned} \tag{4.3}$$

In practice sampling within the interval $(-10, 10)$ or $(-5, 5)$ for w will be sufficient (Oh, 1997).

Output from this model may be used to illustrate the method of calculating marginal likelihoods via the relation

$$\log[P(Y)] = \log[P(Y|\beta)] + \log[P(\beta)] - \log[P(\beta|Y)]$$

as proposed by Chib (1995). The likelihoods and prior ordinates at a high-density point, say the mean $\bar{\beta}$, are readily obtained by substituting the values in $\bar{\beta}$ into the likelihood and the prior densities. To estimate $P(\beta|Y)$ at $\beta = \bar{\beta}$, and so estimate $M(Y) = \log[P(Y)]$ as

$$\hat{M}(Y) = \log[P(Y|\bar{\beta})] + \log[P(\bar{\beta})] - \log[P(\bar{\beta}|Y)]$$

note that

$$P(\beta|Y) = \int P(\beta|Y, W)P(W|Y)\, dW$$

where W is the vector of normally distributed latent data. Assuming a prior $\beta \sim N_p(b, B)$, the posterior $P(\beta|Y, W)$ may be estimated (Albert and Chib, 1993) as

$$\beta|Y, W \sim N(\hat{\beta}_W, V_W)$$

where

$$\hat{\beta}_W = (B^{-1} + X'X)^{-1}(B^{-1}b + X'W)$$

and

$$V_W = (B^{-1} + X'X)^{-1}$$

Given $t = 1, \ldots, T$ MCMC draws of the latent data vectors $W^{(t)}$, a Monte Carlo estimator for $P(\bar{\beta}|Y)$ is therefore provided by

$$\hat{P}(\bar{\beta}|Y) = \sum_{t=1}^{T} \phi(\bar{\beta}|\hat{\beta}_W^{(t)}, V_W)$$

where ϕ is the normal density function.

4.2.1 Other augmented data methods

Several alternatives to the probit may be considered for binary data generated according to

$$Pr(y_i = 1) = Pr(w_i \geq 0) = Pr(\varepsilon_{i0} - \varepsilon_{i1} \leq V_{i1} - V_{i0})$$

The logit link may be sampled directly

$$\begin{aligned} w_i &\sim \text{logistic}(\beta x_i, 1)I(0, \infty) \qquad y_i = 1 \\ w_i &\sim \text{logistic}(\beta x_i, 1)I(-\infty, 0) \qquad y_i = 0 \end{aligned} \tag{4.4}$$

where the logistic density logistic (μ,τ), with mean μ and scale parameter τ has the form

$$f(x) = \tau \exp(\tau[x - \mu])/\{1 + \exp(\tau[x - \mu])\}^2$$

This has variance κ^2/τ^2 where $\kappa^2 = \pi^2/3$. Note that the *standard* logistic density with mean 0 and variance 1 has the form

$$f(x) = \kappa \, \exp(\kappa x)/[1 + \exp(\kappa x)]^2$$

Alternatively the logit link may be approximated by sampling w_i from a Student t with 8 degrees of freedom (Albert and Chib, 1993). This entails constrained normal sampling, as for the probit, but with the precision of one replaced by subject-specific variances sampled from an inverse gamma density with shape and index both equal to four. So

$$
\begin{aligned}
w_i &\sim N(\beta x_i, 1/\lambda_i)I(0, \infty) & y_i &= 1 \\
w_i &\sim N(\beta x_i, 1/\lambda_i)I(-\infty, 0) & y_i &= 0 \\
\lambda_i &\sim \text{Ga}(\nu/2, \nu/2) \\
\nu &= 8
\end{aligned}
\tag{4.5}
$$

Note that the β coefficients under this approach need to be scaled to match those obtained from a logistic regression. Other mixtures are possible, e.g. taking ν as an unknown amounts to a model for the link function; see Chapter 5 and Albert and Chib (1993, p 677). Then various levels of 'heavy tails' are possible as ν varies, with $\nu = 1$ corresponding to the Cauchy density.

Wood and Kohn (1998, p 211) and Oh (1997) propose alternative ways of approximating the logit link by sampling latent w. Following Scott (2003) an augmented data version of the logit link may also be obtained by constrained sampling of $\{w_{i1}, w_{i2}\}$ from exponential densities with means $\lambda_{i1} = 1$ and $\lambda_{i2} = \exp(\beta x_i)$. Then if $D_i = 1$ for 'failure' and $D_i = 2$ for 'success', and if $D_i = \text{argmin}(w_{i1}, w_{i2})$, then $\Pr(D_i = k|x_i) \propto \lambda_{ik}$ as under a logit regression. Another option, suggested by Harvey (1976), is for a heteroscedastic probit with the variance model based on predictors Z_i, such that for $y = 1$

$$
\begin{aligned}
w_i &\sim N(\beta x_i, 1/\lambda_i)I(0, \infty) \\
\lambda_i &= \exp(\gamma Z_i)
\end{aligned}
$$

where Z_i excludes the intercept for identifiability.

Estimated residuals from data augmented models may be used to assess outliers or other aspects of poor fit (Albert and Chib, 1995). Thus for the augmented data probit, the residual

$$\varepsilon_i = w_i - \beta x_i$$

is approximately $N(0,1)$ if the model is appropriate, whereas if the posterior distribution of ε_i is significantly different from $N(0,1)$ then the model conflicts with the observed y. For example, following Chaloner and Brant (1988) one may monitor the probability

$$\Pr(|\varepsilon_i|) > 2 \qquad (4.6a)$$

and compare it will its prior value, which is 0.045. For the augmented data version of the logit as in (4.4), one monitors

$$\Pr(|\varepsilon_i|/\kappa^{0.5}) > 2 \qquad (4.6b)$$

while for the logistic approximation method as in (4.5), one monitors

$$\Pr(|\varepsilon_i|\lambda_i^{0.5}) > 2 \qquad (4.6c)$$

For Poisson data, data augmentation is more complex, especially for overdispersed data. Oh and Lim (2001) mention that the cdf of a Poisson variable y_i with mean μ_i is closely approximated by

$$\Phi(A[y_i, \mu_i]) = \Phi(-3u_i^{-0.5}[(\mu_i u_i)^{1/3} - 1 + u_i/9])$$

where Φ is the standard normal cdf and $u_i = 1/[y_i + 1]$. Define $\eta_i = \log(\mu_i) = \beta x_i$; latent variables underlying the Poisson observations $y_i = 1, 2, 3, \ldots$ may be obtained by sampling

$$w_i \sim N(\mu_i, 1)I(A[y_i - 1, \mu_i] + \eta_i, A[y_i, \mu_i] + \eta_i)$$

For $y_i = 0$ they are obtained as

$$w_i \sim N(\mu_i, 1)I(-\infty, A[0, \mu_i] + \eta_i)$$

Further applications of data augmentation for count and binary regressions occur where the response (e.g. a health behaviour or an economic choice variable) is endogenous with an intervention (e.g. a health intervention or medical advice), especially if important determinants of the intervention are unobserved (Jochmann, 2003). Let $D_i = 1$ or 0 be binary observations according to whether the intervention ('treatment') is taken up by the subject, and Y_i be the observed count or binary outcome. For Y_i a count, one may assume

$$Y_i = (1 - D_i)Y_{i0} + D_iY_{i1}$$

where Y_{i1} and Y_{i0} are unobserved count outcomes with means $\exp(\eta_{i1})$ and $\exp(\eta_{i0})$. Another possibility is to take Y_{i1} and Y_{i0} as gamma variables with the same means. Then assume $\eta_{i1} \sim N(X_i\beta_1, \phi_1)$, $\eta_{i0} \sim N(X_i\beta_0, \phi_0)$

and that take-up of the intervention depends on the response, such that

$$w_i \sim N(\gamma Z_i + \delta_0 \eta_{i0} + \delta_1 \eta_{i1}, 1)I(0, \infty) \qquad D_i = 1$$
$$w_i \sim N(\gamma Z_i + \delta_0 \eta_{i0} + \delta_1 \eta_{i1}, 1)I(-\infty, 0) \qquad D_i = 0$$

where Z_i includes one or more factors not among the X_i. An individual-level treatment effect is provided by $Y_{i1} - Y_{i0}$.

4.3 MODEL ASSESSMENT: OUTLIER DETECTION AND MODEL CHECKS

For detecting unusual cases with binary or binomial data, one may also combine logit or probit links with outlier detection as in Verdinelli and Wasserman (1991) (see section 3.5). Wood and Kohn (1998) consider a shifted location version of the Verdinelli and Wasserman (1991) approach in augmented data sampling of a probit regression. Let outliers be selected by $\delta_i \sim \text{Bern}(\omega)$ where ω is small, say $\omega = 0.05$. Then with inflated variance $\Lambda = 10$ for outliers

$$
\begin{aligned}
w_i|\delta_i = 1 &\sim N(\beta x_i, 10)I(0, \infty) & y_i = 1 \\
w_i|\delta_i = 0 &\sim N(\beta x_i, 1)I(0, \infty) & y_i = 1 \\
w_i|\delta_i = 1 &\sim N(\beta x_i, 10)I(-\infty, 0) & y_i = 0 \\
w_i|\delta_i = 0 &\sim N(\beta x_i, 1)I(-\infty, 0) & y_i = 0
\end{aligned}
\tag{4.7}
$$

An alternative interpretation of outliers when the outcome is binary is that a misclassification has occurred (e.g. Copas, 1988). This might be relevant if y is based on fallible judgement (e.g. $y = 1$ for positive diagnosis under a screening tool with low sensitivity). Let T_i be the true status and φ_i be the probability that $T_i = 1$. Also let ω_1 be the probability that a $T = 1$ is misrecorded as $y = 0$ and ω_0 the probability that $T = 0$ is misrecorded as $y = 1$. Then the probabilities of the actually observed y_i are

$$
\begin{aligned}
\pi_i = P(y_i = 1) &= \Pr(y_i = 1|T_i = 1)p(T_i = 1) \\
&\quad + \Pr(y_i = 1|T_i = 0)\Pr(T_i = 0) \\
&= (1 - \omega_1)\varphi_i + \omega_0(1 - \varphi_i)
\end{aligned}
$$

and similarly

$$
\begin{aligned}
1 - \pi_i = P(y_i = 0) &= \Pr(y_i = 0|T_i = 1)p(T_i = 1) \\
&\quad + \Pr(y_i = 0|T_i = 0)\Pr(T_i = 0) \\
&= \omega_1\varphi_i + (1 - \omega_0)(1 - \varphi_i)
\end{aligned}
$$

Letting $\omega_1 = \omega_0 = \omega$ and allowing the chance of misclassification to vary with $\delta_i \sim \text{Bern}(\omega)$ one can re-express π_i as

$$
\begin{aligned}
\pi_i = P(y_i = 1) &= P(y_i = 1 | T_i = 1)P(T_i = 1) \\
&\quad + \Pr(y_i = 1 | T_i = 0)\Pr(T_i = 0) \qquad (4.8) \\
&= (1 - \delta_i)\varphi_i + \delta_i(1 - \varphi_i)
\end{aligned}
$$

and express the logit or probit of φ_i as a function of predictors (see Example 4.3).

4.3.1 Model assessment: predictive model selection and checks

Model selection methods for binary or count responses involve the same principles, such as predictive cross-validation and choice according to posterior model probability, as for metric outcomes. Applying predictive cross-validation involves relevant adaptations based on the form of deviance appropriate to the response. Thus consider a parameter set $\theta^m = \{\beta^m, \phi^m\}$ including regression parameters β^m. Samples from the posterior predictive density are obtained by sampling from $p(Z|\theta^m)$; for example, $Z_{im} \sim \text{Poi}(\mu_{im})$ in a log-link Poisson regression with $\log(\mu_{im}) = \eta_{im} = \beta^m X_i^m$.

Then for model selection, Carlin and Louis (1996) propose a deviance criterion analogous to the C_m criterion of Laud and Ibrahim (1995), namely

$$
D_m = \mathrm{E}[d(Z_m, y)] \qquad (4.9)
$$

where $d(Z_m, y)$ is the relevant deviance. For y Poisson, for example,

$$
d(Z_m, y) = 2 \sum_{i=1}^{n} [y_i \log(y_i/Z_{im}) - (y_i - Z_{im})]
$$

Gelfand and Ghosh (1998) propose a deviance criterion based on extending to discrete outcomes the predictive loss criterion

$$
\sum_i \varphi_i + \left(\frac{k}{k+1}\right) \sum_i (y_i - \zeta_i)^2 \qquad (4.10)
$$

where ζ_i and φ_i are respectively the posterior mean and variance of Z_i. Thus let τ_i be the posterior average of the deviance term based on the sampled new data at iteration t, Z_i^t. For example, for Poisson distributed count data τ_i is the mean of sampled values of $d(Z_i) = Z_i \log Z_i - Z_i$. The same formula is used for $d(\zeta_i)$ and $d(y_i)$. In practice one would use

$d(y_i) = (y_i + c)\log(y_i + c) - y_i$ and $d(Z_i) = (Z_i + c)\log(Z_i + c) - Z_i$ to avoid logarithms of zero. For binomial data with n_i cases in trial i,

$$d(Z_i) = Z_i/n_i \log(Z_i/n_i) + [(n_i - Z_i)/n_i]\log[(n_i - Z_i)/n_i]$$

Define $\Lambda_i = (\zeta_i + ky_i)/(1 + k)$; then the Gelfand–Ghosh measure is

$$2\sum_i[\tau_i - d(\zeta_i)] + 2(k + 1)\sum_i\{[d(\zeta_i) + ky_i]/[1 + k] - d(\Lambda_i)\} \quad (4.11)$$

From both (4.10) and (4.11), it can be seen that small values of k mean greater stress is put on precision of predictions (so penalizing complex models which produce less precise predictions), whereas increasing k puts more stress on accuracy ('goodness of fit' per se). The first term in both (4.10) and (4.11) is a penalty term for complexity or imprecise predictions, and the second is the goodness-of-fit term (measuring bias in predictions rather than their precision). Meyer and Laud (2002) advocate the measure

$$\sum_i[\varphi_i + (y_i - \zeta_i)^2] \quad (4.12)$$

for y_i as count, binomial or binary data as well as metric data.

For model diagnosis using posterior predictive checks (Gelman *et al.*, 1995), a frequent issue for both binomial and count data is overdispersion (see Chapter 5) and hence a suitable model will replicate the overdispersion present in the data. To assess whether it does so one might compare the ratio $C_Z = \text{Var}(Z)/\bar{Z}$ based on sampled new data with the corresponding ratio H_y for the actual data. This check would be done at each iteration and a satisfactory model will have C_Z exceeding C_y about 50% of the time.

Example 4.2   Model selection, logit regression of nodal involvment
These data have been considered in a number of studies with the issue centred on regressor choice. Congdon (2003) considered Chib's method for obtaining a Bayes factor by marginal likelihood approximation, together with pseudo marginal likelihood estimates and predictive cross-validation. Here we consider the path sampling method for Bayes factor approximation and the parallel sampling method for approximating model probabilities (Chapter 2). The predictors are $x_1 = \log(\text{serum acid}$ phosphate)$, $x_2 = $ result of X-ray ($1 = +$ve, $0 = -$ve), $x_3 = $ size of tumour ($1 = $ large, $0 = $ small) and $x_4 = $ pathological grade of tumour ($1 = $ more serious, $0 = $ less serious). A binary regression model including all four predictors is then $y_i \sim \text{Bern}(\pi_i)$ where

$$\pi_i = \Phi(\beta_0 + \beta_1 x_{1i} + \cdots + \beta_4 x_{4i})$$

or

$$\pi_i = \exp(\beta_0 + \beta_1 x_{1i} + \cdots + \beta_4 x_{4i})/[1 + \exp(\beta_0 + \beta_1 x_{1i} + \cdots + \beta_4 x_{4i})]$$

under probit and logit links respectively. Let M_1 denote the full model (four predictors) and M_0 a reduced model excluding x_4.

The priors used on the regression coefficients are as in Chib (1995). With a logit link, the alternative models are

$$M_1 : \quad \text{logit}(\pi_i) = \beta_0 + \beta_1 x_{1i} + \cdots + \beta_4 x_{4i}$$
$$M_0 : \quad \text{logit}(\pi_i) = \beta_0 + \beta_1 x_{1i} + \cdots + \beta_3 x_{3i}$$

and under the path sampling approach for approximating the Bayes factor, the linking model M_s with $s \in (0,1)$ is

$$M_s : \quad \text{logit}(\pi_{is}) = \beta_0 + \beta_1 x_{1i} + \cdots + \beta_3 x_{3i} + s\beta_4 x_{4i}$$

The log likelihood under M_s is

$$\log P(y_i | \theta, s) = y_i \log(\pi_{is}) + (1 - y_i) \log(1 - \pi_{is})$$

where $\pi_{is} = \exp(\eta_{is})/[1 + \exp(\eta_{is})]$ and $\eta_{is} = \beta_0 + \beta_1 x_{1i} + \cdots + \beta_3 x_{3i} + s\beta_4 x_{4i}$. So, as in section 2.3,

$$R_i(\theta, s) = y_i \beta_4 x_{4i} - \beta_4 x_{4i} \exp(\eta_{is})/[1 + \exp(\eta_{is})]$$

With a grid of 20 intervals the Bayes factor in favour of the smaller model is 2.44. Increasing the number of intervals does not substantially affect this estimate.

Under the parallel sampling approach the posterior probability on M_0 is obtained as 0.79 giving a Bayes factor of 3.76 favouring the simpler model (under equal prior model probabilities). Strictly one should allow for Monte Carlo variation in the posterior model probability estimates $\bar{w}_0$ and $\bar{w}_1$. From a two-chain run of 50 000 iterations the Monte Carlo standard error of both estimates is 0.00116 and a 95% interval on the Bayes factor may be obtained by simulation of the ratio of these two probabilities.

Example 4.3 Latent normal sampling; simulated data   Albert and Chib (1995) present a small simulated sample ($n = 20$) of binary data with a single predictor and discuss the identification of outliers. The data are generated under a logit link, namely

$$y_i \sim \text{Bern}(\pi_i)$$
$$\text{logit}(\pi_i) = \beta_0 + \beta_1 x_i$$

with $\beta_0 = 0$, $\beta_1 = 3$. With the data thus generated, first consider the standard logit and probit regression. Logit regression yields posterior means (and standard deviations) $\beta_0 = 0.03$ (0.6), $\beta_1 = 3.55$ (1.66), while probit regression yields $\beta_0 = 0.03$ (0.34), $\beta_1 = 1.99$ (0.89). Logit regression with a prior misclassification probability $\omega = 0.01$ and misclassification indicators $\delta_i \sim$ Bern(ω) as in (4.8) shows cases 4, 9 and 13 to have posterior probabilities $P(\delta_i = 1)$ of 0.11, 0.11 and 0.07 respectively. The coefficient β_1 is inflated to around 4.1 (s.d. 2.1) under this model.

Table 4.1 shows the probabilities of outliers under different data augmentation options. It also contains the scale weights λ_i under (4.5). Probit regression via augmented data sampling also shows cases 4 and 9 with positive responses ($y = 1$) but negative x to be possible outliers, together with case 13 which has $y = 0$ despite positive x. The posterior

Table 4.1 Outlier detection under data augmentation (DA)

Obs.	y	x	Probability of outlier				
			Probit regression via DA	Logit approximation via DA	DA Logit approximation via scale weights	Logit regression via DA	Probit DA with variance inflation
1	0	−0.740	0.026	0.030	1.022	0.030	0.037
2	0	0.005	0.048	0.051	0.987	0.059	0.053
3	1	0.337	0.036	0.032	1.025	0.035	0.040
4	1	−0.492	0.189	0.180	0.812	0.227	0.155
5	0	−0.268	0.030	0.034	1.027	0.038	0.042
6	1	0.187	0.037	0.036	1.016	0.042	0.042
7	0	−0.211	0.038	0.035	1.022	0.041	0.044
8	0	−0.538	0.030	0.029	1.031	0.032	0.038
9	1	−0.490	0.186	0.180	0.811	0.224	0.151
10	0	−0.377	0.036	0.032	1.026	0.037	0.040
11	0	−0.628	0.029	0.030	1.024	0.030	0.038
12	1	−0.198	0.073	0.076	0.936	0.089	0.071
13	0	0.352	0.155	0.140	0.861	0.176	0.123
14	1	0.366	0.033	0.033	1.023	0.036	0.040
15	0	−0.751	0.031	0.029	1.021	0.031	0.038
16	1	0.359	0.031	0.032	1.025	0.038	0.041
17	0	−0.173	0.039	0.038	1.017	0.043	0.047
18	0	−0.156	0.038	0.038	1.018	0.046	0.044
19	0	−0.432	0.032	0.029	1.030	0.033	0.040
20	1	0.454	0.030	0.031	1.025	0.035	0.041

outlier probabilities of 0.15 or more compare with prior probabilities of 0.05.

Similar results are obtained under the logistic regression schemes (4.4) and (4.5). The scale weights under (4.5) also highlight cases 4, 9 and 13. Probit regression via augmented data sampling with outlier detection in terms of elevated variance, as in (4.7) with $\Lambda = 10$, yields a slightly lower outlier probability on case 13.

4.4 PREDICTOR SELECTION IN BINARY AND COUNT REGRESSION

For regression variable selection in binary or count regression one may adapt the linear regression method of George and McCulloch (1993) and George *et al.* (1996) based on a scale mixture of two normal distributions. As for a metric response, the prior for a regression coefficient when X_j is not certain to be included is

$$\beta_j | \gamma_j \sim \gamma_j N(0, c_j^2 \varsigma_j^2) + (1 - \gamma_j) N(0, \varsigma_j^2) \qquad (4.13)$$

If $\gamma_j = 1$ is chosen at any iteration it corresponds to X_j being included in the model and β_j is then distributed as a normal with mean 0 and large variance $c_j^2 \varsigma_j^2$. If inclusion of X_j is not supported then the prior with a small default variance, namely $N(0, \varsigma_j^2)$, will be selected. c_j should be chosen large enough (and ς_j small enough) to ensure separation of the two densities in the mixture. George *et al.* (1996) recommend taking c_j between 10 and 100, and ς_j to be based on setting a neighbourhood $(-\delta_j, \delta_j)$ around zero within which variation in β_j is unimportant to the outcome. Once δ_j is obtained, $\varsigma_j = \delta_j/2.15$ for $c_j = 10$ and $\varsigma_j = \delta_j/3.04$ for $c_j = 100$ (see George *et al.*, 1996, p 342).

For generalized linear models based on the exponential family density with natural parameter θ, one has

$$f(y_i | \theta_i) = \exp[y_i \theta_i - b(\theta_i) + c(y_i)]$$

where $E(y_i) = b'(\theta_i)$. With $\theta_{0i} = \beta x_i$, the impact on $E(y_i)$ of a change Δx_{ij} in x_j is

$$\Delta y = E(y_{1i}) - E(y_{0i}) = |b'(\theta_{0i} + \beta_j \Delta x_{ij}) - b'(\theta_{0i})|$$

This is approximately equal to $b''(\theta_{0i})|\Delta x_{ij}||\beta_j|$ where $b''(\theta)$ is the variance of y (George *et al.*, 1996). It is required to obtain δ_j so that for $|\beta_j| < \delta_j$, Δy is small, i.e. $\Delta y < \varepsilon$ where ε is small relative to the range in y. Thus one option would be to choose $\delta_j = \varepsilon/(b''(\theta_0)|\Delta x_j|)$, and Δx_j represents a large change in x_j in terms of its dispersion.

Another option is to assess the size that $G_j = |\beta_j \Delta x_j|$ would need to be to ensure that $\Delta y < \varepsilon$, and then take $\delta_j = G_j/|\Delta x_j|$. For example, $E(y) = \Phi(\theta)$ in a binary regression with probit link, with Φ the standard normal cumulative density. Since $E(y)$ can vary only between 0 and 1, a reasonable value for ε is 0.01. Assuming $\theta_0 = 0$ gives $\Phi(\theta_0) = 0.5$, then since $\Phi(\theta_0 + 0.025) = 0.51$, one obtains with $G_j = 0.025$

$$\Delta y = \Phi(\theta_{0i} + G_j) - \Phi(\theta_{0i}) = 0.01$$

and so setting $\delta_j \leq 0.025/|\Delta x_j|$ ensures $|\beta_j \Delta x_j|$ is under 0.025. Δx_j might be the interquartile range of x_j or the difference between the ninth and first decile. For a binary logit regression, $E(y) = L(\theta) = \exp(\theta)/[1 + \exp(\theta)]$ and with $\theta_0 = 0$, $L(\theta_0) = 0.5$. Here $L(0.04) = 0.51$ and one may set $\delta_j = 0.04/|\Delta x_j|$. For count data, one might take Δy as ε times the range of y and with $\theta_0 = 0$, $\exp(\theta_0) = 1$. So then $G_j = \ln(\Delta y + 1)$.

The similar approaches of Smith and Kohn (1996) and Kuo and Mallick (1998) assume binary indicators $\gamma_1, \ldots, \gamma_p$ on regression parameters $\beta_1, \ldots, \beta_p$ such that $\gamma_j = 0$ if $\beta_j = 0$ and $\gamma_j = 1$ if $\beta_j \neq 0$. Suppose the prior on $\rho_j = P(\gamma_j = 1)$ is $\rho_j = 0.5$. Gerlach *et al.* (2002) describe a modification of the Smith and Kohn (1996) approach (see Chapter 3) such that for binary data

$$y_i \sim \text{Bern}(\pi_i)$$
$$\text{logit}(\pi_i) = z_i = \beta X_i + e_i, \quad \text{with } e_i \sim \text{N}(0, \tau)$$

Then $S(\gamma)$ in (3.3) is defined with z replacing y.

Count or logit regression variable selection may also be based on separately running all models and considering predictive summaries or criteria, such as those in (4.9)–(4.12). Meyer and Laud (2002) and Laud and Ibrahim (1995) propose a prior for the regression coefficients β^m in the parameters sets $\theta^m = \{\beta^m, \phi^m\}$ for models m. These are based on eliciting prior guesses for π_{im0} (binary or binomial response), or μ_{im0} (count response), or possibly taking $\text{logit}(\pi_{im0}) = \beta^m X_{im}$, and $\log(\mu_{im0}) = \beta^m X_{im}$ where β^m is the maximum likelihood estimate (Meyer and Laud, 2002, p 864). Strength of belief in the mth prior is represented by the ratio $\gamma_0^m = n_0^m/n$ of prior sample sizes n_0^m to the actual sample size n. For a logistic regression the conjugate prior on the vector β^m is

$$P(\beta^m | \gamma_0^m, \pi_0^m) \propto \exp\left\{ \sum_{i=1}^{n} \gamma_0^m [\pi_{i0}^m X_{im} \beta^m - \log(1 + \exp(X_{im} \beta^m))] \right\}$$

while for a Poisson regression, the conjugate prior is

$$P(\beta^m|\gamma_0^m, \pi_0^m) \propto \exp\left\{\sum_{i=1}^{n} \gamma_0^m [\mu_{i0}^m X_{im}\beta^m - \exp(X_{im}\beta^m)]\right\}$$

Choice between predictors may involves comparing 2^p models against one another and selecting that combination of predictor variables that minimizes predictive criteria such as (4.9). Forward search methods might also be used in tandem with predictive criteria (see also Congdon, 2005, for a parallel sampling method of forward selection).

Example 4.4 Suicides in English local authorities In this example the response is the total male suicides y_i in 1989–1993 in 354 English local authorities. The average suicide count over this five-year period is 52, with a range from 0 (in the Isles of Scilly) to 409 in Birmingham. Expected deaths E_i (based on England-wide age-specific suicide rates in 1991) are used as an offset. There are ten possible predictors, all from the 1991 UK Census. These reflect factors such as isolation and social fragmentation (e.g. high population turnover, one-person households), social class and deprivation (e.g. percentage of economically active in classes IV and V, namely semi- and unskilled manual), and the possible impact of rurality (demonstrated by elevated suicide rates among farmers). The predictors indicative of social fragmentation are X_1 = single, widowed, divorced, X_2 = % of population with different address one year ago, X_3 = one-person households, X_4 = private renting. The remaining predictors are X_5 = unemployment, X_6 = % economically active in classes IV and V, X_7 = agricultural workers, X_8 = renting from local authority/housing association, X_9 = population density (divided by 1000) and X_{10} = non-white ethnicity.

With the mixture selection prior as in (4.13) above, take D as ε times the range of y so that with $\varepsilon = 0.01$,

$$G_j = \ln(\Delta y + 1) = \ln(5.09) = 1.63$$

Then $\delta_j = G_j/|\Delta x_j|, j = 1, \ldots, 10$, where $|\Delta x_j|$ is the difference between the 90th and 10th percentiles for each predictor. Then with $c_j = 10$, $\varsigma_j = \delta_j/2.15$. Here $\varepsilon = 0.01$ and 0.05 are compared in terms of fit using the model averaged Poisson means

$$y_i \sim \text{Po}(E_i\nu_i)$$

where the $\nu_i = \beta X_i$ average over the models defined by which $\gamma_j = 1$ at iteration t. The effect of predictors can be interpreted as measuring changes in the log relative risk, i.e. in $\log(\nu_i)$, using means of $\kappa_j = \gamma_j\beta_j$.

To assess predictions the discrepancy criterion (4.12) is used. A two-chain run of 50 000 iterations (and a 10 000 burn-in) with ς_j based on $\varepsilon = 0.01$ gives a discrepancy of 50 810. The marginal posterior probabilities $\Pr(\gamma_j = 1)$ are above 0.5 for X_3 and X_6 only. For example, the posterior mean of κ_3 is 0.026, so there is a 27% higher suicide risk in areas with 30.5% one-person households (the 90th percentile) than in areas with 21.5% (the 10th percentile). However, the averages of $\kappa_j = \gamma_j \beta_j$ are 'significant' in the sense of having 95% intervals entirely above or below zero for two other predictors, namely X_2 and X_9. The predictors X_2, X_3 and X_6 have positive effects on the suicide rate and the impact of X_9 is negative, perhaps reflecting the impact of rurality as a suicide risk factor.

With ς_j based on $\varepsilon = 0.05$ the discrepancy criterion is lowered slightly to 50 530. With this option X_1 is more apparently significant, with 95% intervals all positive. The predictors X_2, X_3 and X_6 also have positive effects in terms of the 95% intervals for κ_j, whereas X_9 and X_5 have negative effects. Here none of the marginal inclusion probabilities exceed 0.5 so there appear to be a range of plausible models with differing predictors included in them.

4.5 CONTINGENCY TABLES

A specific application of Poisson regression is to contingency tables under which subjects are classified by a number of categorical factors. Thus consider subjects $r = 1, \ldots, N$ with observed values $C_{1r}, C_{2r}, \ldots$ on two or more factors. One might analyse such data at individual level, especially when there are also metric observations $x_{r1}, x_{r2}, \ldots$ and a clear distinction exists between predictors and response(s). However, when the observations are confined to a few categorical variables (which are effectively joint responses) an alternative is to analyse counts accumulated over subjects with the same category values. Suppose there are only two factors $i = 1, \ldots, I$ and $j = 1, \ldots, J$ (i.e. rows and columns of the table) and with associated counts y_{ij}; then $y_{11} + y_{12} + \ldots + y_{21} + y_{22} + \ldots + y_{IJ} = N$ with the total cells in the table being IJ. In general a T-way contingency table involves T cross-classified variables. The totals of the observations over one or more but not all factors (such as the row sums $\sum_j y_{ij}$ in a two-way table) are known as marginal totals.

There are several possible sampling schemes which can produce a contingency table. Poisson sampling is relevant when the total of observations N is random (variants such as negative binomial sampling will be relevant in overdispersed data). Each cell mean in the contingency table is then the mean $\mu_{i_1 i_2 \dots i_T}$ of Poisson distributed counts $y_{i_1 i_2 \dots i_T}$. Other sorts of sampling occur when N is fixed or one or more of the margins are fixed. When N alone is fixed (e.g. the total sample size in a clinical trial or questionnaire survey, or a census population total) one obtains multinomial sampling from a total of N with probabilities of being in a cell defined by categories on the combined factors (see Chapter 6). For instance, in a two-way table, frequencies y_{ij} would be multinomial with probabilities $\pi_{ij} = \text{Pr}(C_1 = i, C_2 = j)$ in relation to $N = \sum_i \sum_j y_{ij}$, with $\sum_i \sum_j \pi_{ij} = 1$. When the margins on one variable are also fixed, for instance in stratified sample designs or when classifiers j, k, etc., relate to subdvisions of populations in units i such as geographic small areas, hospitals or schools, then product multinomial sampling occurs. An example is that considered by Leonard and Hsu (1994) where students in schools $i = 1, \dots, 40$ are classified by grades $j = 1, \dots, 6$ on a mathematics test and the data result from 40 separate multinomial distributions.

Consider Poisson sampling when N is not fixed. Hierarchical log-linear models express the counts in terms of main effects and interactions among the variables, with higher order interactions not being introduced unless lower order ones are present. The simplest model and one the subject of several Bayesian studies (e.g. Epstein and Fienberg, 1991; Albert, 1996; Gunel and Dickey, 1974) is where the variables are mutually independent with main effects only, whereas substantive interest is usually in the form of association between the factors (i.e. in departures from independence). Thus for a two-way table, $y_{ij} \sim \text{Po}(\mu_{ij})$, the independence model

$$\log(\mu_{ij}) = u_0 + u_{1(i)} + u_{2(j)} \tag{4.14}$$

is obtained when parameters $u_{12(ij)}$ representing interactions between the factors are absent. The model with interactions

$$\log(\mu_{ij}) = u_0 + u_{1(i)} + u_{2(j)} + u_{12(ij)} \tag{4.15}$$

is known as a saturated model since the number of possible parameters is $IJ + I + J + 1$, exceeding the number of cells in the table. The usual system of priors to ensure identifiability (number of parameters equal to or less than the number of cells in the table) sets $u_{1(1)} = u_{2(1)} = 0$, and also $u_{12(ij)} = 0$ for all classification pairs $\{i, j\}$ where one or more of the i and j are one. Thus $u_{12(1j)} = 0$ for all j and $u_{12(i1)} = 0$ for all i. The remaining parameters may be taken to be independently distributed fixed effects, with zero means and known large variances, e.g.

$u_{1(2)} \sim N(0, 1000), \ldots, u_{1(I)} \sim N(0, 1000); u_{2(2)} \sim N(0, 1000), \ldots, u_{2(J)} \sim$
$N(0, 1000); u_{12(22)} \sim N(0, 1000), \ldots, u_{12(IJ)} \sim N(0, 1000).$

Possible alternative prior structures include:

1. Two-stage priors where an extra parameter is introduced to model the goodness of fit of the log-linear model and/or account for over-dispersion; Albert (1988) describes a two-stage model where the contingency table results from Poisson sampling and Epstein and Fienberg (1991) do the same when there is multinomial sampling is relation to fixed N (see Chapter 6).
2. Random effects rather than fixed effects priors on interaction parameters, such as $u_{12(ij)}$ in (4.15), that allow a limiting case where the precision on $u_{12(ij)} = 0$ is so high that the independence model becomes plausible (Albert, 1996). Random effects priors that include all the identifiable parameters in a log-linear model are discussed by Knuiman and Speed (1988) and Dellaportas and Forster (1999). These involve covariance matrices which combine projection matrices for each parameter set (e.g. interactions between factors 2 and 3 in a three-way table, or main effects for factor 3) to ensure identifiability, combined with variance parameters which may differ between each parameter set.
3. Models where the interactions are 'structured' for the sake of parsimony (see Chapter 7). For instance, suppose C_1 and C_2 are ordinal and of the same type (parental status and descendant status), so that one can define distance as $|j - i|$ and so model the interaction parameters more parsimoniously as

$$u_{12(ij)} = \kappa |j - i|$$

Despite the reduced parameter count, it is still necessary to set $u_{1(1)} = u_{2(1)} = 0$ in order to separate out the row and column effects from the intercept, but there may be improved precision on derived parameters (e.g. odds ratios) as a result of a more economical model (Albert, 1988).

Raftery (1996) proposes a prior for regression coefficients in general linear models (including contingency table models) and generalizing the prior for linear regression models in Hoeting *et al.* (1996). Assuming standardized covariates, all coefficients β_j except the intercept are assigned independent normal priors with mean 0 and common variance. The prior is then transformed in such a way as to reflect the actual observations (so is a 'data-dependent' prior). The method involves defining a design matrix X which for a contingency table application would be of dimension $N \times p$ where p is the number of independent (identifiable) parameters for the impacts of each level of each categorical factor or factor interaction. N would be the number of cells (e.g. $I \times J \times K$ in a three-way table).

For a regression with p predictors (e.g. a normal linear regression with continuous predictors as well as a Poisson regression with categorical predictors)[1], the prior for β is derived from a prior for γ assuming standardized predictors (Raftery, 1996, p 256). Let $s_j (j = 2, \ldots, p)$ denote standard deviations of predictors in column j of $X_{n \times p}$, and $s_y = [V(y)]^{0.5}$. For preset values of $(\psi, \phi\}$

$$\gamma_1 \sim N(0, \psi^2)$$

and

$$\gamma_j \sim N(0, \phi^2), \quad j = 2, \ldots, p$$

Then $\beta = v + Q\gamma$, where $v_1 = \bar{y}$, and $v_j = 0, j = 2, \ldots, p$. $Q_{p \times p}$ has zero elements except for diagonal elements $(s_y, s_y/s_2, s_y/s_3, \ldots, s_y/s_p)$ and first row $Q_{12} = -s_y \bar{X}_2/s_2, Q_{13} = -s_y \bar{X}_3/s_3, \ldots, Q_{1p} = -s_y \bar{X}_p/s_p$. Values of ϕ should be chosen that have minimal effect on model choice via the Bayes factor and Raftery recommends a range $1 \le \phi \le 5$, with central value $\phi = 1.65$. For non-metric outcomes in general linear models with mean μ_i and link $g, \bar{y}$ and s_y are replaced by $\bar{z}$ and s_Z where $z_i = g(\mu_i) + (y_i - \mu_i)/g'(\mu_i)$.

Priors such as this may be combined with the variable selection methodology of George and McCulloch (1993), Kuo and Mallick (1998), Ntzoufras (1999) and others. Predictor selection for coefficients in log-linear models usually leads to a situation where the number of selection indicators γ_m is less than the number of predictors. For example, in log-linear regression for cross-classified counts $y_{ijk}(i = 1, \ldots, I; j = 1, \ldots, J; k = 1, \ldots, K)$ there would typically be a single inclusion indicator for the entire set of $(I - 1)(J - 1)$ parameters $u_{12(ij)}$, one for $u_{13(ik)}$, one for $u_{23(jk)}$ and another one for $u_{123(ijk)}$. The intercept and main effects are usually included by default. The priors on γ_m would then reflect how many of the possible models involved each parameter set.

Albert (1996) takes the intercept and main effects as included by default but second- and higher order interactions subject to selection.

---

[1] For a log-linear regression 'predictors' are defined by the level of each categorical factor or their interactions. For a $2 \times 2 \times 5$ log-linear model X would be of dimension 20×20 with columns corresponding to an intercept, dummy 'predictors' for each of the six main effects, nine predictors for each of the two-way interactions and four for the three-way interactions.

Thus for a two-way table, the independence model (4.14) may be compared with the model (4.15) including the second-order interaction. The main effects have standard fixed effects priors with corner constraints for identification. However, for interaction terms an exchangeable normal prior $u_{12(ij)} \sim N(0, P)$ is assumed with no corner constraints on the $u_{12(ij)}$. A mixture prior on the inclusion or not of the entire set of interactions can then be set up using the McCulloch and Rossi (1993) method. Let γ_I be a binary selection indicator. Then according as $\gamma_I = 1$ or 0, the variance P or the precision P^{-1} is selected to correspond to the independence model (model 0) or the interaction model (model 1). Model 0 applies as $P \to 0$, so when $\gamma_I = 0, u_{12(ij)} \sim N(0, P_0)$ where P_0 is small, whereas when $\gamma_I = 1$, $u_{12(ij)} \sim N(0, P_1)$ where P_1 is large, allowing non-trivial, non-zero effects to emerge. A robust extension is a scale mixture on the log-linear model parameters subject to doubts on their inclusion. For a two-way table and when $\gamma_I = 1$

$$u_{12ij} \sim N(0, P/\lambda_{ij})$$

where $\lambda_{ij} \sim Ga(0.5\nu, 0.5\nu)$ and ν is small (under five).

Example 4.5   Contraceptive use and heart attack risk   Consider a case–control study of the interdependence between oral contraceptive use and myocardial infarction among women aged 25–49, classified into age bands 25–29, 30–34, 35–39, 40–44 and 45–49 (Raftery, 1996). So the data are of dimension $2 \times 2 \times 5$ arranged as contraceptive use (rows), infarction (columns) and age (slices) (denoted C, M and A for short); see Table 4.2.

Table 4.2   Oral contraceptive use and myocardial infarction

		Control group age					Myocardial infarction age				
		25–29	30–34	35–39	40–44	45–49	25–29	30–34	35–39	40–44	45–49
Contraceptive	No	224	390	330	362	301	2	12	33	65	93
use	Yes	62	33	26	9	5	4	9	4	6	6

Consider first predictor selection with four binary indicators γ_m corresponding to interaction effects (CM), (CA), (MA) and (CMA). There are nine possible models that include intercept and main effects by default:

1. $1 + C + M + A + CM$
2. $1 + C + M + A + CA$

3. $1 + C + M + A + MA$
4. $1 + C + M + A + CM + CA$
5. $1 + C + M + A + CM + MA$
6. $1 + C + M + A + CA + MA$
7. $1 + C + M + A + CA + MA + CM$
8. $1 + C + M + A + CA + MA + CM + CMA$
9. $1 + C + M + A$

CMA appears in only one model, so an appropriate prior is $\delta_4 \sim \text{Bern}(1/9)$. If CMA is not included then each interaction CM, CA and MA appears in four of the remaining eight models. Since non-hierarchical models are not allowed, an appropriate prior for $\delta_1, \delta_2, \delta_3$ is $\delta_j \sim \text{Bern}(\pi)$ where

$$\pi = \delta_4 + 0.5(1 - \delta_4)$$

The choice of one or more $\delta_m = 1$ corresponds to one of the nine models and relative frequencies of selection under MCMC sampling may be monitored. One may also monitor the relative risk of myocardial infarction for oral contraceptive users. This is $\exp(u_{12(22)})$ if CM is included and $\exp(u_{12(22)} + u_{123(22k)})$ if CMA is included. Note that model 7 corresponds to age as a confounder (not a risk factor in itself but the distribution on the true risk factors differs by age), while in model 8 age is a modifier.

   With $N(0, 0.1)$ priors taken on β_j, it appears that model 7 is preferred. The second half of a two-chain run of 100 000 iterations gives a probability exceeding 0.99 on this model.

   Next consider the robust prior model of Raftery (1996), for which in a contingency table application of dimension $2 \times 2 \times 5$ (as in Table 4.2), the model means are given by

$$\log(\mu_{ijk}) = u_0 + u_{1(i)} + u_{2(j)} + u_{3(k)} + u_{12(ij)} + u_{13(ik)} + u_{23(jk)} + u_{123(ijk)} = \beta X$$

where the 20 independent parameters (those not set to zero) are $\beta_1 = u_0$, $\beta_2 = u_{1(2)}$, $\beta_3 = u_{2(2)}$, $\beta_4 = u_{3(2)}$, $\beta_5 = u_{3(3)}$, $\beta_6 = u_{3(4)}$, $\beta_7 = u_{3(5)}$, $\beta_8 = u_{12(22)}$, $\beta_9 = u_{13(22)}$, $\beta_{10} = u_{13(23)}$, $\beta_{11} = u_{13(24)}$, $\beta_{12} = u_{13(25)}$, $\beta_{13} = u_{23(22)}$, $\beta_{14} = u_{23(23)}$, $\beta_{15} = u_{23(24)}$, $\beta_{16} = u_{23(25)}$, $\beta_{17} = u_{123(222)}$, $\beta_{18} = u_{123(223)}$, $\beta_{19} = u_{123(224)}$ and $\beta_{20} = u_{123(225)}$. The design matrix has rows defined by cell entries $\{i, j, k\}$ with k changing most rapidly and i least rapidly. So rows $1, 2, 3, 4, 5, 6, \ldots, 20$ are defined for $y_{111}, y_{112}, y_{113}, y_{114}, y_{115}, y_{121}, \ldots, y_{225}$. The first column relates to β_1 and has ones throughout, the second column relates to β_2 and has entries equal to one for rows 11–20 (where $i = 2$) and zero elsewhere (see Example4_5.xls).

In fact Raftery considers a simplified model (where $p = 17$) under which the relative risk is constant for ages above and below 34. So $\beta_{17} = u_{123(222)}$ for ages $k = 3, 4, 5$ and $\beta_{17} = 0$ for ages $k = 1, 2$, though the full age categorization is retained for lower order effects. Here predictor selection may be used with an indicator only on the inclusion of the redefined MCA, denoted MCA$_2$. Thus $\delta_1 = 1$ if MCA$_2$ is included, $\delta_1 = 0$ for model 7. A prior probability of 0.9 is assumed for $\Pr(\delta_1 = 1)$ and the posterior probability (from the second half of a two-chain run of 25 000 iterations) is 0.767. So the Bayes factor in favour of the simpler model 7 is 2.7, which is not conclusive. A suggested exercise is to obtain the relative risk of myocardial infarction for pill users at ages under 34 under models 7 and 8 and hence a model averaged estimate of the relative risk based on this Bayes factor.

Example 4.6 Thromboembolism data Case–control data first considered by Worcester (1971) has been used in subsequent model choice studies, such as Spiegelhalter and Smith (1982) and Pettit and Young (1990). The data y_{ijk} cross-classify thromboembolism and control patients ($i = 1$ and 2 respectively) by two risk factors: oral contraceptive user ($j = 1$ for user, $j = 2$ for non-user) and smoking ($k = 1$ for smokers, $k = 2$ for non-smokers). The data are in Table 4.3.

Table 4.3 Risks for thromboembolism

Patient type	Smoker contraceptive user		Non-smoker contraceptive user		Totals
	Yes	No	Yes	No	
Thrombolembolism	14	7	12	25	58
Control	2	22	8	84	116

While a design matrix approach to model specification as in Bishop *et al.* (1975) is often used, an economical model notation follows Worcester (1971), with the saturated model (model 1) specified as

$$y_{ijk} \sim \text{Po}(\mu_{ijk})$$

$$\log(\mu_{ijk}) = \beta_0 + \delta_i\beta_1 + \delta_j\beta_2 + \delta_k\beta_3 + \delta_{ij}\beta_{12} + \delta_{ik}\beta_{13} + \delta_{jk}\beta_{23} + \delta_{ijk}\beta_{123}$$

where δ is one only when all the subscripts are one, and is zero otherwise. Thus

$$\delta_1 = \delta_{11} = \delta_{111} = 1$$

but for all other subscript combinations $\delta = 0$. Hence the need to specify corner constraints is eliminated.

For these data a close fit is obtained with a model omitting β_{13} and β_{23} (this forming model 2). Model choice inferences under both methods may be sensitive to the priors on β given the small sample size. Though default priors based on minimal experiments have been suggested (Spiegelhalter and Smith, 1982) one may also draw on substantive epidemiological knowledge. On a log scale the β parameters amount to log relative risks (this is strictly an approximation for case–control data). Relative risks (e.g. of thromboembolism) associated with adverse behavioural risk factors are usually in the range 1–10. A relative risk of 100 amounts to a virtually certain disease inception and allowing the prior on β to include such high relative risks (i.e. virtually certain of disease or death given a risk factor) is itself potentially informative. Instead a normal prior on the β_j ($j \geq 1$) with mean 2 and variance 0.1 is assumed, this being equivalent to a mean relative risk of 7 and 95% point above 1000. For the intercept an N(0,1000) prior is assumed.

A two-chain run of 20 000 iterations is taken, with a 1000 burn-in. With the above prior the Gelfand–Ghosh and DIC criteria both select the smaller model (see Example4_6.xls for calculations). Under the priors used, the Bayes factor estimate is $B_{21} = 23.8$, quite strongly in favour of the smaller model with $\log_e B_{21} = 3.17$. The fact that the reduced model gives a close fit implies that the use of oral contraceptives, particularly among those who smoke, is a risk for thromboembolism, but for smokers who do not take the pill there is no excess risk.

4.6 SEMI-PARAMETRIC AND GENERAL ADDITIVE MODELS FOR BINOMIAL AND COUNT RESPONSES

As for metric outcomes, varying coefficient and general additive models allow possible non-linearity in the impacts of predictors in binomial, binary or count regression but without specifying complex algebraic forms. For instance, suppose y_t is a binary outcome with responses corresponding to ordered values of a regressor x_t

$$x_1 < x_2 < \cdots < x_n \tag{4.16}$$

Assuming $y_t \sim \text{Bern}(\pi_t)$, and predictors $(z_{i1}, \ldots, z_{iq})$ with a conventional linear impact on y, a general additive semi-parametric regression model has the form

$$g[\pi_t] = \beta_0 + s(x_t) + \beta z \tag{4.17}$$

where g is a link function (e.g. the logit). Let $s_t = s(x_t)$ denote the smooth function representing the possibly non-linear impact of x on y as it varies over its range. For count data one might take $y_t \sim \text{Po}(\mu_t)$ or $y_t \sim \text{NB}(\mu_t)$ with

$$\log(\mu_t) = \beta_0 + s(x_t) + \beta z$$

For GAMs based on state-space priors, it is common to assume normal or Student t random walks in the first or second (and occasionally higher order) differences of the s_t. If there are equal gaps between the successive x then a first-order normal random walk prior is

$$s_t \sim \text{N}(s_{t-1}, \tau^2) \tag{4.18}$$

or equivalently

$$s_t = s_{t-1} + u_t \tag{4.19}$$

where $u_t \sim \text{N}(0, \tau^2)$. For greater smoothing, an RW(2) prior may be used with

$$s_t \sim \text{N}(2s_{t-1} - s_{t-2}, \tau^2) \tag{4.20}$$

or

$$s_t = 2s_{t-1} - s_{t-2} + u_t$$

For binary data, especially in small samples, the smoothing function may not be well identified and relatively informative priors on the evolution variance τ^2 may be elicited. With unequal gaps $\delta_t = x_{t+1} - x_t$ between x values, a first-order normal random walk would then be

$$s_t \sim \text{N}(s_{t-1}, \delta_t \tau^2)$$

and a second-order one would be

$$s_t \sim \text{N}(\nu_t, \delta_t \tau^2)$$

where $\nu_t = s_{t-1}(1 + \delta_t/\delta_{t-1}) - s_{t-2}(\delta_t/\delta_{t-1})$.

Since the above priors do not specify a level for s_t, identifiability may be obtained by recentring at each MCMC iteration, namely

$$g[\pi_t] = \beta_0 + S_t + \beta z$$

where $S_t = s_t - \bar{s}$. If there is a smooth in just one predictor then an alternative is to omit the intercept β_0 and then the s_t need not be recentred. Another option is to set the first effect s_1 to zero, rather than a free parameter. Other aspects of identification such as standardizing predictors are also relevant to satisfactory convergence.

Other types of prior include the state-space version of a cubic spline, with bivariate state vector $V_t = (s_t, ds_t/dt)$ evolving via

$$V_t = F_t V_{t-1} + u_t \tag{4.21}$$

where F_t is a 2×2 transition matrix, namely

$$F_t = \begin{bmatrix} 1 & \delta_t \\ 0 & 1 \end{bmatrix}$$

The u_t are bivariate (e.g. MVN) with zero mean and covariance $\tau^2 U_t$, where τ^2 is the smoothing variance and

$$U_t = \begin{bmatrix} \delta_t^3/3 & \delta_t^2/2 \\ \delta_t^2/2 & \delta_t \end{bmatrix}$$

Sometimes a smooth in x interacts with the effect of another predictor z, with

$$g[\pi_t] = \beta_0 + z_t s(x_t)$$

If $x_t = t$ denotes time then a dynamic coefficient model is obtained, with

$$g[\pi_t] = \beta_0 + z_t \beta_{1t}$$

Wood and Kohn (1998) consider general additive regression of binary data using latent continuous data or 'utilities', as in Albert and Chib (1993). So if $y_t = 1$ then the latent response w_t is positive, while if $y_t = 0, w_t$ is constrained to be negative. Equivalence with a probit link involves truncated normal sampling of the w_t with variance 1 and means

$$E(w_t) = \mu_t = \beta_0 + s(x_t) + \beta z \tag{4.22}$$

As mentioned above, this data augmentation approach may have benefits in sampling since full conditionals are the same as for metric (e.g. normal) responses. Residual analysis may also be facilitated.

Spline models based on truncated polynomial or other basis functions can be used, such as the penalized random effects spline of Ruppert $et\ al.$ (2003), with knots $\kappa_k (k = 1, \ldots, K)$ sited at percentiles $(k + 1)/(K + 2)$ of x_t. For binary outcomes modelled via data augmentation, a penalized linear spline would then involve the model

$$w_t | y_t = 1 \sim N(\mu_t, 1) I(0,)$$
$$w_t | y_t = 0 \sim N(\mu_t, 1) I(,0)$$

with means

$$\mu_t = \gamma_0 + \gamma_1 x_t + \sum_{k=1}^{K} \beta_k (x_t - \kappa_k)_+ \tag{4.23}$$

where $\beta_k \sim N(0, \sigma_\beta^2)$. A quadratic spline would have means

$$\mu_t = \gamma_0 + \gamma_1 x_t + \gamma_1 x_t^2 + \sum_{k=1}^{K} \beta_k (x_t - \kappa)_+^2 \qquad (4.24)$$

For count data with $y_t \sim Po(\mu_t)$ a non-linear effect of a single predictor might be combined with an error term to allow for overdispersion (Chapter 5), as in the linear spline model

$$\nu_t = \log(\mu_t) = \gamma_0 + \gamma_1 x_t + \sum_{k=1}^{K} \beta_k (x_t - \kappa_k)_+ + \varepsilon_t$$

With priors $1/\sigma_\beta^2 \sim Ga(a_1, b_1)$, $1/\sigma_\varepsilon^2 \sim Ga(a_2, b_2)$, the full conditionals on $1/\sigma_\beta^2$ and $1/\sigma_\varepsilon^2$ are

$$1/\sigma_\beta^2 \sim Ga\left(a_1 + 0.5K, b_1 + 0.5 \sum_{k=1}^{K} \beta_k^2 \right)$$

$$1/\sigma_\varepsilon^2 \sim Ga\left(a_2 + 0.5n, b_2 + 0.5 \sum_{t=1}^{n} \varepsilon_t^2 \right)$$

where

$$\varepsilon_t = \nu_t - \left[\gamma_0 + \gamma_1 x_t + \sum_{k=1}^{K} \beta_k (x_t - \kappa_k)_+ \right]$$

4.6.1 Robust and adaptive non-parametric regression

To allow for discontinuities or changes in curvature of the smooth or smooths on predictors one may assume that the errors u_t in (4.19) or (4.21) come from a discrete mixture density with L components, e.g. an RW(1) model for univariate u_t would become

$$s_t \sim N(s_{t-1}, \tau_{M_t}^2)$$
$$M_t \sim Categorical(\pi)$$

where $\pi = (\pi_1, \ldots, \pi_L)$. Usually L is two or three at most. One might also (Knorr-Held, 1999) assume a scale mixture (equivalent to a t density with ν degrees of freedom) on the random walk parameters. For example,

$$s_t \sim N(s_{t-1}, \tau^2/\lambda_t) \qquad (4.25)$$

where $\lambda_t \sim Ga(0.5\nu, 0.5\nu)$.

Alternatively a stochastically evolving variance might be considered instead of taking τ^2 constant over time. Thus let $\tau^2 = \exp(h_t)$ and let h_t

itself be an RW(1) or RW(2) series as in (4.18) and (4.20). This provides a dependent (i.e. time structured) prior in the log variances. Alternatively the h_t might be unstructured over time and be a scale mixture just as in (4.24). Despite the extra parameterization involved in such 'spatially adaptive' models there may be a gain in fit (e.g. Jerak and Lang, 2003; Ruppert and Carroll, 2000).

4.6.2   Other approaches to non-linearity

One might also consider approaches analogous to those in section 3.4 for non-linear predictor impacts on discrete outcomes. However, Kay and Little (1987) propose an alternative to the Box–Cox transform

$$x(\lambda) = (x^\lambda - 1)/\lambda$$

of x when y is binary. Specifically they view the predictor as a random variable with density $\{f_j(x), j = 0, 1\}$ according to whether $y = 0$ or $y = 1$ and suppose that

$$\log[f_1(x)/f_0(x)] = \alpha_0 + h(x)$$

where $h(x)$ is a possibly non-linear vector transform of x. They show that for $y_i \sim \text{Bern}(\pi_i)$, the logit transform is then linear in $h(x)$:

$$\text{logit}(\pi_i) = \beta_0 + \beta_1 h(x)$$

Depending on the nature of x (e.g. whether its density may be characterized as normal, gamma, beta, etc.) they define the form $h(x)$ of x such that the logistic model obtains; there may be more than one such h transform variable. For example, if x is gamma then the appropriate $h(x)$ transforms are x and $\log(x)$, while if x is beta with values in $(0,1)$ then the appropriate h variables are $\log(x)$ and $\log(1 - x)$; see Example 4.9. If x is normal then the appropriate h variables are x and x^2.

Example 4.7   Blood cell sedimentation   Everitt and Rabe-Hesketh (2001) present cubic spline smooth estimates obtained via the backfitting algorithm for a sparse binary outcome, i.e. 6 out of 32 observations with $y_t = 1$. The responses relate to the rate at which red blood cells (known as erythrocytes) settle out of suspension in blood plasma, and the response is one if erythrocyte sedimentation (ES) exceeds 20 mm/h, where values below this threshold characterize healthy individuals. The probability of a positive (i.e. morbid) response is related to smooths $s(x)$ in two plasma proteins, fibrinogen (x_{1t}) and gamma-globulin (x_{2t}), which are recorded in gm/l.

Here a direct Bernoulli likelihood approach with logit link is used to obtain smooth functions in the proteins. Because there are ties on both predictors (i.e. two or more subjects with the same values) the smoothing prior is specified in terms of the $n_{u1} = 28$ distinct values of fibrinogen and $n_{u2} = 15$ distinct values of gamma-globulin. Further, the values are unequally spaced, with $\delta_{jt} = x_{j,t+1} - x_{j,t}$ defining the gaps in the ordered predictor values x_{jt}. Since there are two predictors (Table 4.4), the data cannot be uniquely ordered as in (4.16), and it is necessary to record the

Table 4.4 Original data and ranks on predictors

Subject	Fibrinogen (x_1)	Gamma (x_2)	Rank x_1	Rank x_2	y
1	2.52	38	12	11	0
2	2.56	31	14	4	0
3	2.19	33	4	6	0
4	2.18	31	3	4	0
5	3.41	37	25	10	0
6	2.46	36	11	9	0
7	3.22	38	22	11	0
8	2.21	37	5	10	0
9	3.15	39	21	12	0
10	2.6	41	15	13	0
11	2.29	36	8	9	0
12	2.35	29	9	2	0
13	5.06	37	28	10	1
14	3.34	32	24	5	1
15	2.38	37	10	10	1
16	3.15	36	21	9	0
17	3.53	46	26	15	1
18	2.68	34	18	7	0
19	2.6	38	15	11	0
20	2.23	37	6	10	0
21	2.88	30	19	3	0
22	2.65	46	16	15	0
23	2.09	44	1	14	1
24	2.28	36	7	9	0
25	2.67	39	17	12	0
26	2.29	31	8	4	0
27	2.15	31	2	4	0
28	2.54	28	13	1	0
29	3.93	32	27	5	1
30	3.34	30	24	3	0
31	2.99	36	20	9	0
32	3.32	35	23	8	0

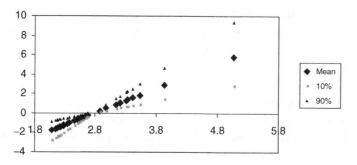

Figure 4.1   Total impact of fibrinogen

ranks r_{1t} and r_{2t} of the tth subject on each predictor. Then $y_t \sim \text{Bern}(\pi_t)$, and assuming a logit link without data augmentation, approach (a) leads to

$$\text{logit}[\pi_t] = \beta_0 + \beta_1 x_{1t} + g_1(x_{1,r1t}) + \beta_2 x_{2t} + g_2(x_{2,r2t})$$

The smoothing prior is as in (4.21) with a $Ga(1, 0.01)$ prior on the smoothing precision $1/\tau^2$.

A two-chain run of 10 000 iterations (convergent after 1000) shows that the total regression effect for fibrinogen, $\beta_1 x_{1t} + s_1(x_{1,r1t})$, is primarily linear in shape, though there is increased variability at larger values (Figure 4.1). The regression effect for gamma-globulin (Figure 4.2) shows some non-linearity, i.e. an attenuation of the effect at middle values. Figure 4.2 replicates that in Everitt and Rabe-Hesketh (2001, p 300).

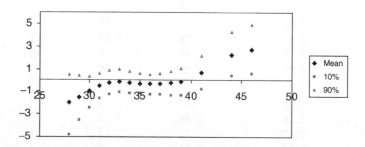

Figure 4.2   Total impact of gamma-globulin

Example 4.8   Union membership and wages   Berndt (1991) considers data from the 1985 US Current Population Survey (CPS) on the relation between union membership and hourly wage rates. Consider a penalized random effects linear spline, as in (4.23), for the impact of wages on the binary outcome. A logit regression without data augmentation is used. $K = 19$ knots are used, starting with the 5th percentile for wages and ending with the 95th percentile. (The code in Program 4.8 allows for a quadratic spline instead.) A Ga(0.5, 0.1) prior is assumed for σ_β^2. A two-chain run of 10 000 iterations (convergent from 1000) shows a clear non-linear effect (Figure 4.3), with the maximum probability occurring at a wage rate of around $14 per hour.

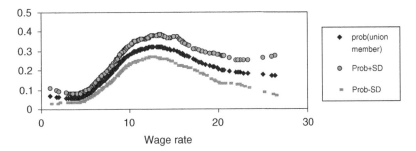

Figure 4.3   Union membership probability and wages per hour

Example 4.9   Age at menarche   Kay and Little (1987) consider binomial data on age at menarche in 3918 Warsaw girls, arranged in 25 age groups. Table 4.5 shows that the ages x range between 9.2 and 17.6 and one possible approach (model A) takes

$$y_i \sim \text{Bin}(n_i, \pi_i)$$
$$\text{logit}(\pi_i) = \beta_0 + \beta_1 x(\lambda)$$

where $x(\lambda) = (x^\lambda - 1)/\lambda$ ($\lambda \neq 0$), $x(\lambda) = \log(x)$ ($\lambda = 0$). Alternatively, by transforming x to range between 0 and 1 according to $v = (x - 9)/18$ it is reasonable to regard v as a beta variable. In this case (model B) the relevant predictors are $\log(v)$ and $\log(1 - v)$, and

$$\text{logit}(\pi_i) = \gamma_0 + \gamma_1 \log(v) + \gamma_2 \log(1 - v)$$

For model A a uniform prior $U(-2, 2)$ is adopted for λ. Should initial runs suggest this not to be wide enough an interval, a prior such as $U(-5, 5)$ might be used. The criterion (4.12) under model A is 496 with

Table 4.5 Age at menarche

Mean age	Total	Number having menstruated	Predictions			
			Box–Cox model		Log (v) and log ($1 - v$), v beta	
			$\overline{Z}$	sd(Z)	$\overline{Z}$	sd(Z)
9.21	376	0	0.2	0.47	0.0	0.00
10.21	200	0	1.2	1.09	0.6	0.81
10.58	93	0	1.1	1.07	1.0	1.03
10.83	120	2	2.5	1.57	2.1	1.51
11.08	90	2	3.0	1.75	2.9	1.76
11.33	88	5	4.8	2.18	4.7	2.23
11.58	105	10	8.7	2.89	9.2	3.02
11.83	111	17	13.9	3.63	14.9	3.96
12.08	100	16	18.4	4.06	19.1	4.22
12.33	93	29	24.0	4.47	25.0	4.37
12.58	100	39	34.9	5.00	35.9	5.00
12.83	108	51	48.6	5.47	49.3	5.74
13.08	99	47	54.5	5.15	54.2	5.34
13.33	106	67	68.7	5.20	67.5	5.32
13.58	105	81	76.5	4.75	75.5	4.60
13.83	117	88	93.5	4.51	92.4	4.77
14.08	98	79	83.5	3.66	82.3	3.71
14.33	97	90	86.6	3.12	86.0	3.27
14.58	120	113	110.7	3.06	110.1	3.19
14.83	102	95	96.4	2.37	96.2	2.45
15.08	122	117	117.1	2.20	117.1	2.32
15.33	111	107	107.8	1.84	108.0	1.93
15.58	94	92	92.1	1.39	92.3	1.39
15.83	114	112	112.3	1.32	112.6	1.22
17.58	1049	1049	1047.0	1.43	1049.0	0.44

$(\beta_0, \beta_1) = (-91, 55)$ and $\lambda = -0.36$ with a 95% interval $(-0.45, -0.24)$. Under model B, the same criterion is 472 with improved predictions at the lowest ages and the highest age; the parameter means from a two-chain run of 50 000 iterations are $(\gamma_0, \gamma_1, \gamma_2) = (2.3, 4.4, -2.3)$. However, the predictions under model B are less precise and using (4.10) with k at one or less shows model B with less advantage.

EXERCISES

1. In Example 4.3 find the outlier probabilities under the variance inflation model when the variance for outliers is five rather than ten.

2. In Example 4.7 use the data augmentation approach equivalent to a logit link to obtain smooths in the proteins. This means sampling according to

$$w_t \sim \text{logistic}(\mu_t, 1)I(A_t, B_t)$$
$$\mu_t = \beta_0 + \beta_1 x_{1t} + s_1(x_{1,r1t}) + \beta_2 x_{2t} + s_2(x_{2,r2t})$$
$$A_t = -\infty, B_t = 0 \quad \text{if } y_t = 0$$
$$A_t = 0, B_t = \infty \quad \text{if } y_t = 1$$

In practice default limits may replace infinities for A_t, B_t. A logit link is also achieved (approximately) by sampling the w_t from a Student t with $\nu = 8$ degrees of freedom, or by scale mixing and normal sampling, namely

$$w_t \sim \text{N}(\mu_t, 1/\lambda_t)I(A_t, B_t)$$

where the λ_t are drawn from a gamma density $G(0.5\nu, 0.5\nu)$. In this case, do the sampled λ_t suggest any outlying cases?

3. There are potentially non-linear impacts of the predictors used in Example 4.4. Use the penalized random effects approach, with either quadratic or linear splines as in (4.23) or (4.24), to investigate possible non-linearity in the impacts of $X_1 = $ population density, $X_2 = \%$ economically active in classes IV and V, and $X_3 = $ private renting. To avoid numerical overflow under the quadratic spline, dividing the predictors by 100 is recommended.

4. In Example 4.8 use an augmented data approach with normal sampling but scale mixing to allow for outliers. Are any apparent? Let the degrees of freedom of the scale mixture be unknown and so assess whether a probit or logit link is more suitable.

5. Using the results in Table 4.5 obtain criterion (4.10) for $k = 1$ and $k = 0.5$. Also add code to Program 4.9 in order to obtain criterion (4.11) for $k = 1000$, $k = 1$ and $k = 0.5$.

6. In Example 4.9 apply the augmented data probit regression model in (4.4). This means adopting a disaggregated model at the level of each girl and using the average age in each band as the predictor. Apply the

$h(x)$ transforms model along with variance inflation as in (4.7) to check for outliers. Alternatively introduce variance scale factors λ_i as in (4.3) with degrees of freedom ν unknown and check for outliers using the λ_i themselves.

REFERENCES

Albert, J. (1988) Bayesian estimation methods for Poisson means using a hierarchical log-linear model. In *Bayesian Statistics 3*, Bernardo, J., DeGroot, M., Lindley, D. and Smith, A. (eds). Oxford University Press, Oxford, 519–531.

Albert, J. (1996) Bayesian selection of log-linear models. *Canadian Journal of Statistics*, **24**, 327–347.

Albert, J. (1999) Criticism of a hierarchical model using Bayes factors. *Statistics in Medicine*, **18**, 287–305.

Albert, J. and Chib, S. (1993) Bayesian analysis of binary and polychotomous response data. *Journal of the American Statistical Association*, **88**, 669–679.

Albert, J. and Chib, S. (1995) Bayesian residual analysis for binary response regression models. *Biometrika*, **82**, 747–759.

Bedrick, E., Christensen, R. and Johnson, W. (1996) A new perspective on priors for generalized linear models. *Journal of the American Statistical Association*, **91**, 1450–1460.

Berndt, E. (1991) *The Practice of Econometrics*. Addison-Wesley: Reading, MA.

Bishop, Y., Fienberg, S. and Holland, P. (1975) *Discrete Multivariate Analysis: Theory and Practice*. MIT Press: Cambridge, MA.

Böhning, D. (1999) *Computer-Assisted Analysis of Mixtures and Applications: Meta-Analysis, Disease Mapping and others*. Chapman and Hall: London/CRC Press: Boca Raton, FL.

Carlin, B. and Louis, T. (1996) *Bayes and Empirical Bayes Methods for Data Analysis*. London: Chapman and Hall.

Chaloner, K. and Brant, R. (1988) A Bayesian approach to outlier detection and residual analysis. *Biometrika*, **75**, 651–659.

Chib, S. (1995) Marginal likelihood from the Gibbs output. *Journal of the American Statistical Association*, **90**, 1313–1321.

Collett, D. (2003) *Modelling Binary Data*, 2nd Edition. Chapman and Hall: London/CRC Press: Baton Roca, FL.

Congdon, P. (2003) *Applied Bayesian Modelling*. John Wiley & Sons: Chichester.

Congdon, P. (2005) Bayesian predictive model comparison via parallel sampling, *Computational Statistics and Data Analysis*, **48**, 735–753.

Copas, J. (1988) Binary regression models for contaminated data. *Journal of the Royal Statistical Society, Series B*, **50**, 225–265.

Dellaportas, P. and Forster, J. (1999) Markov chain Monte Carlo model determination in hierarchical and graphical log-linear models. *Biometrika*, **86**, 615–633.

Epstein, A. and Fienberg, S. (1991) Bayesian estimation in multidimensional contingency tables. In *Computer Science and Statistics: Proceedings of the Twenty-Third Symposium on the Interface*, Keramidas, E (ed.). Interface Foundation of North America: Fairfax, VA, 37–47.

Everitt, B. and Rabe-Hesketh, S. (2001) *Analyzing Medical Data using S-PLUS*. Springer: New York.

Fahrmeier, L. and Knorr-Held, L. (2000) Dynamic and semiparametric models. In *Smoothing and Regression: Approaches, Computation and Application*, Schimek, M. (ed.). John Wiley & Sons: New York, 513–544.

Gelfand, A. and Ghosh, S. (1998) Model choice: A minimum posterior predictive loss approach. *Biometrika*, **85**, 1–11.

Gelman, A., Carlin, J., Stern, H. and Rubin, D. (1995) *Bayesian Data Analysis*, 1st Edition. Chapman and Hall: London.

George, E. and McCulloch, R. (1993) Variable selection via Gibbs sampling, *Journal of the American Statistical Association*, **88**, 881–889.

George, E., McCulloch, R. and Tsay, R. (1996) Two approaches to Bayesian model selection with applications. In *Bayesian Analysis in Statistics and Econometrics: Essays in Honor of Arnold Zellner*, Berry, D., Chaloner, K. and Geweke, J. (eds). John Wiley & Sons: New York.

Gerlach, R., Bird, R. and Hall, A. (2002) Bayesian variable selection in logistic regression: predicting company earnings direction. *Australian & New Zealand Journal of Statistics*, **2**, 155–168.

Gilks, W. and Wild, P. (1992) Adaptive rejection sampling for Gibbs sampling. *Applied Statistics*, **41**, 337–348.

Gunel, E. and Dickey, J. (1974) Bayes factors for independence in contingency tables. *Biometrika*, **61**, 545–557.

Harvey, A. (1976) Estimating regression models with multiplicative heteroscedasticity. *Econometrica*, **44**, 461–465.

Hoeting, J., Raftery, A. and Madigan, D. (1996) A method for simultaneous variable selection and outlier identification in linear regression. *Computational Statistics and Data Analysis*, **22**, 251–270.

Holford, T., White, C. and Kelsey, J. (1978) Multivariate analysis for matched case-control studies. *American Journal of Epidemiology*, **107**, 245–256.

Jerak, A. and Lang, S. (2003) Locally adaptive function estimation for binary regression models. *Discussion Paper* 310, SFB 386, University of Munich.

Jochmann, M. (2003) Semiparametric Bayesian inference for count data treatment models. Department of Economics, University of Konstanz.

Kahn, M. and Raftery, A. (1996) Discharge rates of Medicare stroke patients to skilled nursing facilities: Bayesian logistic regression with unobserved heterogeneity. *Journal of the American Statistical Association*, **91**, 29–41.

Kay, R. and Little, S. (1987) Transformations of the explanatory variables in the logistic regression model for binary data. *Biometrika*, **74**, 495–501.

Knorr-Held, L. (1999) Conditional prior proposal in dynamic models. *Scandinavian Journal of Statistics*, **26**, 129–144.

Knuiman, M. and Speed, T. (1988) Incorporating prior information into the analysis of contingency tables. *Biometrics*, **44**, 1061–1071.

Kuo, L. and Mallick, B. (1998) Variable selection for regression models. *Sankhya*, **60B**, 65–81.

Laud, P. and Ibrahim, J. (1995) Predictive model selection. *Journal of the Royal Statistical Society, Series B*, **57**, 247–262.

Leonard, T. and Hsu, J. (1994) The Bayesian analysis of categorical data – a selective review. In *Aspects of Uncertainty: A Tribute to DV Lindley*, Freeman, P. and Smith, A. (eds). John Wiley & Sons: Chichester.

McCullagh, P. and Nelder, J. (1989) *Generalized Linear Models*, 2nd Edition. Chapman and Hall: London.

Meyer, M. and Laud, P. (2002) Predictive variable selection in generalized linear models. *Journal of the American Statistical Association*, **97**, 859–871.

Ntzoufras, I. (1999) Aspects of Bayesian model and variable selection using MCMC. Department of Statistics, Athens University of Economics and Business.

Oh, M. (1997) A Gibbs sampling approach to Bayesian analysis of generalized linear models for binary data. *Computational Statistics*, **12**, 431–445.

Oh, M. and Lim, Y. (2001) Bayesian analysis of time series Poisson data. *Journal of Applied Statistics*, **28**, 259–271.

Pettit, L. and Young, K. (1990) Measuring the effect of observations on Bayes factors. *Biometrika*, **77**, 455–466.

Raftery, A. (1996) Approximate Bayes factors and accounting for model uncertainty in generalised linear models. *Biometrika*, **83**, 251–266.

Ruppert, D. and Carroll, R. (2000) Spatially-adaptive penalties for spline fitting. *Australian and New Zealand Journal of Statistics*, **42**, 205–223.

Ruppert, D., Wand, M. and Carroll, R. (2003) *Semiparametric Regression*. Cambridge University Press: Cambridge.

Scott, S. (2003) Data augmentation for the bayesian analysis of multinomial logit models. *Proceedings of the American Statistical Association Section on Bayesian Statistical Science*. American Statistical Association: Alexandria, VA.

Smith, M. and Kohn, R. (1996) Nonparametric regression using Bayesian variable selection. *Journal of Econometrics*, **75**, 317–334.

Spiegelhalter, D. and Smith, A. (1982) Bayes factors for linear and log-linear models with vague prior information. *Journal of the Royal Statistical Society, Series B*, **44**, 377–387.

Verdinelli, I. and Wasserman, L. (1991) Bayesian analysis of outlier problems using the Gibbs sampler. *Statistics and Computing*, 1, 105–117.

Wakefield, J. and Best, N. (1999) Accounting for inaccuracies in populations counts and case registration in cancer mapping studies. *Journal of the Royal Statistical Society, Series A*, **162**, 363–382.

Wood, S. and Kohn, R. (1998) A Bayesian approach to robust nonparametric binary regression. *Journal of the American Statistical Association*, **93**, 203–213.

Worcester, J. (1971) The relative odds in the 2^3 contingency table. *American Journal of Epidemiology*, **93**, 145–149.

Zellner, A. and Rossi, P. (1984) Bayesian analysis of dichotomous quantal response models. *Journal of Econometrics*, **25**, 365–393.

CHAPTER 5

Further Questions in Binomial and Count Regression

5.1 GENERALIZING THE POISSON AND BINOMIAL: OVERDISPERSION AND ROBUSTNESS

The Poisson and binomial models assume that the variances of the observations are known functions of the mean parameters. In practice data of both types may be more dispersed than these densities assume. Such overdispersion of excess heterogeneity may reflect a few extreme observations or multiple modes, clustering of high or low rates, clumped or other non-random sampling (Efron, 1986), or variation between units of widely different exposures (e.g. death totals for areas differing widely in population size). Alternatively it may be due to unobserved variations between subjects (or frailties) ϕ_i that are not represented by the observed covariates. Without correction for extra variation the precision of the β parameters will be overstated: their credible intervals will be too narrow (Cameron and Trivedi, 1998). Extra variation may be partly caused by poor fit. For example, standard GLM assumptions may also not hold for the assumed link and a better fit may be obtained with an alternative link. There may also be heterogeneity in the appropriate link.

For metric data an overdispersed alternative to the univariate or multivariate normal is the Student t density, which can be obtained by mixing on the scale parameters (variance or covariance matrix) of the normal. In the same way, the binomial and Poisson densities may need to

---

Bayesian Models for Categorical Data P. Congdon
© 2005 John Wiley & Sons, Ltd

be modified when the observed variance exceeds the form assumed under the density, and this involves a mixture distribution on the binomial probability or the Poisson mean. This may be combined with adoption of alternative link functions or mixing over links.

Mixture and link generalizations can be seen both as providing greater robustness in inferences (Gelman et al., 2003) and as providing a density that is compatible with the data. Another motivation for mixture models is to pool information over units when event counts for each unit may vary considerably; by modelling the rates for individual units in terms of an overall hyperdensity, shrinkage estimates may be obtained for each unit that smooth towards the average. For example, for death rates by area, these estimates would be a weighted combination of the population mean rate and the rate derived by treating each area in isolation (Böhning, 1999). In models with both spatial and unstructured errors pooling may be to local (neighbourhood) means as well as to the global average (Chapter 8). Another application is in 'ecological' inference (e.g. in voting applications) when only the marginals of $R \times C$ tables are observed and a model is required for the internal cells (Rosen et al., 2001; Haneuse and Wakefield, 2004) – see Chapter 11.

For a set of occurrences/trials data $[\{y_1, n\}, \{y_2, n\}, \ldots, \{y_g, n\}]$ conforming to the standard binomial density with common probability π and $Be(a, b)$ prior, the posterior mean of y is that of a $Be(a + \sum_i y_i, a + b + gn)$ density, with $E(\pi|y) = (a + \sum_i y_i)/(a + b + gn)$ and variance $V(y) = nE(\pi|y)[1 - E(\pi|y)]$. Overdispersion would mean $V(y)$ exceeding $nE(\pi|y)[1 - E(\pi|y)]$. For Poisson count data with prior $\mu \sim Ga(c, d)$ and observations $\{y_1, \ldots, y_n\}$ the posterior density is $Ga(c + \sum_i y_i, d + n)$ with mean $(c + \sum_i y_i)/(d + n)$ and the variance equals this mean. For data subject to overdispersion the observed variance exceeds the mean. It is also possible, though less frequent, for data to be underdispersed relative to the standard Poisson assumption. Similarly in Poisson or binomial regression it is often found that the conditional variance $V(y_i|x_i)$ exceeds its nominal value under Poisson or binomial sampling. For Poisson data this means $V(y_i|x_i)$ exceeds $\mu_i = E(y_i|x_i)$ where for instance $\mu_i = \exp(\beta x_i)$.

In many cases data y_i are not simply overdispersed but contain too many zeros to be Poisson or binomial. Such systematic departures from Poisson or binomial sampling require a different method that explicitly models the excess frequency of zero observations. Retaining the usual sampling models, albeit mixed with gamma or beta frailties, will lead to underprediction of the percentage of zero counts (Hall, 2000; Lambert,

1992). Models explicitly approaching excess zeros in relation to the Poisson or binomial models are considered in section 5.4.

5.2   CONTINUOUS MIXTURE MODELS

While other analyses are possible and commonly used in more complex applications (e.g. general linear mixed models for multilevel and panel applications), the conjugate mixture distributions are often used in simple applications, especially in non-regression situations. The conjugate model for count data that is often used to account for extra variation replaces the model $y_i \sim \text{Po}(\mu_i)$ by a model with heterogeneous means following a gamma density. Thus

$$y_i \sim \text{Po}(\mu_i)$$
$$\mu_i \sim \text{Ga}(\delta, \gamma)$$

Parameterizing the mean of the μ_i as $\lambda = \delta/\gamma$, one obtains $\text{E}(\mu) = \lambda$ and $V(\mu) = \delta/\gamma^2 = \lambda^2/\delta$. Then

$$V(y) = \text{E}[V(y|\mu)] + \text{Var}[\text{E}(y|\mu)] = \lambda + \lambda^2/\delta = \lambda(1 + \lambda/\delta)$$

so that overdispersion applies if $\alpha > 0$ where $\alpha = 1/\delta$.

To represent unobserved heterogeneity for count data in a regression analysis, a multiplicative frailty with log-link would mean

$$y_i \sim \text{Po}(\nu_i)$$
$$\nu_i = \mu_i\phi_i = \exp(\beta X_i)\phi_i$$

Assuming a gamma frailty model $\phi_i \sim \text{Ga}(\delta, \gamma)$, conjugate to the Poisson density, then

$$P(y_i|X_i, \phi_i)P(\phi_i) = \{\exp(-\phi_i\mu_i)[\phi_i\mu_i]^{y_i}/y_i!\}\{\delta^\gamma \phi_i^{\delta-1} \exp(-\gamma\phi_i)/\Gamma(\delta)\}$$

Integrating out ϕ_i, as in

$$P(y_i|X_i) = \int P(y_i|X_i, \phi_i)P(\phi_i)d\phi_i$$

leads to a marginal negative binomial density for y_i. The identifiability constraint $\delta = \gamma$ is frequently used, the alternative being to exclude an intercept in βX_i. With this constraint $V(\phi_i) = 1/\delta$ and the negative binomial $\text{NB}(\mu_i, \delta)$ has the form

$$P(y_i|X_i) = \Gamma(\delta + y_i)/\{\Gamma(\delta)\Gamma(y_i + 1)\}\left(\frac{\delta}{\delta + \mu_i}\right)^\delta \left(\frac{\mu_i}{\mu_i + \delta}\right)^{y_i} \quad (5.1)$$

With estimation by repeated sampling it is straightforward to analyse using either the negative binomial likelihood or the mixed Poisson–gamma likelihood. Alternative gamma mixtures are possible (Albert, 1999), namely

$$y_i \sim \text{Po}(\nu_i)$$
$$\nu_i \sim \text{Ga}(\mu_i \alpha, \alpha) \tag{5.2}$$

giving $\text{Var}(\nu_i) = \mu_i/\alpha$ and

$$V(y_i|X_i) = \mu_i + \mu_i/\alpha$$

Conditional on α, the ν_i are $\text{Ga}(y_i + \alpha\mu_i, 1 + \alpha)$ with mean

$$(1 - S)y_i + S\mu_i$$

where $S = \alpha/(\alpha + 1)$ is a shrinkage factor. If the observations result from different exposures E_i with $y_i \sim \text{Po}(\nu_i E_i)$, then the underlying rates $\nu_i|\alpha$ are $\text{Ga}(y_i + \alpha\mu_i, E_i + \alpha)$ with posterior mean

$$(1 - S_i)y_i/E_i + S_i\mu_i$$

where $S_i = \alpha/(\alpha + E_i)$. S_i is the proportion of shrinkage from the crude unpooled estimate towards the hierarchical model mean μ_i. Larger values of α mean greater shrinkage, and the non-hierarchical Poisson corresponds to $\alpha \to \infty$. Alternatively

$$\nu_i \sim \text{G}(\alpha, \alpha/\mu_i) \tag{5.3}$$

gives

$$V(y_i|X_i) = \mu_i + \mu_i^2/\alpha$$

and $\nu_i|\alpha \sim \text{Ga}(y_i + \alpha_i, E_i + \alpha/\mu_i)$ with mean

$$(1 - S_i)y_i + S_i\mu_i$$

where $S_i = \alpha/(\alpha + E_i\mu_i)$. These are known as the NB1 and NB2 models respectively (Cameron and Trivedi, 1998). $P(y|X)$ as in (5.1) corresponds to the NB2 form, since $V(\nu_i) = \mu_i^2/\delta$.

In a generalization of the negative binomial, Winkelmann and Zimmermann (1995) propose a variance function

$$V(y_i|X_i) = \mu_i + \alpha\mu_i^{\kappa+1}$$

where $\alpha > 0$ and $\kappa \geq -1$. The Poisson is the limiting case when α tends to zero. This variance function is obtained by

$$y_i \sim \text{Po}(\nu_i)$$
$$\nu_i \sim \text{G}(\mu_i^{1-\kappa}/\alpha, \mu_i^{-\kappa}/\alpha) \tag{5.4}$$

The value $\kappa = 0$ means the conditional variance is $(1 + \alpha)\mu_i$, while $\kappa = 1$ gives $V(y_i|X_i) = \mu_i + \alpha\mu_i^2$. So this mixture includes the NB1 and NB2 models and the form (5.4) may be used to decide which is more appropriate to the data.

A further development in the negative binomial or Poisson–gamma approach is to let the variance parameter α vary by subjects (e.g. Dey et al., 1997; Efron, 1986), so that

$$y_i \sim \text{Po}(\nu_i)$$
$$\nu_i \sim \text{G}(1/\alpha_i, 1/[\alpha_i\mu_i])$$

with $\mu_i = \exp(\beta X_i)$, and $\alpha_i = \exp(\gamma Z_i)$, where X_i and Z_i may overlap. For instance, in count data with varying exposures E_i it may be that variances depend on E_i, so that

$$\log(\alpha_i) = \gamma_0 + \gamma_1 E_i$$

The main alternative to a conjugate frailty is an additive random error in $g(\mu_i)$ so that

$$y_i \sim \text{Po}(\mu_i)$$
$$g(\mu_i) = \beta_0 + \beta X_i + \varepsilon_i \tag{5.5}$$

where ε_i may follow a parametric density (e.g. normal or Student t). As noted by Engel and Keen (1994, p 12) if the log-link is assumed and $\varepsilon \sim \text{N}(0, \alpha)$ then to a close approximation, $V(y_i|X_i) = \mu_i + \alpha\mu_i^2$. To avoid possible identifiability problems with regard to β_0 and the average of ε_i, one may instead take

$$g(\mu_i) = \beta_{0i} + \beta X_i$$

where β_{0i} is randomly distributed.

For occurrences–trials data the conjugate frailty model assumes π to vary from trial to trial according to a beta density. Thus instead of $y_i \sim \text{Bin}(n_i, \pi)$ it is assumed that

$$y_i \sim \text{Bin}(n_i, \pi_i)$$
$$\pi_i \sim \text{Beta}(\kappa_1, \kappa_2)$$

A reparameterized beta density, setting $\gamma = \kappa_1 + \kappa_2$, $\rho = \kappa_1/\gamma$, is often preferred since ρ and γ are closer to being orthogonal, so that

$$\pi_i \sim \text{Beta}(\gamma\rho, \gamma(1 - \rho)) \tag{5.6}$$

The variance of y_i is then given by

$$V(y_i) = n_i\pi_i(1 - \pi_i)\left(\frac{n_i + \gamma}{\gamma + 1}\right)$$

and the correlation between individual responses within each trial is

$$1/(1 + \kappa_1 + \kappa_2) = 1/(1 + \gamma) \qquad (5.7)$$

The beta–binomial is not in the exponential family even for known γ. If covariates are available, one may (Kahn and Raftery, 1996) take ρ to vary over cases with

$$g(\rho_i) = \beta X_i$$

where g is an appropriate link, such as $\text{logit}(\rho_i) = \beta X_i$.

The GLMM alternative is to take $y_i \sim \text{Bin}(n_i, \pi_i)$ as above but assume an additive random error in $g(\pi_i)$. For example,

$$\text{logit}(\pi_i) = \beta X_i + \varepsilon_i \qquad (5.8)$$

where ε_i may follow a parametric density or a non-parametric mixture. Taking ε to be normal gives the logistic normal model (Pierce and Sands, 1975). Engel and Keen (1994) show that, to a close approximation, if $\varepsilon_i \sim N(0, \phi)$ and $\mu_i = n_i \pi_i$ then

$$V(y_i | X_i) = \mu_i \left(1 - \frac{\mu_i}{n_i} \right) \left[1 + \phi \mu_i \left(1 - \frac{\mu_i}{n_i} \right) \left(\frac{n_i - 1}{n_i} \right) \right]$$

Example 5.1   Male suicides in England   To illustrate count overdispersion, consider (as in Example 4.4) male suicides y_i in 1989–1993 in 354 English local authorities (Example5_1.xls). Four predictors are used to predict suicide relative risks. These are $X_1 =$ single, widowed, divorced, $X_2 =$ one-person households, $X_3 = \%$ economically active in classes IV and V, and $X_4 =$ population density (divided by 1000). On the basis of Example 4.4, X_2–X_4 would be expected to be the most important of these. However, other aspects of the data need to be considered, e.g. the adequacy of Poisson regression in modelling the dispersion in the data. The variance of the y_i is 1567 compared with the mean of 52 and the predictions of an adequate model would reflect this overdispersion.

In particular, if z_i are samples of new data from the model being applied, a posterior predictive check for an adequate model would show overdispersion in the z compatible with that in the y. So one might compare the ratio of variance with mean between z and y, by calculating

$$C_z^{(t)} = V(z^{(t)})/\bar{z}^{(t)}$$

for the sampled z at iteration t and comparing it with $C_y = V(y)/\bar{y} = 1567/52 = 30.1$

A basic assessment of how far overdispersion is adequately modelled involves comparing the average of the sampled deviances, namely $D^{(t)}$ where

$$D^{(t)} = 2\left\{\sum(y_i + 0.5)\log[(y_i + 0.5)/(\mu_i^{(t)} + 0.5)] - (y_i - \mu_i^{(t)})\right\}$$

with the available degrees of freedom, namely $n - p$ or $n - p_e$. On this basis, a simple Poisson regression (model A), sampling for which converges from 1000 iterations of a two-chain run of 20 000 iterations, seems reasonably adequate with $\bar{D} = 526$. However, the predictive check shows an average probability $P_c = 0.96$ that $C_z^{(t)}$ exceeds C_y, whereas acceptable bounds for P_c are between 0.10 and 0.90. Predictors $\{X_1–X_4\}$ under model A are 'significant' in the sense of 95% intervals above zero for X_1, X_2 and X_3 and below zero for X_4. The 95% interval for X_1 is from 0.0003 to 0.0082.

For the NB1 Poisson–gamma mixture as in (5.2), a two-chain run of 20 000 iterations (convergent from 10 000) gives an improved deviance $\bar{D} = 361$, and a more compatible predictive check with $P(C_z^{(t)} > C_y) = 0.86$. The dispersion parameter α is estimated at 0.47. The impact of X_1 remains significant under this model with 95% interval $\{0.0004, 0.0024\}$.

A further improved fit and satisfactory predictive check are achieved by the generalized gamma mixture (5.4) with $\bar{D} = 337$ and $P(C_z^{(t)} > C_y) = 0.73$. Here $\kappa = 1.05$ and $\alpha = 0.02$. Estimates of CPOs under this model are obtained from (2.15). The 16 areas with scaled CPOs under 0.01 (Table 5.1) may suggest excluded predictors; for example, 3 of the 16 are coastal resorts with higher suicide rates than predicted under the model, perhaps reflecting the transient populations attracted to such areas.

5.2.1 Modified exponential families

As well as GLMM and conjugate mixture models, more fundamental extensions of the GLM allow modelling of the variance (and hence over-dispersion) as well as regression for the mean. Efron (1986) pointed out the limitation of the binomial variance assumption in which $\text{Var}(y_i)$ is inversely proportional to the sample size n_i, especially when samples of subjects are subject to clumping. Efron discussed the family of double-exponential models that include double-binomial and double-Poisson models. Instead of the standard binomial and Poisson densities that contain only mean parameters, their double versions include mean and variance parameters μ_i and φ_i that may be related to predictors (including actual sample size in the case of φ_i under the binomial). For example, the double Poisson takes the form

$$\log[P(y|\mu, \varphi)] = c(y) + 0.5\log(\varphi) - \varphi\mu + \varphi y\log(e\mu/y)$$

Table 5.1 Diagnostics for another areas

Area	Log (CPO)	Ratio of CPO to maximum CPO	$\hat{Y}$ (posterior mean of μ_i)	SMR	Crude SMR	Y (actual deaths)	Expected deaths
Lambeth	−6.39	0.00289	130.0	135.1	151	146	96.2
Sheffield	−6.38	0.00294	194.6	94.7	87	179	205.5
Wycombe	−6.20	0.00351	62.5	106.1	130	77	58.9
Burnley	−6.18	0.00357	42.3	131.2	172	56	32.3
Taunton Deane	−6.16	0.00365	43.4	123.6	162	57	35.1
Bedford	−6.07	0.00401	60.6	120.2	148	75	50.4
Bournemouth	−5.90	0.00473	81.5	135.8	159	96	60.0
Reigate & Banstead	−5.63	0.00623	48.5	110.5	139	61	43.9
Waltham Forest	−5.43	0.00757	70.3	88.9	74	58	79.0
Bassetlaw	−5.37	0.00808	47.0	119.0	149	59	39.5
Torbay	−5.34	0.00829	56.8	131.8	160	69	43.1
Manchester	−5.27	0.00889	224.5	143.3	150	235	156.7
Sunderland	−5.24	0.00912	95.4	90.3	79	83	105.6
East Riding of Yorkshire	−5.20	0.00959	97.7	87.6	77	86	111.6
Derwentside	−5.19	0.00962	31.3	99.1	67	21	31.6
Hastings	−5.17	0.00984	39.9	130.1	165	51	30.7

SMR = Standard Mortality Ratio

where $c(y)$ is a function of y alone. Under the double binomial, the variance of y_i is approximately $\pi_i(1 - \pi_i)/[n_i\varphi_i]$ rather than $\pi_i(1 - \pi_i)/n_i$. Efron's double-binomial analysis of toxoplasmosis data for 34 cities in El Salvador took the variance function to be a quadratic in the sample sizes

$$\varphi_i = M/[1 + \exp(-\kappa_i)]$$
$$\kappa_i = \gamma_0 + \gamma_1 n_i + \gamma_2 n_i^2$$

It may be noted that modelling of binomial data by Bernoulli disaggregation and then data augmentation, as in (4.3), has a similar impact to the double exponential if combined with scale mixing as in (4.5). Thus for $i = 1, \ldots, g$,

$$w_{ij} \sim N(\beta X_i, 1/\lambda_i)I(0, \infty) \quad j = 1, y_i$$
$$w_{ij} \sim N(\beta X_i, 1/\lambda_i)I(-\infty, 0) \quad j = n_i - y_i + 1, n_i$$

where $\log(\lambda_i)$ may be modelled as a function of n_i. The double exponential also has affinities with the scaled exponential families proposed by West (1985) and the exponential dispersion models of Jorgensen (1987). Gelfand and Dalal (1990) and Dey *et al.*, (1997) consider overdispersed GLMs based on a two-parameter exponential family of the form

$$\log[P(y|\theta, \tau)] = c(y) + y\theta + \tau T(y) - \rho(\theta, \tau) \qquad (5.9)$$

where the usual exponential family of Chapter 1 occurs when $\tau = 0$. Including a variance function $a(\phi)$ in (5.9) and setting $\rho(\theta, 0) = b(\theta)$, the usual exponential family is

$$\log[P(y|\phi, \theta)] = c(y, \phi) + [y\theta - b(\theta)]/a(\phi)$$

Often $a(\phi) = \phi/n$ and another way of modelling overdispersion is to take ϕ to vary over subjects according to $\log(\phi_i) = \gamma Z_i$ (Ganio and Schafer, 1992) The usual one-parameter family is obtained only for given ϕ_i.

5.3   DISCRETE MIXTURES

While continuous mixtures such as the gamma–Poisson and beta–binomial model have advantages of conjugacy, they are not necessarily always the most suitable way of modelling overdispersion. Finite mixture models as discussed in Chapter 3 may have benefit in representing overdispersion while simultaneously providing robust inferences (see e.g. Aitkin, 1996; Aitkin, 2001; Hurn *et al.*, 2003). For count data y_i without covariates one might assume a mixture over Poisson or negative binomial densities

$$P(y_i|\pi, \mu) = \sum_{j=1}^{J} \pi_j Po(y_i|\mu_j)$$

$$P(y_i|\pi, \mu, \delta) = \sum_{j=1}^{J} \pi_j NB(y_i|\mu_j, \delta_j)$$

As in Chapter 3 a mixture of Poisson or negative binomial densities can be formulated as a hierarchical model using the latent variable G_i that allocates subject i to one of the components. So a Poisson mixture can be represented as

$$P(y_i|G_i = j) = Po(\mu_j)$$

With predictors X_i one might consider mixtures of Poisson regressions such that

$$E(y_i|X_i) = \sum_{j=1}^{J} \pi_j \exp(\beta_j X_i) \qquad (5.10a)$$

or mixtures of logistic regressions for binary data, with

$$E(y_i|X_i) = \sum_{j=1}^{J} \pi_j \{\exp(\beta_j X_i)/[1 + \exp(\beta_j X_i)]\} \qquad (5.10b)$$

A discrete mixture may also be relevant in modelling overdispersion. Thus for Poisson overdispersion, one might assume

$$y_i \sim \text{Po}(\mu_i)$$
$$g(\mu_i) = \beta_0 + \beta_1 X_i + \varepsilon_i = \beta_{0i} + \beta_1 X_i$$

where the prior for β_{0i} may follow a finite mixture of normal densities with differing means and variances

$$\beta_{0i} \sim \sum_{j=1}^{J} \pi_j N(B_j, \tau_j)$$

One might also take discrete mixtures of the conjugate Poisson–gamma or beta–binomial models (Moore *et al.*, 2001). For example, with binomial data, let

$$y_i \sim \text{Bin}(n_i, \kappa_i)$$
$$\kappa_i \sim \sum_{j=1}^{J} \pi_j \, \text{Beta}(\alpha_{ij}, \beta_{ij}) \qquad (5.11)$$

A reparameterization of the beta in terms of $\alpha_{ij} = \rho_{ij}\gamma_j$ and $\beta_{ij} = (1 - \rho_{ij})\gamma_j$ facilitates regression modelling (e.g. a logit regression for predicting ρ_{ij} using predictors X_i). It also permits simple identifiability constraints (e.g. $\rho_1 > \rho_2 > \ldots > \rho_J$) when predictors are not used, and $\alpha_{ij} = \rho_j\gamma_j$, $\beta_{ij} = (1 - \rho_j)\gamma_j$.

Another type of mixture strategy involves outlier detection models. For example, in a conjugate Poisson model, with $y_i \sim \text{Po}(\nu_i)$, and $\nu_i \sim \text{Ga}(\alpha, K\alpha/\mu_i)$. The parameter α is a precision parameter (as α tends to infinity the Poisson is approached). For outlier resistance one may assume

$$\nu_i \sim \pi \text{Ga}(K\alpha, K\alpha/\mu_i) + (1 - \pi) \, \text{Ga}(\alpha, \alpha/\mu_i)$$

where π is small (e.g. $\pi = 0.05$) and $0 < K < 1$ (e.g. $K = 0.25$). The first component is 'precision deflated'. In a GLMM Poisson model with

$$\log(\nu_i) = \mu_i + u_i$$

one might take

$$u_i \sim \pi N(0, K\zeta) + (1 - \pi) N(0, \zeta)$$

where $K > 1$ (e.g. $K = 5$ or $K = 10$).

Finally, as in Dey and Ravishanker (2000) and Kleinman and Ibrahim (1998), one may adopt Dirichlet process mixing. Here, for $y_i \sim EF(\mu_i)$,

$$g(\mu_i) = \beta_0 + \beta_1 X_i + \varepsilon_i \qquad (5.12)$$
$$\varepsilon_i \sim DP(\kappa H_0)$$

where κ is the concentration parameter and H_0 is a base density. H_0 is often taken to be normal with variance τ an extra parameter. The preset concentration parameter, or the prior on it (if not preset), includes beliefs regarding how close the true density of ε_i is reflected by H_0. Low values of κ allow more non-normality (and fewer clusters) than larger values. The number of potential clusters K^* under (5.12) may equal n but in practice may be set at a value such as 10 or 20 $(\leq n)$ to encompass realistic amounts of possible error clustering. The same prior might be applied to varying slopes

$$g(\mu_i | G_i = j) = \beta_0 + \beta_{1j} X_i$$

where $G_i = j$ is the cluster selected for observation i. In a model with varying intercepts and slopes, namely

$$g(\mu_i | G_i = j) = \beta_{0j} + \beta_{1j} X_i$$

one might adopt different DPPs for the intercepts and slopes, allowing for more clustering in one than the other.

Example 5.2 Sole egg hatchings Lindsey (1975) presents binomial data on the effects of salinity (X_1) and temperature (X_2) in an egg hatching experiment for English sole. Specifically the observations are the numbers of eggs hatching y_i out of n_i total eggs in $i = 1, \ldots, 72$ tanks. He observed that extra variation was substantial after allowing for these known influences. The average deviance for a simple binomial model is over $12\,000$. Here two types of mixture analysis are applied: the conjugate beta–binomial model and a non-parametric mixture.

Under the conjugate mixture approach of (5.6)

$$y_i | \pi_i \sim Bin(n_i, \pi_i)$$
$$\pi_i \sim Beta(\gamma \rho_i, \gamma(1 - \rho_i))$$

with the impact of the predictors modelled via

$$\text{logit}(\rho_i) = \beta X_i$$

with the prior for γ assumed to be a Ga(1, 0.001). A two-chain run of 5000 iterations for this model shows early convergence, with the average deviance reduced to around 71.9. With 70.3 effective parameters the DIC is 142.2. γ stands at around 2.9, implying a within-cluster correlation, from (5.7), of $1/(1 + 2.9) = 0.26$. Of the two predictors, only temperature appears to have a coefficient different from zero, with a 95% interval (0.17, 0.36).

A non-parametric mixture involves a Dirichlet process with a G(0.01, 0.1) prior on the Dirichlet concentration parameter κ, and an N(0, 1) baseline density assumed for varying intercepts β_{0j} in the model

$$y_i \sim \text{Bin}(n_i, \pi_i)$$
$$\text{logit}(\pi_i) = \beta_{0G_i} + \beta x_i$$

An upper limit of 20 clusters for the cluster index G_i is assumed. Convergence is obtained from around 1500 iterations (with centred covariates) and the deviance averages 87. The average number of clusters is 15.4 and the mean of κ is 8.4. The effective parameter count is slightly smaller (66) than for the beta–binomial model but the DIC higher at 153. The interval for β_2 is slightly less favourable to high values with a 95% interval (0.18, 0.29).

Example 5.3   Protozoa mortality   Data from Follman and Lambert (1989) record deaths y_i from n_i exposed to poison dose x_i for $g = 8$ groups of protozoa. Thus $y = \{0, 8, 18, 18, 22, 37, 47, 50\}$, $n = \{55, 49, 60, 55, 53, 53, 51, 50\}$ with dosages $x = \{4.7, 4.8, 4.9, 5, 5.1, 5.2, 5.3, 5.4\}$. Follman and Lambert try a logistic regression mixture with $K = 2$ groups as in (5.10), with varying intercepts but homogeneous effect of dose:

$$y_i \sim \text{Bin}(n_i, \rho_i)$$
$$G_i \sim \text{Cat}(\pi_1, \pi_2)$$
$$\text{logit}(\rho_i) = \beta_{0,G_i} + \beta_1(x_i - \bar{x})$$

Follman and Lambert (1989) obtain a deviance of 21.3 at the maximum likelihood parameter estimate; see also Agresti (2002, p 546). The irregularity making a robust mixture necessary is the flat mortality rate at medium doses; a simple binomial regression adequately models overdispersion as a predictive check will show.

Here a two-group regression with varying intercepts and slopes is used, namely

$$\text{logit}(\rho_i) = \beta_{0,G_i} + \beta_{1,G_i}(x_i - \bar{x})$$

N(0,10 000) priors are assumed on β_{0j} and β_{1j} with an identifiability constraint $\beta_{12} > \beta_{11}$ on the slopes, and a Dirichlet D(5, 5) prior on $\pi = (\pi_1, \pi_2)$. The posterior classification rates from a two-chain run of 100 000 iterations are $\pi = (0.58, 0.42)$ with $\beta_0 = (-0.2, -5.2)$, $\beta_1 = (6.5, 91)$. Posterior probabilities $P(G_i = 2)$ are above 0.5 only for the first and last observations. The average deviance is now 10.6, indicating an improved model.

5.4 HURDLE AND ZERO-INFLATED MODELS

Hurdle and zero-inflated models are specialized instances of discrete mixture models commonly used for count or binomial data with excess zeros. In the hurdle model non-zero observations (counts of one, two or more) occur from crossing a threshold or hurdle (Mullahy, 1986). The probability of crossing this hurdle involves a binary sampling model, while the sampling of non-zero counts involves a truncated Poisson or binomial (sampling confined to values of y above zero). Explanatory variables X_i can be introduced in both parts of the model. In economic applications, an interpretation in terms of a two-stage decision process may apply, with the first binary stage involving a selection mechanism.

Let f_1 and f_2 be probability densities appropriate to integer data. For count observations y_i, f_1 might be Bernoulli and f_2 Poisson or negative binomial. Then the probabilities of the two stages are given by

$$
\begin{aligned}
P(y_i = 0) &= f_1(0) \\
\Pr(y_i = j) &= \{[1 - f_1(0)]/[1 - f_2[0]]\}f_2[j] \qquad j > 0 \\
&= \omega f_2[j]
\end{aligned}
$$

where $\omega = [1 - f_1(0)]/[1 - f_2[0]]$. The correction factor $1 - f_2[0]$ is needed to account for the truncated sampling at stage 2 (i.e. ensure the model probabilities sum to unity). Note that for count data f_1 might also be the Poisson or negative binomial density truncated at zero (Cameron and Trivedi, 1998, p 124).

Suppose count observations are arranged so that cases $i = 1, \ldots, n_1$ have zero outcomes, while for $i = n_1 + 1, \ldots, n$, all responses are non-zero. Thus if f_1 were Bernoulli with $f_1(1) = \pi_i$, $f_1(0) = (1 - \pi_i)$

and f_2 Poisson with mean μ_i, with $f_2[0] = \exp(-\mu_i)$, the likelihood would be defined by

$$y_i \sim \text{Bern}(\pi_i) \quad i = 1, \ldots, n_1$$

$$\Pr(y_i = j) = [(1 - \pi_i)/(1 - \exp(-\mu_i))] \exp(-\mu_i)\mu_i^{y_i}/y_i! \quad i = n_1 + 1, \ldots, n$$

π_i and μ_i may be functions of possibly overlapping covariates W_i and X_i. Setting $f_1 = f_2$ or equivalently $\omega = 1$ reduces to the standard model (Cameron and Trivedi, 1998). The factor ω can be seen to vary by obervation. In terms of this factor, the expectation and variance are (Winkelmann, 2000)

$$E(y) = \omega \sum_{j=1}^{\infty} j f_2(j)$$

$$V(y) = \sum_{j=1}^{\infty} j^2 f_2(j) - \left[\omega \sum_{j=1}^{\infty} j f_2(j)\right]^2$$

which reduces to the standard variance-mean relation when $\omega = 1$. The range $0 < \omega < 1$ yields overdispersion with excess zeros, while $\omega > 1$ yields underdispersion (subject to the variance being defined) with zeros less frequent than under the standard Poisson. These properties are defined at the individual level in terms of means μ_i, such that

$$V(y_i|X_i) = \mu_i + \frac{1 - \omega_i}{\omega_i} \mu_i^2$$

The larger the ratio of $1 - \omega_i$ to ω_i the more an observation is contributing to excess dispersion.

An alternative zero inflated Poisson (ZIP) mechanism is proposed by Lambert (1992) whereby zero observations may occur either by definition (e.g. when a manufacturing process is under control) or by a chance mechanism (when the manufacturing process is producing some defective items but sometimes yields zero defectives). Let $G_i = 1$ or 2 according to which latent state or regime is operating. Then

$$\Pr(y_i = 0) = \Pr(G_i = 1) + P(y_i = 0|G_i = 2)\Pr(G_i = 2)$$

$$\Pr(y_i = j) = P(y_i = j|G_i = 2)\Pr(G_i = 2) \quad j = 1, 2, \ldots$$

If the probability that $G_i = 1$ is denoted ξ_i then the overall density is

$$\Pr(y_i = j) = \xi_i(1 - g_i) + (1 - \xi_i)P(y_i)$$

where $g_i = \min(y_i, 1)$ and $P(y_i)$ is a standard density for count data. A logit model with covariates W_i might be used to model the ξ_i. Under a

Poisson model with mean $\mu_i = E(y_i|X_i)$

$$P(y_i = 0) = \xi_i + (1 - \xi_i)\exp(-\mu_i)$$
$$P(y_i = j) = (1 - \xi_i)\exp(-\mu_i)\mu_i^{y_i}/y_i! \quad j = 1, 2, \ldots$$

The variance is then

$$V(y_i|X_i) = (1 - \xi_i)[\mu_i + \xi_i\mu_i^2] > \mu_i(1 - \xi_i) = E(y_i|X_i)$$

and again the modelling of excess zeros implies overdispersion. The zero inflated model can also be used with $P(y)$ as negative binomial in case the Poisson option does not produce predictions that account both for overdispersion and excess zeros.

Example 5.4 Major derogatory reports This example uses data from Greene (2000) on y_i, the number of derogatory credit reports, for 100 subjects in terms of age, dollar income (divided by 10 000), monthly credit card expenditure (divided by 100) and home ownership vs. renting (1 for owners). The covariates W_i used to explain the hurdle parameters π_i and the ZIP parameters ξ_i are age, income and home ownership, while age, income and expenditure are used in the second-stage truncated Poisson model. Simple analysis of average expenditure and income by counts of derogatory reports shows that average expenditure is highest for zero reports but that average income peaks for $y = 2$ reports. The second stage of the hurdle model (applicable only to the 18 subjects with $y_i > 0$) may be achieved by truncated Poisson sampling,

$$y_i \sim Po(\mu_i)I(1,)$$

The fit of the hurdle and ZIP models is compared by sampling new data. The new data z are obtained at each iteration t from the relevant Poisson density (defined by $G_i^{(t)} = 1$ or 2) for the inflated model, and from the second stage of the hurdle model for all subjects, even those with $y_i = 0$. The Poisson sampling for z under the hurdle model allows zero counts. The predictive criterion (2.10) is then used to assess fit.

The coefficients of the ZIP model show that higher income people are less likely to be in the zero-risk group (those for whom $G_i = 1$), while the frequency of reports declines with average expenditure (compare Greene, 2000, p 892). As may be verified by estimating CPO statistics (see 2.15), the fit of the hurdle model is worsened by subjects 85 and 91 who, despite large expenditures, have a derogatory report (Table 5.2). The ZIP model is much less affected by these subjects, and under criterion (2.10) its fit is markedly better.

Table 5.2 Derogatory reports

	Hurdle selection model			ZIP zero group		
	Mean	2.5%	97.5%	Mean	2.5%	97.5%
Intercept	3.27	0.92	5.59	5.56	1.44	10.24
Age	−0.043	−0.110	0.027	−0.040	−0.128	0.048
Income	−0.16	−0.49	0.18	−1.22	−2.56	−0.30
Own/rent	0.91	−0.36	2.28	0.47	−1.28	2.30
	Frequency model			Standard Poisson group		
Intercept	1.31	−0.09	2.73	1.33	−0.39	2.70
Age	−0.01	−0.05	0.03	−0.01	−0.05	0.04
Income	−0.06	−0.30	0.16	−0.12	−0.44	0.10
Expenditure	−0.31	−0.68	−0.01	−0.60	−1.00	−0.29

Example 5.5   Recreation trips   To illustrate the use of negative bino-mial and NB hurdle regression, especially in a setting with excess zeros, consider the recreation trips data of Cameron and Trivedi (1998). These data relate to 659 respondents to a survey on boating trips to Lake Somerville, Texas, during 1980. Of the 659, 50 respondents report ten or more trips (with a maximum of 88) but 417 respondents report zero trips. Potential models should reflect both overdispersion and the excess zeros.

Predictors include a subjective quality rating (X_1 = SQ), a binary index for involvement in water-skiing at the lake (X_2 = SKI), annual household income in \$1000 (X_3 = INC), and a binary index for season fee payment (X_4 = FEE). The other predictors are dollar expenditures (TEXP) by lakes: Conroe (X_5 = TEXP1), Somerville (X_6 = TEXP2) and Houston (X_7 = TEXP3). Cameron and Trivedi consider several mixture models, including a hurdle version of the negative binomial.

Here a binary model with the above seven predictors is used to model the first stage of the NB hurdle, i.e. the distinction between zero-trip res-pondents ($d_i = 0$) and those with at least one trip ($d_i = 1$). Thus

$$d_i \sim \text{Bern}(\pi_i)$$
$$\text{logit}(\pi_i) = \beta X_i$$

The second stage is negative binomial with these predictors and with truncated sampling (confined to values 1 and above). Thus

$$y_i \sim \text{NB}(\mu_i, \delta)I(1,)$$
$$\log(\mu_i) = \gamma X_i$$

A Ga(1, 0.1) prior is set on the δ parameter in (5.1). There are only 13 subjects with FEE = 1 and none of them have $d_i = 0$. The parameter β_4 on FEE is therefore poorly identified by the data. Cameron and Trivedi give a maximum likelihood estimate for β_4 of 9.4 (s.e. $= 0.6$) and a suitably downweighted (e.g. precision reduced by 100) version of this estimate might be used to provide an informative prior. Hence a prior $\beta_4 \sim (9.4, 36)$ is assumed.

A two-chain run of 100 000 iterations converges after 40 000 iterations. The results (based on iterations 50 000–100 000) are similar to those of Cameron and Trivedi (1998, p 214). Among differences are the less precise estimate of β_4, i.e. 8.2 with a 95% interval (2, 15) but a significant parameter γ_4. δ is estimated at 1.64 with a 95% interval (1.31, 2.03).

5.5   MODELLING THE LINK FUNCTION

Under the GLM, a known link function g connects the mean of the density to a systematic linear regression term. Thus for count data, $y_i \sim \text{Po}(\mu_i)$ and $g(\mu_i) = X_i\beta$ where $\log(\mu_i)$ is the canonical link. Binary regression models assume that y_i is related to predictors X_i via

$$\Pr(y_i = 1|x_i) = F(X_i\beta)$$

with F being the inverse link function. For binary regression the range of F is within [0,1] and F is therefore typically a known cumulative distribution function. F as the normal cdf is equivalent to the probit link, and F as the cdf of the standard logistic function corresponds to a logit link, which is the canonical GLM link. Sometimes standard choices of link lead to poor fit, and one may instead allow for a choice of the link function based on the data. Selection of the link usually leads to a form of model averaging and may be combined with predictor selection (Ntzoufras *et al.*, 2003).

For example, Basu and Mukhopadhyay (2000) propose normal scale mixtures in the probit regression model

$$\Pr(y_i = 1|X_i) = \int \Phi(X_i\beta/\sigma)\mathrm{d}h(\sigma) \tag{5.13}$$

where h is a discrete mixture such as that formed by a Dirichlet process; see also Basu and Chib (2003). Newton *et al.*, (1996) also consider non-parametric estimation of the binary link function by assuming a Dirichlet process mixture on the mean, with a logistic or normal distribution function as baseline. Mallick and Gelfand (1994) propose mixture models for link functions g, in particular discrete mixtures of beta densities. A baseline link such as $g_0(\eta_i) = g_0(X_i\beta) = \exp(X_i\beta)$ in a Poisson regression provides the basis for mixture modelling of the cumulative density function defined by $J(\eta) = ag(\eta)/[ag(\eta) + b]$, where $Y_i \sim \text{Po}(g(\eta_i))$. Another approach takes the link as belonging to a parametric class of link transformations. For binary and binomial data, one- and two-parameter link families have been suggested by Aranda-Ordaz (1981), Stukel (1988) and Czado (1994).

5.5.1 Discrete (DPP mixture)

The model in (5.13) may be operationalized by using the Albert and Chib (1993) data augmentation approach. Letting λ_i denote precisions for subjects i, the scale mixture generalization of the probit is

$$W_i \sim N(X_i\beta, 1/\lambda_i)I(L_i, U_i)$$

where the λ_i are drawn from a density confined to positive values and with mean 1 and $\{L_i, U_i\}$ are defined by the observed binary y_i. A conjugate density for λ_i is provided by a gamma density, leading to a t density with ν degrees of freedom (see Chapter 3). Thus

$$\lambda_i \sim \text{Gamma}(\nu/2, \nu/2) \tag{5.14}$$

Instead of assuming n distinct values of λ_i (i.e. one for each observation), Basu and Mukhopadhyay (2000) propose a DPP whereby λ_i values fall into a relatively small number of clusters (perhaps $K^* = 5$ to $K^* = 20$ would be a typical number). The baseline density of the DPP would be a gamma as above, with the degrees of freedom ν taken as either known or an extra parameter, and the concentration parameter κ of the DPP may also be taken as known or left as a free parameter.

In many situations, including Example 5.6 below, the data are presented as binomial but result from repeated binary observations at covariate level X_i, so the actual observations are repetitions within clusters, of the form y_{ij}, $i = 1, \ldots, g$, $j = 1, n_i$ where n_i is the total cluster size and $r_i = \sum y_{ij}$ is the number of 'successes' when the covariate has value X_i. Thus at the lowest dose in Table 5.3 below there are six cases where $y_{ij} = 1$ and 53 cases where $y_{ij} = 0$. The augmented data sampling would be at this disaggregated data level but the regression model would be at the cluster level. So the probit model would involve

$$W_{ij} \sim N(X_i\beta, 1)I(L_{ij}, U_{ij})$$

with truncation below or above by zero according as y_{ij} were 1 or 0.

5.5.2 Parametric link transformations

For models where the link belongs to a parametric class of link transformations, let ψ be the shape parameter of the link density and $\eta = X\beta$.
Then possible families of densities include

$$h(\eta, \psi) = (1 + \eta\psi)^{1/\psi} \tag{5.15}$$

$$h(\eta, \psi) = \log(1 + \eta\psi)/\psi \tag{5.16}$$

$$h(\eta, \psi) = [\exp(\eta\psi) - 1]/\psi \tag{5.17}$$

and

$$h(\eta, \psi) = [(\eta + 1)^\psi - 1]/\psi \tag{5.18}$$

Czado (1994) uses the last of these in single parameter link functions
which are appropriate to left and right tails of the link F. In particular,
taking $F[h(\eta, \psi)] = \Phi[h(\eta, \psi)]$ where Φ is the standard normal cdf, the
option

$$h(\eta, \psi) = \begin{cases} \eta & \text{if } \eta \geq 0 \\ -[(-\eta + 1)^\psi - 1]/\psi & \text{otherwise} \end{cases}$$

is used the modify the left tail and

$$h(\eta, \psi) = \begin{cases} [(\eta + 1)^\psi - 1]/\psi & \text{if } \eta \geq 0 \\ \eta & \text{otherwise} \end{cases}$$

allows for modification of the right tail. Then under either option $\psi = 1$
corresponds to the usual probit link. Czado and Raftery (2001) consider
the choice between tail-modified models using the Bayes factor methods
of Raftery (1996).

As for the Basu–Mukhopadhyay model, augmented data sampling
may be used to apply these modified link models. With $\eta_i = X_i\beta$, a left
modified tail approach to a probit link would mean sampling

$$W_i \sim \text{N}(\eta_i, 1) \qquad\qquad\qquad \text{if } \eta_i \geq 0 \tag{5.19}$$

$$W_i \sim \text{N}(-[1 - \eta_i]^\psi - 1]/\psi, 1) \quad \text{otherwise} \tag{5.20}$$

with sampling additionally constrained according to whether y_i were 1
or 0. For example, if $y_i = 1$ then sampling would be truncated below at
zero and switch between (5.19) and (5.20) according to whether η_i were
positive or negative. It is possible to combine the tail modifications
(5.15)–(5.18) with scale mixing as in (5.14); see Example 5.7.

5.5.3 Beta mixture on cumulative densities

A parametric approach is also proposed by Mallick and Gelfand (1994) and Gelfand and Mallick (1995). The basis of their approach is that a discrete mixture of beta densities can closely approximate any continuous density on (0, 1). In particular the modelling of the cumulative density

$$J(\eta) = ag(\eta)/[ag(\eta) + b]$$

or equivalently the inverse link g with

$$g(\eta) = bJ(\eta)/[a - aJ(\eta)]$$

involves a discrete mixture (with K components, with K usually in small integers) over the cumulative densities associated with beta densities Beta(c, d). The cumulative densities are

$$B(\eta, c, d) = \frac{\Gamma(c+d)}{\Gamma(c)\Gamma(d)} \int_0^{J_0(\eta)} u^{c-1}(1 - u)^{d-1} du$$

where $J_0(\eta) = ag_0(\eta)/[ag_0(\eta) + b]$ is the cumulative density function associated with the baseline inverse link g_0. Thus

$$J(\eta) = \sum_k \pi_k B(\eta, c_k, d_k)$$

where π_k are mixture proportions. Mallick and Gelfand (1994) argue that taking π_k unknown and $\{c_k, d_k\}$ known is more tractable than taking $\{c_k, d_k\}$ as parameters. Gelfand and Mallick (1995) propose generating c_k and d_k by the scheme

$$c_k = \lambda k$$
$$d_k = (K + 1 - k)\lambda$$

in order to provide a set of densities that fully cover (0,1), with λ governing how peaked the beta densities are. Priors on parameters $\{a, b, \lambda)$ might be considered but defaults such as $a = b = \lambda = 1$ are often sufficient.

Example 5.6 Beetle mortality Data from Bliss (1935) record the number of beetles dead after five hours' exposure to carbon disulphide (transformed to $\log_{10}$ dose in Table 5.3).

The standard probit gives a coefficient on the standardized dose variable of 1.34 and mean deviance of 38.3. The predictions of the death rates are most discrepant at the four lowest doses, with underpredictions at the two lowest doses and overpredictions at the following two. Various analyses of the data have suggested alternative links (e.g. complementary log–log) or a generalized logistic model (Stukel, 1988).

Table 5.3   Flour beetle deaths

Dose	Log_{10}dose	Deaths	Exposed	Death rate (%)
49.1	1.6907	6	59	10.2
53.0	1.7242	13	60	21.7
56.9	1.7552	18	62	29.0
60.8	1.7842	28	56	50.0
64.8	1.8113	52	63	82.5
68.7	1.8369	53	59	89.8
72.6	1.861	61	62	98.4
76.5	1.8839	60	60	100.0

Here a left tail modification was found to give a significant gain in fit (cf. Czado, 1994) with mean deviance reduced to 31.6 (with effective parameters $p_e = 2.5$). The predictions of the first four death rates, i.e. 0.12, 0.19, 0.31 and 0.54, are much improved. The posterior mean of the link parameter ψ is close to zero (Figure 5.1) with a 95% interval $(-0.67, 0.62)$ from a two-chain run of 10 000 iterations.

The impression from Table 5.3 is that the death rate becomes much higher for doses over 60, implying a non-constant coefficient on dose. To assess whether this option provides an alternative way of improving fit a discrete mixture model is applied for the dose coefficient. Specifically a

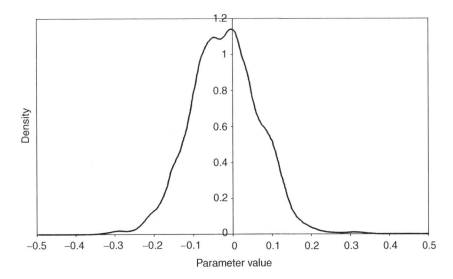

Figure 5.1   Kernel plot, link parameter

DPP is used with concentration parameter of one and up to eight possible clusters on the dose coefficient. The average coefficient at each dose level is monitored and the lowest dose coefficients (around 1.27) are found at doses 1.691 and 1.724, whereas the highest is 1.44 at dose 1.884. This model does improve the average deviance (to around 36.8) but not as much as the left tail modification. Comparison with the deviance at the mean of the parameters shows there are around 3.7 effective parameters. A quadratic in dose is another option that has been mentioned in the literature for these data.

Finally, a data augmentation approach is adopted with scale mixing as in Basu and Mukhopadhyay (2000), and left tail modification, as in Czado (1994). A logit link is used as it is more robust to extreme linear predictor values. A Dirichlet concentration parameter κ of one is assumed and ν in (5.14) set at ten. Up to 12 clusters are assumed in line with there being two possible responses at six doses. The augmented data have the form W_{ij} where $i = 1,6$ denotes the dose levels and $j = 1, \ldots, n_i$ denote repetitions at each dose. A two-chain run shows convergence from around 5000 iterations and an average deviance 35.0 with six effective parameters. The highest residuals are for those dying at the lowest dose and for the survivors at 1.837 and 1.861. Evidence for variation in scales is not pronounced with the lowest precision λ_i being 0.88 for the lone survivor at dose 1.861. Note that the marginal likelihood and Bayes factors for this type of model can be obtained as in Basu and Chib (2003).

Example 5.7   Unknown link for simulated Poisson data   To illustrate the beta mixture modelling of the cumulative density $J(\eta)$ associated with the inverse link $g(\eta)$, consider 20 Poisson observations generated from a model with $\eta_i = \beta_0 + \beta_1 x_i = -2 + 3x_i$ where the x_i are standard normal variates and $g(\eta) = \exp(\eta)$. It is assumed that $a = b = 1$, but a prior $\lambda \sim G(1, 1)$ is assumed in the scheme $c_k = \lambda k$, $d_k = (K + 1 - k)\lambda$ where $K = 4$. A simple Poisson regression with these simulated data shows posterior standard deviations on β_0 and β_1 of 0.66 and 0.5.

The precisions of the priors on these coefficients in the beta mixture link model then downweight those from the Poisson regression by ten. A two-chain run of 40 000 iterations for this link mixture model converges at around 22 500 iterations and the second half of the run shows $\lambda = 1.27$ and $\{\beta_0, \beta_1\} = \{-2.1, 3.7\}$ so that the covariate effect is amplified. Figure 5.2 shows the relation between the estimated $g(\eta) = J(\eta)/[1 - J(\eta)]$ and the baseline $g_0(\eta) = \exp(\eta)$. The EPD for this model obtained by sampling 'new data' from the model (see 2.11) in fact

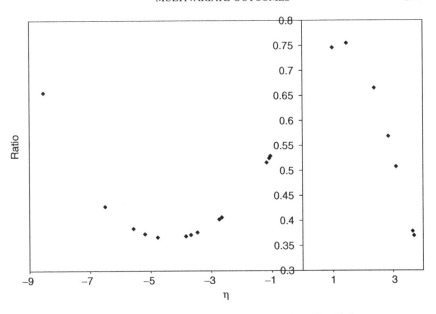

Figure 5.2 Ratio of estimated link to baseline link

improves on that obtained with a simple Poisson regression, i.e. 20.0 vs. 23.5, with closer predictions of the zero observed counts.

5.6   MULTIVARIATE OUTCOMES

For K multivariate outcomes of binary, proportion or count data, similar themes to those above recur. These include direct modelling of the outcome as against modelling of the latent continuous scale producing the outcome, the choice between different links, and the possible necessity to account for overdispersion on one or more of the outcomes. There are also choices over how to use random effects to model the correlations between the outcomes.

Consider the case of K binary outcomes, $Y_i = \{Y_{i1}, Y_{i2}, \ldots, Y_{iK}\}$. Among possible frameworks for such data are:

1. K separate Bernoulli likelihoods (i.e. direct modelling of the outcomes) with correlations between outcomes modelled in the link function, for instance by additive multivariate normal errors ε_{ij} in the logit or other link. The correlations between responses are obtained from the estimated covariance matrix Σ of $\varepsilon_i = (\varepsilon_{i1}, \varepsilon_{i2}, \ldots, \varepsilon_{iK})$.

2. A multivariate probit model which may be estimated directly (by multivariate integration) or by augmenting the data with K underlying latent continuous values $\{W_{i1}, W_{i2}, \ldots, W_{iK}\}$ (Chib and Greenberg, 1998). The correlations between responses may be modelled by assuming a multivariate normal error of dimension K, or scale mixtures of multivariate normal errors (e.g. leading to multivariate Student t errors). A multivariate logit regression may also be achieved with suitable mixing strategies (Chen and Dey, 2003; O'Brien and Dunson, 2004).

3. Either separate Bernoulli likelihoods assumed for each outcome as in option 1, or separate latent scale sampling of each outcome as in option 2, but with correlations between outcomes modelled by $Q < K$ factors.

Under option 1 random effects over subjects or subjects–outcomes may be modelled parametrically (e.g. MVN for ε_i), though the parameters of the covariance matrix may be weakly identified in small samples. A possible alternative is a Bernoulli–beta mixture with beta frailties ϕ_i for subjects to model the correlation between outcomes. Thus, with a logit link

$$y_{ij} \sim \text{Bern}(\pi_{ij})$$
$$\pi_{ij} = \{\exp(\beta_j x_{ij})/[1 + \exp(\beta_j x_{ij})]\}\phi_i$$
$$\phi_i \sim \text{Beta}(\kappa, \kappa_2)$$

Under the multivariate version of the probit model (option 2) the known conditional variance of one in the univariate probit, as in (4.2), is replaced by a correlation matrix $R = [r_{jk}]$ for the density $Y|X$ with known diagonal elements but correlations as free parameters. There will also be outcome-specific regression parameter vectors β_k of dimension p, assuming that a common regression vector $x_i = (1, x_{i2}, x_{i3}, \ldots, x_{ip})$ is used to predict all outcomes. The probability of a particular pattern $y_i = \{y_{i1}, y_{i2}, \ldots, y_{iK}\}$ is, with $\theta = \{\beta_1, \ldots, \beta_K, R\}$,

$$\text{Prob}(Y_i = y_i|\theta) = \int_{D_{i1}} \int_{D_{i2}} \cdots \int_{D_{iK}} \phi_K(u|0, R)du$$

with the regions of integration D_{ik} defined according to whether $y_{ik} = 1$ or $y_{ik} = 0$. Thus D_{ik} is between $-\infty$ and $x_i\beta_k$ when $y_{ik} = 0$, but between $x_i\beta_k$ and ∞ when $y_{ik} = 1$.

As for the univariate probit, a tractable formulation of the multivariate probit involves augmenting the data with the latent normal variables

$W_i = \{W_{i1}, W_{i2}, \ldots, W_{iK}\}$. Then W_i is truncated MVN with mean $\mu_i = \{\mu_{i1}, \mu_{i2}, \ldots, \mu_{iK}\}$ where $\mu_{ik} = x_i \beta_k$ and dispersion (correlation) matrix R. Sampling of the constituent W_{ik} of W_i is confined to values above zero when $y_{ik} = 1$ and to values below zero when $y_{ik} = 0$. One may generalize the multivariate probit models to multivariate t (MVT) or other models by scale mixing, which amounts to dividing the correlation matrix R by a weighting factor λ_i

$$W_i \sim TN_K(\mu_i, R/\lambda_i)$$
$$\lambda_i \sim \text{Ga}(\nu/2, \nu/2)$$

(5.21)

5.6.1   The factor analysis approach

To continue the multivariate binary example, the factor analysis perspective (option 3), accounting for both the interdependence among the y_{ik} and the different regression effects β_k on each outcome, might involve either direct Bernoulli likelihoods or truncated univariate normal sampling. Under this option $Q < K$ factors $F_{i1}, F_{i2}, \ldots, F_{iQ}$ are introduced to account for the correlations between the observations. As noted by Cox and Wermuth (2002), choosing latent scale sampling instead of the direct Bernoulli likelihood involves two levels of latent variables. Thus for the kth outcome for subject i

$$W_{ik} \sim N(\mu_{ik}, 1)I(L_{ik}, U_{ik})$$

(5.22)

$$\mu_{ik} = x_i \beta_k + \lambda_{k1} F_{i1} + \lambda_{k2} F_{i2} + \cdots + \lambda_{kQ} F_{iQ}$$

(5.23)

where for $y_{ik} = 0$, U_{ik} specifies an upper sampling limit of zero, and for $y_{ik} = 1$, L_{ik} specifies a lower sampling limit of zero. However, conditional on the factor scores F_{iq}, the observations $\{y_{ik}\}$ are usually taken to be independent. This is the conditional independence assumption, under which the F_{iq} account for the correlations between the y_{ik}.

Regression and factor loading parameters for the kth variable may be standardized (Bock and Gibbons, 1996) by dividing by

$$d_k = \left(1 + \sum_{m=1}^{Q} \lambda_{km}^2\right)^{0.5}$$

The correlations between the latent scales on different outcomes W_j and W_k (conditional on x) are then obtained under the factor model as

$$\rho_{jk} = \sum_{q=1}^{Q} \lambda_{jq} \lambda_{kq}$$

where the λ are standardized. Note that a factor structure might also be used with direct sampling of binary items (Bartholomew and Knott, 1999), as in

$$y_{ik} \sim \text{Bern}(\pi_{ik}) \tag{5.24}$$

$$\text{logit}(\pi_{ik}) = x_i\beta_k + \lambda_{k1}F_{i1} + \lambda_{k2}F_{i2} + \cdots + \lambda_{kQ}F_{iQ} \tag{5.25}$$

For multivariate binary responses, identifiability requires the factors F_{iq} in (5.23) or (5.25) to have mean 0 and variance 1 and to be uncorrelated.

Additional restrictions are also needed on the $K \times M$ matrix Λ of factor loadings with multiple binary outcomes. These either are equality constraints or involve setting loadings to fixed values; thus for $q > k$ one might set $\lambda_{kq} = 1$ or $\lambda_{kq} = 0$ (Bartholomew, 1987, chapter 6; Bock and Gibbons, 1996). So for $Q = 2$ one might set $\lambda_{12} = 1$ or 0. Different fixed values might be tried (e.g. $\lambda_{12} = 1$ instead of $\lambda_{12} = 0$) to find which is better supported by the data. If all observed items are anticipated to be positive measures of an underlying construct, as when K items are all binary measures of mathematical aptitude in the case $Q = 2$ (and $y_{ik} = 1$ for greater aptitude), then the setting of a loading such as λ_{12} to one acts to prevent label switching on the remaining loadings, namely $\{\lambda_{22}, \lambda_{32}, \ldots, \lambda_{K2}\}$.

5.6.2 Specific modelling of marginal and odds ratios

Multivariate binary data may also be analysed in terms of the marginal outcomes $\Pr(y_{ij} = 1)$, joint pairwise outcomes $\Pr(y_{ij} = 1, y_{ik} = 1)$, or conditional outcomes $\Pr(y_{ij} = 1|y_{ik} = 1)$. For larger K this means selecting certain aspects of the total likelihood (i.e. the joint density of $y_1, \ldots, y_K$) for consideration. As discussed by Liang *et al.*, (1992), the joint distribution of a K vector of binary responses $y = (y_1, y_2, \ldots, y_K)$ can be represented by a saturated log-linear model

$$\log[\text{P}(y)] = u_0 + \sum_{k=1}^{K} u_k y_k + \sum_{j<k}^{K} u_{jk} y_j y_k + \cdots + u_{12\ldots K} y_1 y_2 \cdots y_K$$

with $2^K - 1$ parameters. These parameters express conditional probabilities, for instance:

$$u_k = \text{logit}[\Pr(y_k = 1|y_j = 0, j \neq k)]$$

and

$$u_{jk} = \log[\text{OR}(y_j, y_k|y_m = 0, m \neq j, m \neq k)]$$

where the odds ratio (OR) is defined by

$$OR(y_j, y_k) = [\Pr(y_j = 1, y_k = 1)\Pr(y_j = 0, y_k = 0)]/$$
$$[\Pr(y_j = 1, y_k = 0)\Pr(y_j = 0, y_k = 1)]$$

Usually, more parsimonious models are applied which seek to represent substantively more important questions, as represented by the marginal outcomes and pairwise odds ratios, or the impact that positives on y_k ($k \neq j$) have on the occurrence of $y_j = 1$.

For example, if the y_{ik} are all positive disease indicators ($y = 1$ for disease present), then the conditional probability that $y_{ik} = 1$ may be related to the number of positive responses on the other outcomes (Connolly and Liang, 1988). Let $Y_{i[k]}$ denote the indicators for subject i excluding y_{ik}, namely $\{y_{i1}, y_{i2}, \ldots, y_{i,k-1}, y_{i,k+1}, \ldots, y_{iK}\}$. Then one form for this type of conditional logistic model specifies

$$y_{ik}|Y_{i[k]} \sim \text{Bern}(\pi_{ik})$$
$$\text{logit}(\pi_{ik}) = \log\left[\frac{\theta_1 + S_k\theta_2}{1 - \theta_1 + (K - 1 - S_k)\theta_2}\right] + \beta_k x_i$$

where S_k is the number of indicators in $Y_{i[k]}$ which have value 1. The alternating logistic model of Carey *et al.*, (1993) is somewhat similar with model means for marginal and joint outcomes

$$\mu_{ij} = E(y_{ij})$$
$$\nu_{ijk} = E(y_{ij}, y_{ik})$$

contributing to a model for the conditional outcome, namely

$$\Pr(y_{ij} = 1|y_{ik} = 1) = g^{-1}[\alpha y_{ik} + \log(R_{ijk})]$$

where

$$R_{ijk} = (\mu_{ij} - \nu_{ijk})/(1 - \mu_{ij} - \mu_{ik} + \nu_{ijk})$$

McCullagh and Nelder (1989) propose a multivariate logit model in which marginal and joint associations (pairwise and higher) are modelled together; see also Glonek and McCullagh (1995) and Liang *et al.* (1992). Pairwise associations between binary outcomes are generally stated in odds ratio form but can be modelled via a multinomial likelihood (Chapter 6). For example, $K = 2$ binary responses y_{i1} and y_{i2} define a multinomial of dimension 4 formed by combining the levels of y_1 and y_2. Let $\pi_{11}, \pi_{10}, \pi_{01}, \pi_{00}$ be combined response categories ($1 = \text{Yes}, 0 = \text{No}$) with $\sum_i \sum_j \pi_{ij} = 1$. Independence between the responses may be assessed via the odds ratio

$$(\pi_{11}\pi_{00})/(\pi_{10}\pi_{01})$$

with null value 1, or log odds ratio

$$\log[(\pi_{11}\pi_{00})/(\pi_{10}\pi_{01})]$$

with null value 0. Consider the data in aggregate form with n_{11} being the number of subjects with $y_1 = 1$, $y_2 = 1$, n_{10} being the total with $y_1 = 1, y_2 = 0$, and so on. Suppose additionally there is a predictor with M levels (e.g. grouped drug exposure) with value z_m for subtable n_{jkm} with probabilities $\pi_{11m}, \pi_{10m}, \pi_{01m}, \pi_{00m}$ defining the joint outcomes in that subtable. Then the multivariate logit model of McCullagh and Nelder for $K = 2$ involves marginal logits to predict $P(y_1 = 1)$ and $Pr(y_2 = 1)$, while to model the odds ratio, a multinomial logit regression can estimate how joint response changes with the predictor, via

$$\pi_{ijm} = \phi_{ijm} / \sum_j \sum_k \phi_{jkm}$$

where

$$\phi_{11m} = \exp(\alpha x_m + \beta x_m + \gamma x_m)$$
$$\phi_{10m} = \exp(\alpha x_m)$$
$$\phi_{01m} = \exp(\beta x_m) \tag{5.26}$$
$$\phi_{00m} = 1$$

The predictor vector has the form $x_m = (1, z_m)$. Independence within each subtable of the predictor using odds ratios defined by relations among the ϕ_{ijm} (and π_{ijm}) is given by

$$\log[(\phi_{11m}\phi_{00m})/(\phi_{10m}\phi_{01m})] = \log[(\pi_{11m}\pi_{00m})/(\pi_{10m}\pi_{01m})] = \gamma x_m$$

This analysis can also be performed with disaggregated data using the marginal responses y_{i1}, y_{i2} and the multinomial variable D_i formed by the joint values of (y_{i1}, y_{i2}), where $D_i = 1$ if $y_1 = 1, y_2 = 1$, $D_i = 2$ if $y_1 = 1, y_2 = 0$, etc.

5.6.3 Multivariate Poisson data

For multivariate count data, option 1 above no longer poses identifiability issues (see Chib and Winkelmann, 2001), and option 3 applies with a direct likelihood approach (e.g. Poisson or NB). However, the augmented data approach in option 2 is not applicable. Thus under option 1

$$y_{ij} \sim Po(\mu_{ij})$$
$$\log(\mu_{ij}) = \beta_j x_{ij} + \varepsilon_{ij}$$

where the multivariate error $\varepsilon_i = \{\varepsilon_{i1}, \varepsilon_{i2}, \ldots, \varepsilon_{iK}\}$ might follow an MVN, or a mixture of MVNs, such as

$$\varepsilon_i \sim \pi N_K(0, \Sigma_1) + (1 - \pi) N_K(0, \Sigma_2)$$

where $\Sigma_2 = \gamma \Sigma_1$ with scalars π and γ possibly both taking preset values. Other options are a scale mixture of MVNs

$$\varepsilon_i \sim N_K(0, \Sigma / \lambda_i)$$

with $\lambda_i \sim Ga(\nu/2, \nu/2)$.

The negative multinomial model (Guo, 1996) may also be relevant to account for both correlated outcomes and overdispersion. Thus

$$y_{ij} \sim Po(\mu_{ij})$$
$$\mu_{ij} = \exp(\beta_j x_{ij}) \phi_i$$
$$\phi_i \sim Ga(\alpha_1, \alpha_2)$$

where $\alpha_1 = \alpha_2$ for identifiability.

A factor model may also be used to model the correlations between errors. For both binary and count data an alternative factor model approach is to the regression component (systematic part) of the model, assuming that a common set of predictors x_i is used across the responses. This is relevant if the observed responses are indicators of an underlying construct. Thus with a single factor

$$y_{ij} \sim Po(\mu_{ij})$$
$$\log(\mu_{ij}) = \beta_0 + \lambda_{1j} F_i$$
$$F_i = \beta x_i + u_i$$

Example 5.10 considers multivariate count data with $K = 2$ when a single unobservable construct plausibly underlies both outcomes. If $Q = 2$ (with $K > 2$) then

$$\log(\mu_{ij}) = \beta_0 + \lambda_{1j} F_{i1} + \lambda_{2j} F_{i2}$$
$$F_{i1} = \beta_1 x_i + u_{i1}$$
$$F_{i2} = \beta_2 x_i + u_{i2}$$

The number of free factor loadings depends on the assumptions about the conditional variances $V_j = Var(u_{ij})$ of the factor scores $F_{ij}, j = 1, \ldots, Q$. If $Q = 1$, for example, then taking $V_1 = 1$ means $\lambda_{11}, \ldots, \lambda_{1K}$ may be estimated whereas taking V_1 unknown requires one of the λ_{1j} to be fixed.

Muthen (1979) proposes this model for binary data when there are multiple indicators of one or possibly more underlying constructs. His analysis focuses on a survey among married women of attitudes to sex typing. The survey questions yield binary indicators $Y_1, \ldots, Y_K$ of an underlying construct F 'the propensity to reject sex typing' and predictors x_i include length of education and length of time married.

Example 5.8 Pneumoconiosis symptoms   Consider the data analysed by Bock and Gibbons (1996) on five binary pneumoconiosis symptom variables, namely cough ($1 = $ Yes, $0 = $ No), phlegm, breathlessness, wheeze and weather-related symptoms. The single covariate is $x_{i2} = $ age with $x_{i1} = 1$. Age takes the values 22.5, 27.5, etc., up to 62.5. Here a 10% sample is obtained from the full set of 29 720 observations.

An initial analysis considers just the first two outcomes and takes a single factor ($Q = 1$). The constraint $\lambda_{11} \sim N(1, 1)I(0, \ )$ is adopted, i.e. the positive first loading of cough on the single factor amounts to constraining this factor to be a positive morbidity measure. The prior on λ_{21} is taken as $N(0, 1)$ though an $N(1, 1)$ prior as in Johnson and Albert (1999) is a possibility in line with expectations of a unipolar morbidity factor. A two-chain run of this model shows early convergence (at around 500 iterations) and gives a steeper age gradient for cough than phlegm, namely 0.036 (s.d. $= 0.0022$) as against 0.031 (0.0021). The correlation between W_1 and W_2 is obtained as 0.85 – based on posterior means of standardized loading λ_{11} and λ_{21}, which are both 0.92. The DIC is 3955 with 1160 effective parameters.

The bivariate probit is estimated under the assumption of a $U(-1, 1)$ prior for r_{12} and $N(0, 100)$ priors for β_{kg}, where $k = 1, 2$ for the two outcomes and $g = 1$ for the constant and $g = 2$ for age. With convergence at around 800 iterations in a 2000-iteration run, the age gradients are found to be similar to the factor model, though more precisely estimated, i.e. 0.035 (0.0020) for cough and 0.030 (0.0020) for phlegm. The correlation r_{12} is estimated at 0.90. However, the fit measured by DIC deteriorates to 6253. The DIC is obtained by monitoring the mean of the multivariate binary deviance and comparing it with the deviance obtained using posterior averages of $\mu_{ik} = \beta_{k1} + \beta_{k2}x_{i2}$.

Example 5.9 Use of health care   As an illustration of options for modelling correlations in a bivariate count outcome (Y_1, Y_2), consider joint modelling of discrete outcomes relating to demand for medical care by the elderly in the USA (Munkin and Trivedi, 1999; Chib and Winkelmann, 2001). The two variables considered, emergency room visits and number of hospitalizations (EMR and HOSP) for $n = 4406$ subjects, are obtained from the National Medical Expenditure Surveys in 1987 and

1988. These are related to categorical predictors (health status, region, type of medical insurance, ethnicity, sex, employed or not, difficulties with activities in daily living), a count predictor for illness conditions, and to three continuous predictors, age (divided by 10), years of education, and relative family income.

The predictors are the same for both outcomes, so $x_{ij} = x_i$. Both variables have relatively low means $\bar{y}$, namely 0.26 for EMR and 0.30 for HOSP, and high proportions of zero counts. So one aspect of model checking is whether this high zero proportion is replicated by the model predictions. There is also some overdispersion with $V(y_1) = 0.7$ and $V(y_2) = 0.75$.

The first model (model A) is the bivariate Poisson lognormal mixture, with unrestricted dispersion structure, namely

$$y_{ij} \sim \text{Po}(\mu_{ij}), \quad i = 1, \ldots, N, j = 1, K \qquad (5.27)$$

where $K = 2$,

$$\log(\mu_{ij}) = \beta_j x_i + \varepsilon_{ij} \qquad (5.28)$$

and the ε_{ij} are bivariate normal. If the ε_{ij} are highly correlated, a one-factor model might be considered as an alternative. Under this

$$\log(\mu_{ij}) = \beta_j x_i + \lambda_{1j} F_i \qquad (5.29)$$

where $F_i \sim \text{N}(0, \tau_F)$. The variance parameter τ_F may be preset (when both λ_{1j} are unknown) or estimated (if one of the λ_{1j} is preset, typically to one). Since both outcomes may be taken as proxies for an underlying morbidity construct, a further possible model is

$$y_{ij} \sim \text{Po}(\mu_{ij})$$
$$\log(\mu_{ij}) = \beta_0 + \lambda_{1j} F_i \qquad (5.30)$$
$$F_i = \beta x_i + u_i$$

The applicability of this model would be greater when β_1 and β_2 in (5.28) are similar.

A Poisson lognormal model (model A) as in (5.27)–(5.28) has an average deviance $\bar{D}$ of 2256 for EMR and 2320 for HOSP with deviances $D(\bar{\mu}_{ij})$ at models means of 1643 and 1648 (from the second half of a two-chain run of 5000 iterations). This gives a DIC of 5859 with $p_e = 1284$. Predictor effects are given in Table 5.4 with a correlation between the errors ε_{i1} and ε_{i2} of over 0.9. There is some doubt on how well this model accounts for the excess zeros: in fact the model replicates seem to have more zeros than are present in the data, since a predictive check comparing the zero proportion among replicates z_{ij} with that among the y_{ij} averages 0.92.

Table 5.4   Health demand: bivariate Poisson

	Mean	S.d.	2.5%	Median	97.5%
$V(\varepsilon_1)$	2.23	0.55	1.80	2.02	4.11
$\mathrm{Cov}(\varepsilon_1,\varepsilon_2)$	2.07	0.44	1.71	1.89	3.38
$V(\varepsilon_2)$	2.37	0.77	1.79	2.02	4.56
$\mathrm{Corr}(\varepsilon_1,\varepsilon_2)$	0.92	0.05	0.77	0.93	0.96
EMR visits					
Constant	−4.050	0.465	−4.735	−4.014	−3.042
Excellent health	−0.676	0.219	−1.138	−0.668	−0.282
Poor health	0.626	0.158	0.377	0.610	0.995
Number chronic conditions	0.243	0.030	0.187	0.240	0.307
ADL difficulties	0.281	0.180	−0.104	0.307	0.572
N East	−0.009	0.118	−0.253	−0.004	0.205
Mid West	−0.095	0.128	−0.335	−0.104	0.172
West	0.149	0.124	−0.109	0.151	0.414
Age	0.163	0.070	0.006	0.179	0.244
Black	0.156	0.123	−0.085	0.151	0.413
Male	0.090	0.104	−0.112	0.092	0.297
Married	−0.157	0.135	−0.458	−0.135	0.057
Years education	−0.022	0.016	−0.056	−0.022	0.007
Family income	−0.001	0.016	−0.033	−0.001	0.030
Employed	0.220	0.132	−0.044	0.220	0.473
Private insurance	0.058	0.117	−0.155	0.054	0.280
MedicAid	0.199	0.181	−0.119	0.197	0.593
Hospitalizations					
Constant	−5.436	0.355	−6.413	−5.358	−4.916
Excellent health	−0.750	0.229	−1.222	−0.733	−0.351
Poor health	0.730	0.249	0.407	0.669	1.419
Number chronic conditions	0.271	0.043	0.172	0.275	0.358
ADL difficulties	0.146	0.263	−0.533	0.229	0.482
N East	−0.005	0.119	−0.237	−0.003	0.223
Mid West	−0.004	0.143	−0.297	0.008	0.243
West	0.130	0.127	−0.112	0.127	0.402
Age	0.300	0.034	0.239	0.295	0.374
Black	0.085	0.134	−0.155	0.072	0.390
Male	0.164	0.094	−0.022	0.171	0.332
Married	−0.040	0.131	−0.349	−0.013	0.159
Years education	−0.003	0.015	−0.033	−0.002	0.025
Family income	−0.002	0.017	−0.033	−0.003	0.032
Employed	0.104	0.138	−0.186	0.102	0.387
Private insurance	0.198	0.128	−0.056	0.200	0.437
MedicAid	0.235	0.186	−0.161	0.236	0.610

It follows from the high correlation that a single factor model for the errors as in (5.29) is a viable alternative. The variance τ_F is preset at one so both loadings λ_j are free parameters. These are assigned N(1, 1) priors with λ_1 additionally constrained to be positive, with the goal of avoiding label switching. The label switching problem occurs because F_i may be either a morbidity effect (with both λ positive) or a good health effect (with both λ negative). In fact, the posterior means of the λ_j are both around 1.28. Although $\bar{D}$ worsens to 2277 for EMR and 2349 for HOSP, the gap between $\bar{D}$ and $D(\bar{\mu})$ is reduced because the model has reduced complexity (in fact $p_e = 1180$ with a DIC of 5806). The predictive check P_c on zero counts is still on the boundary of acceptability, i.e. $P_c = 0.89$.

Estimates of the model (5.27)–(5.28) show a similarity in the profile of effects β_{jv}, $v = 1, 17$ (see Table 5.4). From this it follows that a one-factor regression model as in (5.30) might be considered as a further alternative. The loadings have priors as in the factor error model so one would expect F_i to be a morbidity index, i.e. an index of patient propensity to make more EMR and HOSP visits. This model leads to generally similar influences of the 17 predictors on the single 'needs factor' as were obtained in the bivariate Poisson model. As for the factor error model, the fit is slightly improved over model A with a DIC of 5808 and $p_e = 1157$. The effects on the outcomes (via the factor) of difficulties in activities of daily living (ADL) and of male gender are clarified (Table 5.5).

Table 5.5 Regression effects under factor regression model

	Mean	S.d.	2.5%	Median	97.5%
Excellent health	−0.553	0.133	−0.812	−0.550	−0.302
Poor health	0.427	0.074	0.286	0.426	0.572
Number chronic conditions	0.215	0.020	0.175	0.215	0.253
ADL difficulties	0.317	0.068	0.181	0.318	0.447
N East	0.032	0.074	−0.114	0.032	0.177
Mid West	0.066	0.068	−0.071	0.067	0.196
West	0.116	0.076	−0.038	0.116	0.270
Age	0.110	0.053	0.033	0.102	0.236
Black	0.112	0.085	−0.054	0.113	0.278
Male	0.123	0.059	0.006	0.123	0.239
Married	−0.063	0.063	−0.187	−0.062	0.062
Years education	−0.005	0.008	−0.020	−0.005	0.011
Family income	−0.002	0.010	−0.021	−0.001	0.018
Employed	0.106	0.093	−0.084	0.108	0.286
Private insurance	0.099	0.073	−0.041	0.098	0.242
MedicAid	0.158	0.099	−0.038	0.158	0.349

Both observed variables have similar loadings on the factor. The check on zero counts is improved with $P_c = 0.83$. Further model simplification, perhaps via regressor selection to eliminate predictors of marginal significance, might enhance the impacts of those regressors which distinguish patients with non-zero visits – this would improve this predictive check further.

Example 5.10   Visual impairment by eye   Liang *et al.* (1992) consider bivariate binary data for visual impairment (vision less than 20/60) among 5199 subjects under the State of Maryland driving regulations. Thus $Y_{i1} = 1$ if sight in the left eye is impaired and $Y_{i2} = 1$ if there is impairment in the right eye. They consider the predictors age (in bands 40–50, 50–60, 60–70 and 70+), race (1 for blacks, 0 for whites) and education (Table 5.6). Here only age and race are considered and age is in the form $A_i - 2.5$ where $A_i = 1$ for persons aged 40–50, $A_i = 2$ for persons 50–60, etc. (i.e. age centred at 60 and in units of ten years).

Table 5.6   Visual impairment by age and race

Left eye	Right eye	Black 40–50	50–60	60–70	70+	Subtotal	White 40–50	50–60	60–70	70+	Sub-total	Grand total
No	No	729	551	452	307	2039	602	541	752	606	2501	4540
No	Yes	21	23	21	37	102	15	16	37	67	135	237
Yes	No	19	24	22	29	94	11	15	31	60	117	211
Yes	Yes	10	14	28	56	108	4	9	11	79	103	211
Total		779	612	523	429	2343	632	581	831	812	2856	5199

As in section 5.4.2, one may model the marginal models and odds ratio separately. Here, however, the marginal models $\eta_a = \text{logit}(\Pr(Y_1 = 1))$ and $\eta_b = \text{logit}(\Pr(Y_2 = 1))$ are assumed to be the same so that $\eta = \eta_a = \eta_b$ is an epidemiological model for visual impairment (VI) regardless of eye affected or whether sight is impaired in one or both eyes. Then following Liang *et al.*, the predictors for the marginal model are age, age squared, race, and an interaction between age squared and race. Liang *et al.*, also include an interaction between race and age itself but this appears to be insignificant (Liang *et al.*, 1992, Table 4). The model for the odds ratio, reflecting whether VI in one eye is associated with VI in the other, is

$$\log[\text{OR}(Y_1, Y_2)] = \log[(\pi_{11}\pi_{00})/(\pi_{10}\pi_{01})] = \alpha_0 + \alpha_1 \text{Race}$$

Here disgaggregated data for the 5199 subjects is used for estimation.

Liang *et al.* report in their GEE1 analysis a coefficient α_1 of 0.54 with t ratio 2.2. Here the multinomial logit approach of (5.26) (after a two-chain run of 10 000 iterations, convergent from 1000) shows α_1 to straddle zero, with mean 0.35 and 95% interval $\{-0.13, 0.81\}$. The effects of age, age squared, race (all positive) and of race by age squared (negative) are, by contrast, all well defined. The coefficients in the marginal model are higher than reported by Liang *et al.*, (1992), with the race coefficient in η being 0.6 compared with 0.33.

As one possible model check the proportion of blacks impaired in both eyes among those over 70 is compared with the proportion among whites over 70 so impaired. Thus the risk difference based on sampling replicate data from the multinomial model is compared with the observed risk difference. The observed rates are 13.05% (56/429) among blacks and 9.73% among whites, giving a difference of 3.32%. The model tends to underestimate this difference $(P_c = 0.05)$ predicting an average difference of around 1.1%. This suggests that age or an age–race interaction might be added to the model for the odds ratio. In fact the observed risk differences between races are similar for the two age bands under 60 and the two bands over 60, so the interaction model might be quite simple.

EXERCISES

1. In Example 5.1 try the Poisson–lognormal mixture as in (5.5), with $\varepsilon_i \sim N(0, \sigma^2)$ and also a non-parametric mixture with

$$y_i \sim Po(\mu_i)$$
$$\log(\mu_i) = \beta x_i + \varepsilon_i$$
$$\varepsilon_i \sim DP(\nu G_0)$$

   where G_0 is also a normal density with mean 0 and unknown variance, where ν is preset at five, and the maximum number of potential clusters is 20. How far does either approach modify inferences on the predictor effects or affect goodness of fit?

2. The toxoplasmosis data used by Efron (1986) and others record positive cases y_i in total samples of size n_i in $i = 1, \ldots, 34$ cities in El Salvador in relation to rainfall totals x_i. Consider the beta-binomial model

$$y_i \sim Bin(n_i, p_i)$$
$$p_i \sim Be(\delta_i \alpha_i, \delta_i[1 - \alpha_i])$$

where $\text{logit}(\alpha_i) = \beta x$ and $\log(\delta_i) = \gamma z$. Using standardized predictors $X_i = (x_i - \bar{x})/s_x$ and $N_i = (n_i - \bar{n})/s_n$, consider the model with fixed variance parameters $\delta_i = \delta$ against models allowing them to vary linearly and quadratically with N_i. Thus the simplest model has $\log(\delta_i) = \gamma_0$ and the most complex has $\log(\delta_i) = \gamma_0 + \gamma_1 N_i + \gamma_2 N_i^2$. The model for $E(p_i)$ in each case is

$$\text{logit}(\alpha_i) = \beta_0 + \beta_1 X_i + \beta_2 X_i^2 + \beta_3 X_i^3$$

Obtain predictions of y_i via replicate sampling $z_i \sim \text{Bin}(n_i, p_i)$, and assess via a predictive loss criterion such as (2.10) how far their bias and precision are affected by different variance models.

City	Rainfall	p(MLE)	n_i	y_i	$n_i - y_i$
1	1735	0.5	4	2	2
2	1800	0.6	5	3	2
3	2050	0.292	24	7	17
4	1770	0.545	11	6	5
5	1756	0.167	12	2	10
6	1871	0.438	16	7	9
7	1780	0.615	13	8	5
8	1936	0.3	10	3	7
9	1750	0.25	8	2	6
10	1830	0	1	0	1
11	1920	0	1	0	1
12	1650	0	1	0	1
13	2063	0.561	82	46	36
14	1900	0.3	10	3	7
15	2000	0.2	5	1	4
16	2077	0.368	19	7	12
17	1650	0.5	30	15	15
18	1770	0.611	54	33	21
19	2250	0.727	11	8	3
20	2100	0.692	13	9	4
21	1976	0.167	6	1	5
22	1973	0.3	10	3	7
23	1920	0.5	6	3	3
24	2200	0.182	22	4	18
25	2240	0.444	9	4	5
26	1796	0.532	77	41	36
27	1918	0.535	43	23	20
28	2292	0.622	37	23	14

(Continue)

(*Continued*)

City	Rainfall	p(MLE)	n_i	y_i	$n_i - y_i$
29	1750	1	2	2	0
30	1800	0.8	10	8	2
31	2000	0	1	0	1
32	1620	0.278	18	5	13
33	1890	0.471	51	24	27
34	1834	0.707	75	53	22

3. Fit a DPP mixture with varying intercepts β_{0j}, but uniform slope, namely

$$y_i \sim \text{Bin}(n_i, \pi_i)$$
$$\text{logit}(\pi_i) = \beta_{0G_i} + \beta_1 x_i$$

to the data in Example 5.3 from Follman and Lambert (1989). For example, one might try taking the maximum potential clusters as $J = 5$, and the Dirichlet concentration parameter as an unknown (e.g. updated as in section 3.5.3) or preset at $\kappa = 5$.

4. Fit the Poisson ZIP and hurdle models to the fetal lamb movement data which contains 182 zeros, 41 ones, 12 twos, 2 threes, 2 fours, and one seven (Leroux and Puterman, 1992). Thus for the ZIP model the unknowns are ξ and μ in the mixture model

$$\Pr(y_i = 0) = \xi + (1 - \xi)\exp(-\mu)$$
$$\Pr(y_i = j) = (1 - \xi)\exp(-\mu)\mu^{y_i}/y_i!$$

Include a predictive check of the zero counts from the ZIP model. Note that these data have a time series aspect that is not allowed for in these mixture models.

5. In Example 5.6 try the scale mixture model but with a right tail rather than left tail modification. How does the fit compare with the left tail modification (as in Model D of Program 5.6) and for which cases does an improvement/deterioration occur?

6. In Example 5.7 how is the EPD affected by setting λ at one rather than taking it as a free parameter, and taking $K = 3$ rather than $K = 4$?

7. In Example 5.8 try one- and two-factor models with all five outcomes and compare their fits using DIC or EPD measures. Also try the bivariate probit model applied to breathlessness and wheeze (Y_3 and Y_4) in the data given in the example; compare with the results (based on 29 720 subjects) given by Bock and Gibbons (1996).

8. In Example 5.9 try the negative multinomial model and compare its fit with the factor model and bivariate Poisson–lognormal model.

9. In Example 5.9 include a predictive check on how far replicate data from the model reproduces the observed overdispersion. Also apply a selection procedure on the regressors to assess whether the predictive check for zero counts is thereby improved.

10. In Example 5.10 add an age–race interaction to the model for the odds ratio and assess whether the predictive check on both eye VI by race improves.

REFERENCES

Agresti, A. (2002) *Categorical Data Analysis*, 2nd edition. John Wiley & Sons: New York.

Aitkin, M. (1996) A general maximum likelihood analysis of overdispersion in generalized linear models. *Statiststics and Computing*, **6**, 251–262.

Aitkin, M. (2001) Likelihood and Bayesian analysis of mixtures. *Statistical Modelling*, **1**, 287–304.

Albert, J. (1999) Criticism of a hierarchical model using Bayes factors. *Statistics in Medicine*, **18**, 287–305.

Albert, J. and Chib, S. (1993) Bayesian analysis of binary and polychotomous response data. *Journal of the American Statistical Association*, **88**, 669–679.

Aranda-Ordaz, F. (1981) On two families of transformations to additivity for binary response data. *Biometrika*, **68**, 357–363.

Bartholomew, D. (1987) *Latent Variable Models and Factor Analysis*. Charles Griffin: London.

Bartholomew, D. and Knott, M. (1999) *Latent Variable Models and Factor Analysis*, Kendall's Library of Statistics, 7. Arnold: London.

Basu, S. and Chib, S. (2003) Marginal likelihood and Bayes factors for Dirichlet process mixture models. *Journal of the American Statistical Association*, **98**, 224–235.

Basu, S. and Mukhopadhyay, S. (2000) Binary response regression with normal scale mixture links. In *Generalized Linear Models: A Bayesian Perspective*, Dey, D., Ghosh, S. and Mallick, B. (eds). Marcel Dekker: New York.

Bliss, C. (1935) The calculation of the dosage-mortality curve. *Annals of Applied Biology*, 22.

Bock, R. and Gibbons, R. (1996) High dimensional multivariate probit analysis. *Biometrics*, **52**, 1183–1194.

Böhning, D. (1999) *Computer-Assisted Analysis of Mixtures and Applications: Meta-Analysis, Disease Mapping and Others*. Chapman and Hall: London/CRC Press: Boca Raton, FL.

Cameron, C. and Trivedi, P. (1998) *Regression Analysis of Count Data*, Econometric Society Monograph No. 30, Cambridge University Press: Cambridge.

Carey, V., Zeger, S. and Diggle, P. (1993) Modelling multivariate binary data with alternating logistic regressions. *Biometrika*, **80**, 517–526.

Chen, M. and Dey, D. (2003) Variable selection for multivariate logistic regression models. *Journal of Statistical Planning and Inference*, **111**, 37–55.

Chib, S. and Greenberg, E. (1998) Analysis of multivariate probit models. *Biometrika*, **85**, 347–361.

Chib, S. and Winkelmann, R. (2001) Markov chain Monte Carlo analysis of correlated count data. *Journal of Business and Economic Statistics*, **19**, 428–435.

Connolly, M. and Liang, K. (1988) Conditional logistic regression models for correlated binary data. *Biometrika*, **75**, 501–506.

Cox, D. and Wermuth, N. (2002) On some models for multivariate binary variables parallel in complexity with the multivariate Gaussian distribution. *Biometrika*, **89**, 462–469.

Czado, C. (1994) Bayesian inference of binary regression models with parametric link. *Journal of Statistical Planning and Inference*, **41**, 121–140.

Czado, C. and Raftery, A. (2001) Choosing the link function and accounting for link uncertainty in generalized linear models using Bayes factors. *Discussion Paper* 262, SFB 386, University of Munich.

Dey, D. and Ravishanker, N. (2000) Bayesian approaches for overdispersion in *Generalized Linear Models*. In *Generalized Linear Models: A Bayesian Perspective*, Dey, D., Ghosh, S. and Mallick, B. (eds). Marcel Dekker: New York.

Dey, D., Peng, F. and Gelfand, A. (1997) Overdispersed generalized linear models. *Journal of Statistical Planning and Inference*, **64**, 93–107.

Efron, B. (1986) Double exponential families and their use in generalized linear regression. *Journal of the American Statistical Association*, **81**, 709–721.

Engel, B. and Keen, A. (1994) A simple approach for the analysis of generalized linear mixed models. *Statistica Neerlandica*, **48**, 1–22.

Follman, D. and Lambert, D. (1989) Generalizing logistic regression by nonparametric mixing. *Journal of the American Statistical Association*, **84**, 295–300.

Ganio, L. and Schafer, D. (1992) Diagnostics for overdispersion. *Journal of the American Statistical Association*, **87**, 795–804.

Gelfand, A. and Dalal, S. (1990) A note on overdispersed exponential families. *Biometrika*, **77**, 55–64.

Gelfand, A. and Mallick, B. (1995) Bayesian analysis of semiparametric proportional hazards models. *Biometrics*, **51**, 843–852.

Gelman A., Carlin J., Stern, H. and Rubin, D. (2003) *Bayesian Data Analysis*, 2nd Edition. CRC Press: Boca Raton, FL.

Glonek, G. and McCullagh, P. (1995) Multivariate logistic models. *Journal of the Royal Statistical Society, Series B*, **53**, 533–546.

Greene, W. (2000) *Econometric Analysis*, 4th edition, Prentice Hall: Englewood Cliffs, NJ.

Guo, G. (1996) Negative multinomial regression models for clustered event counts. *Sociological Methodology*, **26**, 113–132.

Hall, D. (2000) Zero-inflated Poisson and binomial regression with random effects: a case study. *Biometrics*, **56**, 1030–1039.

Haneuse, S. and Wakefield, J. (2004) Ecological inference incorporating spatial dependence. Chapter 13 in *Ecological Inference: New Methodological Strategies*, King, G., Rosen, O. and Tanner, M. (eds). Cambridge University Press: Cambridge.

Hurn, M., Justel, A. and Robert, C. (2003) Estimating mixtures of regressions. *Journal of Computational and Graphical Statistics*, **12**, 1–25.

Johnson, V. and Albert, J. (1999) *Ordinal Data Models*. Springer: New York.

Jorgensen, B. (1987) Exponential dispersion models. *Journal of the Royal Statistical Society, Series B*, **49**, 127–162.

Kahn, M. and Raftery, A. (1996) Discharge rates of Medicare stroke patients to skilled nursing facilities: Bayesian logistic regression with unobserved heterogeneity. *Journal of the American Statistical Association*, **91**, 29–41.

Kleinman, K. and Ibrahim, J. (1998) A semiparametric Bayesian approach to the random effects model. *Biometrics*, **54**, 921–938.

Lambert, D. (1992) Zero-inflated Poisson regression, with an application to defects in manufacturing. *Technometrics*, **34**, 1–14.

Leroux, B. and Puterman, M. (1992) Maximum penalized likelihood estimation for independent and Markov-dependent Poisson mixtures. *Biometrics*, **48**, 545–558.

Liang, K.-Y., Zeger, S. and Qaqish, B. (1992) Multivariate regression analyses for categorical data. *Journal of the Royal Statistical Society, Series B*, **54**, 3–40.

Lindsey, J. (1975) Likelihood analysis and test for binary data. *Applied Statistics*, **24**, 1–16.

Mallick, B. and Gelfand, A. (1994) Generalized linear models with unknown link function. *Biometrika*, **81**, 237–245.

McCullagh, P. and Nelder, J. (1989) *Generalized Linear Models*, 2nd Edition. Chapman and Hall: London.

Moore, D., Park, C. and Smith, W. (2001) Exploring extra-binomial variation in teratology data using continuous mixtures. *Biometrics*, **57**, 490–494.

Mullahy, J. (1986) Specification and testing of some modified count data models. *Journal of Econometrics*, **33**, 341–365.

Munkin, M. and Trivedi, P. (1999) Simulated maximum likelihood estimation of multivariate mixed-Poisson regression models, with application. *Econometrics Journal*, **2**, 29–48.

Muthen, B. (1979) A structural probit model with latent variables. *Journal of the American Statistical Association*, **24**, 807–811.

Newton, M., Czado, C. and Chappell, R. (1996) Bayesian inference for semi-parametric binary regression. *Journal of the American Statistical Association*, **91**, 142–153.

Ntzoufras, I., Dellaportas, P. and Forster, J. (2003) Bayesian variable and link determination for generalised linear models. *Journal of Statistical Planning and Inference*, **111**, 165–180.

O'Brien, S. and Dunson, D. (2004) Bayesian multivariate logistic regression. *Biometrics*, **60**, 739–746.

Pierce, D. and Sands, B. (1975) Extra-Bernouilli variation in binary data. *Technical Report* No. 46, Dept. of Statistics, Oregon State University.

Raftery, A. (1996) Approximate Bayes factors and accounting for model uncertainty in generalised linear models. *Biometrika*, **83**, 251–266.

Rosen, O., Jiang, W., King, G. and Tanner, M. (2001) Bayesian and frequentist inference for ecological inference: the R×C case. *Statistica Neerlandica*, **55**, 134–156.

Stukel, T. (1988) Generalized logistic models. *Journal of the American Statistical Association*, **83**, 426–431.

West, M. (1985) Generalised linear models: scale parameters, outlier accommodation and prior distributions (with discussion). In *Bayesian Statistics 2*, Bernardo, J., DeGroot, M., Lindley, D. and Smith, A. (eds). North-Holland: Amsterdam.

Winkelmann, R. (2000) *Econometric Analysis of Count Data*. Springer: New York.

Winkelmann, R. and Zimmermann, K. (1995) Recent developments in count data modeling. *Journal of Economic Surveys*, **9**, 1–24.

CHAPTER 6

Random Effect and Latent Variable Models for Multicategory Outcomes

6.1 MULTICATEGORY DATA: LEVEL OF OBSERVATION AND RELATIONS BETWEEN CATEGORIES

Suppose the observed data are polytomous, when for individual data level one among the elements of the vector $y_i = (y_{i1}, y_{i2}, \ldots, y_{iJ})$ takes the value 1. Equivalently this type of outcome may be represented by a categorical indicator $D_i = j$ if and only if $y_{ij} = 1$. Much survey and census data regarding attitudes or demographic attributes are of this kind, as are choice data in econometrics. Usually such data are defined in terms of mutually exclusive alternatives: if the jth outcome of the J possible outcomes occurs then $y_{ij} = 1$, and the others are zero ($y_{ik} = 0$ for $k \neq j$). Aggregated multinomial (or 'compositional') data accumulate such responses over subjects within clusters i and will be in the form $f_i = (f_{i1}, f_{i2}, \ldots, f_{iJ})$ specifying counts in response categories $j = 1, \ldots, J$ for clusters i. Such categories may be ranked (e.g. attainment levels j by individual pupils or schools i) or unranked. This chapter considers unranked or nominal categorizations, whereas Chapter 7 tackles ordinally ranked multinomial data. In each situation a model is required for the probabilities π_{ij} of choice or allocation j for the ith unit or individual.

As for binary data the 'revealed' choice or allocation may be regarded as reflecting the operation of an underlying latent utility or frailty (Albert

Bayesian Models for Categorical Data   P. Congdon
© 2005 John Wiley & Sons, Ltd

and Chib, 1993; Scott, 2003) and data augmentation at the level of the individual may be introduced to facilitate estimation of the π_{ij}. While a standard multinomial likelihood, or data augmentation corresponding indirectly to the standard likelihood, may be appropriate in many cases, whether for individual or aggregate data, interdependence between choice categories and other sorts of heterogeneity (leading to multinomial overdispersion) may need to be considered.

As for binomial and count data, representations of heterogeneity in choice modelling may be classed as discrete or continuous (Wedel *et al.*, 1999), with discrete approaches exemplified by finite mixture models that may be taken to correspond to subpopulations or market segments (e.g. in consumer choice modelling). However, the underlying distribution of preferences or utilities may in fact be continuous, and continuous mixture models may outperform discrete mixture models in out-of-sample predictions (Lenk *et al.*, 1996). The conjugate approach for such heterogeneity is the multinomial–Dirichlet mixture where the Dirichlet is the multivariate generalization of the beta density. However, as for binary logit and Poisson data it is often easier to model random interdependent choices and heterogeneity within the regression link, as in random effects MNL models (see section 6.6) (Hensher and Greene, 2001), or via probit models (Hausman and Wise, 1978).

6.2 MULTINOMIAL MODELS FOR INDIVIDUAL DATA: MODELLING CHOICES

A number of modelling schemes have been proposed for multiple category variables observed at individual level, including multinomial generalizations of the logit and probit models for binary outcomes. Examples of multicategory modelling include human choice applications, for instance in transportation (Greene, 2000; Nobile *et al.*, 1997), in marketing (Rossi and Allenby, 2003), party choice in political science (Glasgow, 2001) and in occupational and residential mobility. They may involve unordered or ranked categories between which a choice is made, and predictors may relate to subject characteristics (e.g. class or income), attributes of different choices (e.g. price), or be subject-specific measures of attractiveness of different choices.

The revealed choice is often taken (e.g. in econometrics) to reflect an underlying latent continuous utility scale. In some applications (e.g. political affiliation or religious belief) a single underlying scale may be less substantively justified. Assume a subject chooses the option with the highest utility and that the utility function is written as a sum of an error term (summarizing unknown influences) and a systematic component for the utility of known predictors. Different sorts of model follow according to the density assumed for the errors: the extreme value distribution of McFadden (1974) leads to logit models (e.g. the multinomial logit and categorical logit models discussed below), while a normal error leads to multiple and categorical probit models (McCulloch and Rossi, 1994). Other forms for ε may be used, such as scale mixtures of multivariate normals (Chen and Kuo, 2002). Discrete mixtures of normal or other densities are another possibility.

Predictors of choice or group membership may include social and demographic characteristics of individuals i, X_i, namely 'characteristics of the chooser' in choice models (Hoffman and Duncan, 1988). However, they may also be attributes of the different choices j, A_j, or attributes of choices specific to individuals C_{ij} (e.g. individual costs attached to different transport modes). In a choice application the underlying utility takes the form

$$U_{ij} = V_{ij} + \varepsilon_{ij} = X_i\beta_j + C_{ij}\gamma + A_j\varphi + \varepsilon_{ij} \qquad (6.1)$$

where $\eta_{ij} = X_i\beta_j + C_{ij}\gamma + A_j\varphi$ reflects the impact of known aspects of utility measured by X_i, C_{ij} and A_j. What is actually observed is the choice j which is taken to correspond to a maximum utility decision with $U_{ij} = \max(U_{i1}, U_{i2}, \ldots, U_{ij})$, so that

$$\pi_{ij} = \text{Prob}(U_{ij} > U_{i[j]})$$

with $[j]$ denoting the set of choices apart from the jth. Subject to identifiability constraints, data augmentation provides a way to simulate the underlying utilities (Albert and Chib, 1993; McCulloch and Rossi, 1994), and may lead to simplified sampling by converting the analysis to normal linear multivariate regression. However, direct estimation using the multinomial likelihood remains an option.

Different model frameworks are defined according to the type of regressor variable. If the predictors include only individual characteristics then what may be termed a pure multinomial logit or probit is appropriate, usually with choice-specific parameters β_j and with parameters for a baseline choice category set to zero to avoid model indeterminacy. If the predictors include attributes C_{ij} of the jth alternative specific for individual i, then an appropriate model is the conditional logit or probit model. With $X_{i0} = 1$ for a constant term included in a $1 \times p$ vector X_i of predictors for subjects $i = 1, \ldots, n$, and with the last among $j = 1, \ldots, J$

categories as baseline or reference, the pure multinomial logit model has
the form

$$\Pr(Y_{ij} = 1) = \pi_{ij} = \exp(X_i\beta_j) / \left[1 + \sum_{k=1}^{J-1} \exp(X_i\beta_k) \right]$$

$$j = 1, \ldots, J - 1 \tag{6.2a}$$

$$\Pr(Y_{iJ} = 1) = \pi_{iJ} = 1 / \left[1 + \sum_{k=1}^{J-1} \exp(X_i\beta_k) \right] \tag{6.2b}$$

where $\beta_j = (\beta_{j0}, \ldots, \beta_{jp})'$ describes how the subject attributes affect
different choices. The effect of X_{im} (the mth attribute for subject i) on
the logit of choice j versus choice k is obtained as the contrast $\beta_{jm} - \beta_{km}$.
An alternative parameterization for identifiability is to specify a regression parameter prior γ_j for $j = 1, \ldots, J$ and then centre the parameters as
$\gamma_j^c = \gamma_j - \bar{\gamma}$, with

$$\pi_{ij} = \exp(X_i\gamma_j^c) / \sum_{k=1}^{J} \exp(X_i\gamma_k^c) \tag{6.2c}$$

One point to note is that one may define the variable regression
coefficients in terms of subsets of the choices, rather than all the β_j
necessarily being different. For example, taking β_{1m} as a free parameter
for predictor m but other β_{km} as fixed at zero (for $k = 2, \ldots, J - 1$)
amounts to comparing the impact of a covariate between choice 1 against
all other choices combined. So in the data considered in Example 6.2
below, Greene (2000) compares the impact of household income on travel
by air as against travel by train, car or bus.

By contrast to the multinomial logit model, the conditional logit
considers choice attribute data C_{ij} specific to individuals, and typically
involves regression coefficients γ constant over alternatives j

$$\pi_{ij} = \exp(C_{ij}\gamma) / \sum_{k=1}^{K} \exp(C_{ik}\gamma) \tag{6.3}$$

Dividing by the numerator, as in

$$\pi_{ij} = 1 / \sum_{k=1}^{K} \exp([C_{ik} - C_{ij}]\gamma)$$

shows that choice probabilities are then determined by differences in
attribute values between alternatives. When equality over categories in
the γ coefficients is not assumed, it is necessary to constrain one
category's parameter to zero (e.g. $\eta_{ij} = \beta_{j0} + C_{1ij}\gamma_{j1} + C_{2ij}\gamma_{j2}$, for $j =
1, \ldots, J - 1$ and $\eta_{iJ} = 0$).

With data containing both chooser attributes X and attributes C specific to choices and individuals, coefficients on intercepts or chooser attributes replicate the format in (6.2) but the coefficients for C variables replicate (6.3). So with a single variable of each type, X_i and $C_{ij}, J = 3$ choices, and taking the third category as reference,

$$\eta_{i1} = \alpha_1 + X_i\beta_1 + C_{i1}\gamma$$
$$\eta_{i2} = \alpha_2 + X_i\beta_2 + C_{i2}\gamma$$
$$\eta_{i3} = C_{i3}\gamma \qquad\qquad (6.4)$$
$$\pi_{ij} = \exp(\eta_{ij}) / \sum_{k=1}^{K} \exp(\eta_{ik})$$

This type of mixed attribute model is generally also called the multi-nomial logit model or MNL model for short, and this is the terminology subsequently adopted. However, it is worth being aware of the distinction sometimes drawn between conditional logit and logit models with chooser attributes only (e.g. Greene, 2000, section 19.7.1). While any choice-specific fixed regression effects (e.g. the intercepts α_j) are set to zero in the reference category model, note that choice–subject predictors C_{ij} with constant coefficients over choices are included.

In the multinomial and conditional logit models, the ratio π_{ij}/π_{ik}, i.e. the probability of choosing the jth alternative compared with the kth, can be seen to be independent of the presence or characteristics of other alternatives. This is known as the independence of irrelevant alternatives (IIA) assumption or axiom. However, assuming this property may be inconsistent with observed choice behaviour in that utilities over different alternatives may be correlated (e.g. there may be sets of similar alternatives with similar utilities between which substitution may be made). One way to correct for clustering is to assume subject or subject–choice errors in the generalized logit link, leading to mixed logit models or mixed MNL (MMNL) models (section 6.4).

Another option is the use of multinomial probit models (section 6.4) since these are not restricted to the IIA axiom. Their estimation by classical methods is complicated by the need to evaluate multidimensional normal integrals. However, MCMC sampling using data augmentation is relatively easy computationally. Other options to tackle departures from IIA include nested logit models (e.g. Lahiri and Gao, 2002) which group the choices into subsets such that error variances differ between subsets.

6.3   MULTINOMIAL MODELS FOR AGGREGATED DATA: MODELLING CONTINGENCY TABLES

Strictly, individual choice categorical regression is only necessary when one or more predictors are continuous. When all predictors are categorical, equivalences with contingency table analysis may be useful to exploit. As mentioned in Chapter 4, contingency tables (e.g. f_{ij} for a two-way table) can be taken to follow a multinomial distribution if the total count (e.g. f_{++} for a two-way table) is fixed or if row totals are fixed. The latter case is known as product multinomial sampling since there are I separate multinomial densities, one for the data in each row, and the total likelihood is a product over these. Authors such as Lindley (1964) and Good (1976) set out the Bayesian analysis for $I \times J$ two-way tables with multinomial sampling in relation to the total count f_{++} with cell probabilities π_{ij} subject to $\sum_i \sum_i \pi_{ij} = 1$. This involves a Dirichlet prior[1] on $\pi = (\pi_{11}, \pi_{12}, \ldots, \pi_{IJ})$ such that

$$P(\pi \mid \alpha) \propto \prod_{ij} \pi_{ij}^{\alpha_{ij}-1}$$

The α_{ij} can be considered as prior counts for the cell (i,j) and the total α_{++} as a prior sample size. The posterior is also Dirichlet with elements $y_{ij} + \alpha_{ij}$ and total sample size $f_{++} + \alpha_{++}$. Epstein and Fienberg (1991) consider a profile of fits using an extended prior allowing for varying strengths of belief in independence. Thus $(f_{11}, f_{12}, \ldots, f_{IJ}) \sim M(y_{++}; \pi_{11}, \pi_{12}, \ldots, \pi_{IJ})$ and $(\pi_{11}, \pi_{12}, \ldots, \pi_{IJ}) \sim \mathrm{Dir}(K, \alpha)$, namely

$$P(\pi \mid K, \alpha) \propto \prod_{ij} \pi_{ij}^{K\alpha_{ij}-1}$$

A multiple logit model is specified for α_{ij} based on a log-linear model

$$\alpha_{ij} = \exp(\delta_0 + \delta_{1i} + \delta_{2j}) / \sum_i \sum_j \exp(\delta_0 + \delta_{1i} + \delta_{2j})$$

with constraint

$$\delta_0 = -\log \left\{ \sum_{ij} [\exp(\delta_{1i} + \delta_{2j})] \right\}$$

[1] It is worthwhile knowing that a sample $(y_1, \ldots, y_j)$ from a Dirichlet with parameters $(\alpha_1, \ldots, \alpha_j)$ may be obtained by sampling x_j from J independent gamma densities $x_j \sim \mathrm{Ga}(\alpha_j, 1)$ and then setting $y_j = x_j / \sum_j x$. Alternatively select y_1 from a beta density $\mathrm{Be}(\alpha_1, \alpha_{++} - \alpha_1)$ and $x_j (j = 2, \ldots, J-1)$ from $\mathrm{Be}(\alpha_j, \sum_{k=j+1}^{J} \alpha_k)$. Then

$$y_j = \left(1 - \sum_{k=i}^{j-1} y_k\right) x_j \quad (j = 2, \ldots, J-1) \quad \text{and} \quad y_k = \left(1 - \sum_{k=1}^{J-1} y_k\right)$$

K is preset and taking large values ($K \to \infty$) gives estimates π_{ij} corresponding to independence, whereas $K = 0$ corresponds to a saturated model with estimated π_{ij} close to the observed proportions. Below, a prior on K is proposed as a way of assessing independence.

Alternatively, as in Leonard and Hsu (1994, p 290), a model can be framed directly in terms of the multiple logits

$$\pi_{ij} = \exp(\eta_{ij}) / \sum_i \sum_j \exp(\eta_{ij})$$

where

$$\eta_{ij} = \delta_i^A + \delta_j^B + \delta_{ij}^{AB} \tag{6.5}$$

subject to the identification constraints

$$\sum_i \delta_i^A = \sum_j \delta_j^B = \sum_i \delta_{ij}^{AB} = \sum_j \delta_{ij}^{AB} = 0$$

Via MCMC sampling one may assess significance for omnibus association criteria based on models such as (6.5) for $I = J = 2$

$$\phi = \log[(\pi_{11}\pi_{22})/(\pi_{12}\pi_{21})] = \eta_{11} + \eta_{22} - \eta_{12} - \eta_{21}$$

Thus if variables A and B are positively associated then sampled values $\phi^{(t)}$ will be predominantly positive.

6.3.1 Conditional contingency tables: histogram smoothing and multinomial logit via Poisson regression

For product multinomial analysis with row totals fixed, the analysis conditions on known y_{i+}. So with probabilities ρ_{ij} of variable B taking values $j = 1, \ldots, J$ given variable $A = i$, and with $\sum_j \rho_{ij} = 1$, the sampling model is

$$f_{ij} \sim M(y_{i+}; \rho_{i1}, \ldots \rho_{iJ}) \quad i = 1, \ldots, I$$

The probabilities defining sampling in each row may be obtained as

$$\rho_{ij} = \pi_{j|i} = \Pr(B = j \mid A = i) = \Pr(A = i, B = j)/\Pr(A = i) = \pi_{ij}/\pi_{i+}$$

with π_{ij} obtained by multinomial sampling in relation to y_{++} as above. Gunel and Dickey (1974) show how a two-step prior may be specified for product multinomial data, by first sampling y_{i+} from y_{++}, and then y_{ij} from y_{i+}. A product multinomial also applies if column totals are fixed.

However, the Dirichlet has a restricted covariance structure when there are dependencies between the 'response' categories j within rows i. For example, for I constituencies and J political parties one may expect both negative and positive correlations between ρ_{ij} for different parties. Instead consider the logit form

$$\rho_{ij} = \exp(\eta_{ij}) / \sum_k \exp(\eta_{ik})$$

and assume the parameters of the I multinomial densities under fixed row totals are exchangeable under a hyperdensity. For example, setting $\eta_i = (\eta_{i1}, \ldots, \eta_{iJ})$, with $\eta_{iJ} = 0$ for identifiabilty, one may assume

$$\eta_{i,1:J-1} \sim N_{J-1}(\mu, \Sigma)$$

or

$$\eta_{i,1:J-1} \sim N_{J-1}(\mu, \Sigma/\lambda_i)$$

where the λ_i are positive variables with mean 1. Alternatively one may take

$$\eta_i \sim N_J(\mu, \Sigma)$$

where the μ parameters include a constraint such as $\mu_J = 0$ for identification and the identified centred means $\mu_j^c = \mu_j - \bar{\mu}(j = 1, \ldots, J)$ may then be obtained.

The raw percentages f_{ij}/f_{i+}, providing estimates of ρ_{ij} based on a separate row analysis without pooling strength under a hyperdensity, are likely to show erratic features, whereas prior structures such as the above both lead to frequency smoothing and provide for a model of interdependencies between categories. This model may also be fitted by Poisson regression using the fact that the multinomial is equivalent to a Poisson distribution conditional on a fixed total; this involves defining I fixed effects predictors ξ_i to ensure the row totals are maintained. Thus

$$f_{ij} \sim Po(\mu_{ij})$$
$$\log(\mu_{ij}) = \xi_i + \gamma_{ij}$$

where ξ_i would typically be assigned vague priors, e.g. $\xi_i \sim N(0, 10\,000)$.

Product multinomial sampling may be extended to combinations of conditioning factors. In particular, when all predictors are categorical, an MNL model may be fitted by Poisson regression again using the multinomial–Poisson equivalence conditional on a fixed total. Consider four variables A, B, C and D in categorical form with I, J, K and M levels respectively. Assume a single multinomial for the contingency table f_{ijkm} such that the joint probabilities $\Pr(A = i, B = j, C = k, D = m) = \pi_{ijkm}$ sum to one, namely $\sum\sum\sum\sum \pi_{ijkm} = \pi_{++++} = 1$.

Let D be the dependent variable and A, B and C be predictors. So the goal is to model conditional probabilities that D is at level m given the predictors at various combinations of levels,

$$\pi_{m|ijk} = \Pr(D = m \mid A = i, B = j, C = k)$$
$$= \Pr(A = i, B = j, C = k, D = m)/\Pr(A = i, B = j, C = k)$$
$$= \pi_{ijkm}/\pi_{ijk+}$$

The distribution of D given $(A = i, B = j, C = k)$ is multinomial with denominator f_{ijk+} and probabilities $\pi_{m|ijk}$. So via multinomial–Poisson equivalence this is the same as a Poisson model with fixed marginal totals f_{ijk+}. The Poisson model would need to include dummy terms for all main effects and interactions among the predictors, in order to ensure the marginal totals f_{ijk+} are preserved by the model. However, this reduces to including a set of IJK fixed effects ξ_{ijk}.

Thus an MNL approach with main effect impacts of the categorical predictors A, B and C on the response D involves the model

$$f_{ijkm} \sim \text{Po}(\mu_{ijkm})$$
$$\log(\mu_{ijkm}) = \xi_{ijk} + \alpha_m + \beta_{im} + \gamma_{jm} + \delta_{km} \tag{6.6}$$

where the usual corner constraints apply to the intercept terms (e.g. α_J must be set to zero and only $\alpha_1, \ldots, \alpha_{J-1}$ are free) and to the impacts of A, B and C. The predicted probabilities are then $\pi_{m|ijk} = \mu_{ijkm}/f_{ijk+}$. The model may be extended to interactions between predictors, e.g. η_{ijm} models an interactive impact of A and B on D.

Example 6.1 Occupational attainment This example illustrates an application where the pure multinomial logit model of equation (6.2) rather than conditional logit of equation (6.3) is relevant: the only predictors pertaining to the allocation of subjects between a set of five occupational categories are subject attributes. The example illustrates how summaries may be obtained for regression parameter contrasts that occur in discrete choice models, and how 'Bayesian significance tests' can be applied to such contrasts. It also considers an alternative Poisson log-linear regression model that may be applied when predictors are available in (or converted to) categorical form.

The data are from the 1982 US General Social Survey and also considered by Long (1997), among others. The occupations are 'menial' ($j = 1$), blue collar ($j = 2$), craft ($j = 3$), white collar ($j = 4$) and professional ($j = 5$). The three predictors are X_1 = ethnicity (1 = white, 0 = non-white), X_2 = years of education, and X_3 = years of work experience. Low-skill manual occupations are taken as the reference category, so $\beta_{10} = \cdots = \beta_{13} = 0$ with $\beta_{j0}, \beta_{j1}, \beta_{j2}, \beta_{j3}$ being the intercept, ethnicity, education and experience effects for occupations $j = 2, \ldots, 5$. This is the corner constraint parameterization.

The IIA property of the MNL without random effects means that the odds of being in occupation j as against occupation k, namely

$$\Pr(y_{ij} = 1)/\Pr(y_{ik} = 1) \equiv \Pr(D_i = j)/\Pr(D_i = k)$$

are given by $\exp([\beta_j - \beta_k]X_i)$, so the contrast $\beta_j - \beta_k$ defines the effect of all covariates X_i on the logit of occupation j versus occupation k. The effect of covariate X_{im} (e.g. years of education) on the logit of occupation j vs. k is $\beta_{jm} - \beta_{km}$. Sometimes such contrasts may be significant (i.e. positive or negative with a probability exceeding 0.95 or higher), even when the original coefficients are not significantly different from zero. In particular, one may consider the probabilities

$$\Pr[\beta_{jm} - \beta_{km} > 0]$$

of positive contrasts between choice j and k for predictor m. It may be noted that maximum likelihood estimation (MLE) shows significant positive effects (t statistics > 1.96) of (a) white ethnicity on the odds (relative to menial jobs) of being in professional jobs; of (b) education on being in white-collar and professional jobs; and of (c) experience on being in professional jobs.

With N(0,10 000) priors on the free parameters and a two-chain run of 20 000 iterations, convergence is obtained from around 5000 iterations. By contrast to the MLE results, the Bayesian estimation shows more significant coefficients (Table 6.1); for example, white ethnicity is advantageous for entry to both blue-collar jobs and white-collar jobs as opposed to menial jobs. It has been noted that maximum likelihood analysis of logistic models for binary outcomes may have drawbacks with assessing significant predictor effects in small samples (Zellner and Rossi, 1984) and the present analysis may illustrate similar issues in multiple logit analysis.

Two of the experience effects are significant in the sense that 95% intervals are entirely positive or negative, though that for white-collar jobs is on the margin. Some of the experience contrasts $\beta_{j3} - \beta_{k3}$ are, however, clearly significant. For example, the contrast between β_{43} and β_{23} is positive with a probability of 0.98, showing an advantage of experience in white-collar as against blue-collar jobs.

Examination of the CPOs estimated via (2.15) shows subjects 336 and 295 with the lowest log CPOs (-9.1 and -7). These are individuals with

Table 6.1 Multinomial logit occupational choice (effects relative to menial jobs); Bayes estimates

	Blue collar	Craft	White collar	Professional
Constant (β_{j0})	−0.6	0.6	−1.4	−1.1
2.5%	−2.0	−0.6	−2.8	−2.0
97.5%	0.3	1.3	−0.5	−0.2
White (β_{j1})	1.36	0.63	2.12	2.24
2.5%	0.16	−0.34	1.05	1.35
97.5%	2.77	1.88	3.43	3.21
Education (β_{j2})	−0.11	0.09	0.37	0.82
2.5%	−0.30	−0.07	0.16	0.61
97.5%	0.06	0.26	0.59	1.05
Experience (β_{j3})	0.005	0.030	0.037	0.039
2.5%	−0.029	−0.002	0.000	0.004
97.5%	0.041	0.065	0.077	0.077

under ten years of education but in professional jobs. Low CPOs also occur for individuals with 16 years of education who are in blue-collar jobs. Such findings suggest some heterogeneity in the impact of education on occupational allocation.

As will be considered below, discrete mixture models offer one approach to such heterogeneity. For example, assume

$$\Pr(y_{ij} = 1 \mid L_i) = \exp(\beta_{jL_i} x_i) / \left[1 + \sum_{k=2}^{J} \exp(\beta_{kL_i} x_i) \right] \quad j > 1$$

where L_i is latent group for subject i and predictor effects $\beta_{jg} = (\beta_{jg0}, \dots, \beta_{jgp})$ differ between groups $g = 1, \dots, G$. However, in this application discrete mixture models with $G = 2$ subpopulations differing either in all predictor effects $\beta_{jgm}, m = 1, 3$, or just in the effect of education β_{jg2} (experience and ethnicity not differing by subpopulation) did not yield a clearly interpretable pattern. One drawback with these data is the relative sparsity in terms of measuring the impact of ethnicity (only 28 of 337 subjects are non-white). Other approaches to choice heterogeneity (e.g. effects of attributes on allocation varying between subjects) are considered in section 6.5.

To illustrate the Poisson regression approach to the MNL model (section 6.3.1), education was grouped into five categories (under

12 years, 12 years, 13–14 years, 15–16 years and 17 + years) while experience was formed into seven bands (under 6 years, 6–10, 11–15, 16–20, 21–25, 26–35, over 35 years). So the joint predictor distribution has $IJK = 2 \times 5 \times 7 = 70$ cells and 70 fixed effects ξ_{ijk} as in (6.6) must be included to ensure equivalence to a multiple logistic. Note that a multiple logistic using multinomial sampling remains an option and has the benefit of not involving dummy predictors for each combination of predictor (which inflate the parameter count). The substantive interest under either approach is in the parameters β_{im}, γ_{jm} and δ_{km} relating to the impact of ethnicity, education and experience respectively.

With a Poisson likelihood, the posterior means of the white ethnicity parameters $\{\beta_{im}\}$, namely $\{\beta_{22}, \beta_{23}, \beta_{24}, \beta_{25}\}$, are similar to those in Table 6.1 but less precisely estimated (i.e. wider 95% credible intervals): 1.2 (−0.1,2.8), 0.5 (−0.7,1.7), 1.8 (0.1,3.9) and 1.95 (0.5, 3.6). The effects of education suggest some non-linearity: the coefficients γ_{j5} for entry to professional jobs for education levels 2, 3, 4 and 5 are respectively 1.1 (with a 95% interval from −0.4 to 2.7), 1.8 (0.3,3.3), 5.3 (3.1,8.4) and 10.6 (4.6,17.9). While losing continuity in predictor values, this approach may be useful to detect non-linearity as part of an exploratory data analysis. However, the estimates obtained depend on whatever grouping of the originally continuous variable is used. Example 6.4 considers an alternative approach to the non-linear impact of education for these data.

Example 6.2 Exam grade frequencies Leonard and Hsu (1994) present totals f_{ij} of students allocated to six grades in mathematics for 40 London schools. The frequency smoothing model described in section 6.3.1 is fitted with multinomial sampling conditioning on the school totals f_{i+}, and with parameters ρ_{ij}, μ and Σ, as in

$$f_{ij} \sim M(f_{i+}, \rho_{ij}) \quad i = 1, \ldots, I$$

$$\rho_{ij} = \exp(\eta_{ij}) / \sum_i \sum_j \exp(\eta_{ij})$$

$$\eta_i \sim N_J(\mu, \Sigma)$$

where the μ parameters include a constraint $\mu_1 = 0$ for identification. The centred means $\mu_j^c = \mu_j - \bar{\mu}(j = 1, \ldots, 6)$ may then be obtained.

A Wishart prior on Σ is adopted with 6 degrees of freedom and identity scale matrix, and $N(0,10^3)$ priors on the μ_j. The second half of a two-chain run of 10 000 iterations results in a correlation matrix

$$R = \begin{bmatrix} 1 & 0.71 & 0.51 & -0.02 & -0.40 & -0.69 \\ 0.71 & 1 & 0.56 & 0.14 & -0.30 & -0.54 \\ 0.51 & 0.56 & 1 & 0.29 & -0.10 & -0.36 \\ -0.02 & 0.14 & 0.29 & 1 & 0.24 & 0.19 \\ -0.40 & -0.30 & -0.10 & 0.24 & 1 & 0.47 \\ -0.69 & -0.54 & -0.36 & 0.19 & 0.47 & 1 \end{bmatrix}$$

showing positive correlations for adjacent grades, especially the first and second. The centred means have posterior means (with standard deviations) similar to those cited by Leonard and Hsu, namely -0.47 (0.2), 0.46 (0.1), 0.61 (0.1), 0.61 (0.1), -0.90 (0.17) and -0.31 (0.17). The smoothing is illustrated by school 22 with the lowest student total (14) students and $f_{22} = (2, 3, 6, 3, 0, 0)$ but with posterior mean $\rho_{22} = (0.132, 0.239, 0.317, 0.221, 0.039, 0.052)$.

Using chi-square deviations to measure discrepancies, the more unusual schools in terms of the model are those either with high proportions in grade 1 or those with high proportions in grades 5 or 6. A predictive posterior check on model adequacy may be used comparing the usual chi-square statistic $C(f, \hat{f})$ (where $\hat{f}_{ij} = \rho_{ij}f_{i+}$) with that based on sampling replicate data, namely $C(f_{\text{new}}, \hat{f})$. The value 0.67 so obtained indicates an adequate model.

Example 6.3 Parental style Epstein and Fienberg (1991) consider a two-way table (Table 6.2) relating to parental decision making (authoritarian vs. democratic) and political affiliation (SDS vs. YAF).

They specify N(0.5, 4) priors for both δ_{1i} and δ_{2j} under the independence prior for the Dirichlet means, namely

$$\alpha_{ij} = \exp(\delta_0 + \delta_{1i} + \delta_{2j}) / \sum_i \sum_j \exp(\delta_0 + \delta_{1i} + \delta_{2j})$$

with constraint

$$\delta_0 = -\log\left\{ \sum_{ij} [\exp(\delta_{1i} + \delta_{2j})] \right\}$$

Table 6.2 Parental style and political affiliation

	SDS	YAF	All
Authoritarian	29	33	62
Democratic	131	78	209
All	160	111	271

Table 6.3   Contingency table probabilities for different K

K	0	100	200	400	600	1000	2000	∞
π_{11}	0.107	0.115	0.119	0.125	0.130	0.126	0.135	0.133
π_{12}	0.122	0.115	0.110	0.105	0.102	0.098	0.100	0.093
π_{21}	0.483	0.474	0.471	0.469	0.468	0.463	0.453	0.459
π_{22}	0.288	0.296	0.300	0.301	0.300	0.313	0.312	0.315

They take various values of K and compare the resulting estimates of π_{ij} (see Table 6.3).

Here a prior on K is added, obtained as $K = 1/L$ where L is uniform between zero and a large number (e.g. 10^7). A two-chain run of 50 000 iterations (with K converging after 10 000) gives mean K of 254 and median 70. This suggests a model intermediate between the independence and saturated models. π is estimated as (0.115, 0.115, 0.475, 0.295). The DIC is 21.6 with 2.2 effective parameters. A predictive check probability of 0.41 based on a chi-square calculation for actual and replicate data suggests that this approach provides an adequate model.

6.4   THE MULTINOMIAL PROBIT

Independence between choices is often not appropriate, as Example 6.2 illustrates. Among the limitations of the MNL forms considered in section 6.2 in analysing individual choice data are inflexibility in the face of correlated choices (and substitutability between choices) and heterogeneity in predictor effects. The multinomial probit (MNP) model seeks especially to reflect interdependent choices, but may be extended to reflect heterogeneity (Glasgow, 2001). It starts with the random utility model of (6.1) but involves reframing it to obtain identifiability. Thus in a generalization of the utility comparison underlying dichotomous choice (Chapter 4), the utility for choice j has systematic and stochastic components according to

$$U_{ij} = \eta_{ij} + \varepsilon_{ij} = X_i\beta_j^* + C_{ij}\gamma^* + A_j\varphi^* + \varepsilon_{ij}$$

where $y_{ij} = 1$ if $U_{ij} = \max(U_{i1}, U_{i2}, \ldots, U_{iJ})$. Since the density $y \mid X, C, A$ is unchanged by adding a scalar random variable to U_{ij}, identifiability in terms of location requires differencing against the utility of a reference category, such as the Jth (Geweke *et al.*, 1994). So the latent utilities which are modelled are differences

$$W_{ij} = U_{ij} - U_{iJ}, \quad j = 1, \ldots, J - 1$$

giving $J - 1$ latent variables. So $y_{ij} = 1$ $(j = 1, \ldots, J - 1)$ if both $W_{ij} = \max(W_{i1}, W_{i2}, \ldots, W_{iJ-1})$ and $W_{ij} > 0$. If the observed choice is J $(y_{iJ} = 1)$ then all the W_{ij} are negative. Note that this redefinition entails taking differences in the predictors C_{ij} and A_j; so for instance the model now involves differenced predictors $c_{ij} = C_{ij} - C_{iJ}, j = 1, \ldots, J - 1$.

The augmented data W_{ij} may be sampled in the same way that the latent scale is sampled in binary regression except that they will be correlated (Albert and Chib, 1993, p 673). Under the multinomial probit, an MVN linear regression with correlated errors (as in Chapter 3) is assumed for the latent W_{ij}. For example, a regression with both chooser characteristics and subject-specific choice measures has the form

$$W_{ij} = \beta_j X_i + \gamma c_{ij} + \varepsilon_{ij} \tag{6.7}$$

where

$$\varepsilon_i \sim N_{J-1}(0, \Sigma)$$

and $\varepsilon_i = (\varepsilon_{i1}, \varepsilon_{i2}, \ldots, \varepsilon_{i, J-1})$. Unlike the MNL models considered in section 6.2, the correlation among the choices induced by the error structure in (6.7) means that the restrictive IIA no longer holds. So one may assess how far voters view political parties as similar or different (Glasgow, 2001), or allow for substitution between travel modes (Daganzo, 1979). Scale mixing may also be used, as in the simple probit (see equation 4.5) to allow robustness to aberrant observations. Then

$$\varepsilon_i \sim N_{J-1}(0, \Sigma/\lambda_i)$$

where $\lambda_i \sim Ga(\nu/2, \nu/2)$ and ν is a degrees of freedom parameter (Linardakis and Dellaportas, 2003).

Despite differencing via $W_{ij} = U_{ij} - U_{iJ}$, there is still an issue of unique scale, since multiplying (6.7) through by a constant leaves the likelihood unchanged. In general one element of Σ must be preset, leaving $[J(J-1)/2] - 1$ free parameters (Glasgow, 2001; Albert and Chib, 1993). Setting the first variance term in Σ to one is the most common strategy, though this leads to more complex sampling than an unrestricted Σ. Let this first variance term be denoted σ_{11}, the variance for ε_{i1}, with $\sigma_1 = \sigma_{11}^{0.5}$.

McCulloch and Rossi (1994) propose a Gibbs sampling scheme that involves an unrestricted Σ but monitors the identified parameters, such as the regression parameters, $\tilde{\beta}_j = \beta_j/\sigma_1$ and $\tilde{\gamma} = \gamma/\sigma_1$, the scaled covariances

$$\tilde{\Sigma}_{jk} = \Sigma_{jk}/\sigma_1$$

and hence the correlations between the errors. Yu (2000) reports the success of such a strategy in an application to psychometric rankings. Specifying a prior on β_j, γ and the unrestricted error covariance matrix means that the prior on the identified parameters is induced. To ascertain the form of the induced prior might involve simulation using the observations without actual updating with the response data y. Problems may occur with informative priors on the unidentified parameters, though Nobile (1998) suggests a strategy to improve convergence under the unrestricted Σ approach.

Schemes with fully identified prior for the MNP have also been proposed (McCulloch *et al.*, 2000; McCulloch and Rossi, 2000; Lahiri and Gao, 2002). Thus from the properties of the multivariate normal Σ may be written as a partitioned matrix

$$\Sigma = \begin{bmatrix} \sigma_{11} & \omega \\ \omega & \Phi \end{bmatrix} \tag{6.8}$$

where σ_{11} is the variance of ε_{i1}, and the $J - 2$ parameters ω define the covariance between ε_{i1} and the errors $\eta_i = (\varepsilon_{i2}, \varepsilon_{i3}, \ldots, \varepsilon_{i,J-1})$. Then for sampled ε_{i1}, the η_i are N_{J-2} with covariance Φ and means $(\omega/\sigma_{11})\varepsilon_{i1}$. Taking $\sigma_{11} = 1$ reduces the unknown parameters to a $(J - 2)$-dimensional 'regression' parameter ω and the $(J - 2)(J - 1)/2$ parameters of Φ.

Another method proposed by Chib *et al.* (1998) also sets $\sigma_{11} = 1$ and uses a prior based on a Choleski decomposition to represent the free elements in Σ. Thus let

$$\Sigma = HH'$$

where H is a $(J - 1) \times (J - 1)$ lower triangular matrix with $h_{11} = 1$. For example, with $J = 3$, H would be a 2×2 matrix with first row (1 0), and with second row (h_{21}, h_{22}), so that $\sigma_{11} = 1, \sigma_{12} = \sigma_{21} = h_{21}$ and $\sigma_{22} = h_{21}^2 + h_{22}^2$. Letting $\psi = (h_{21}, \log h_{22}, h_{31}, h_{32}, \log h_{33}, \ldots, h_{J-1,1}, \ldots, \log h_{J-1,J-1})$ priors may be in the form of unrestricted normal densities on ψ_{jk}.

Heterogeneity in the multinomial probit may be introduced via

$$W_{ij} = \beta_j X_i + \gamma c_{ij} + \varphi_i Z_{ij} + \varepsilon_{ij}$$

where Z_{ij} is a vector of predictors the impact of which varies over subjects according to the subject-level random effect φ_i. If Z_{ij} contains just the intercept then a nested error structure is obtained. The parameters of the MNP, especially its heterogeneous version, may be weakly identified by the data. The W_{ij} are unobserved, and if themselves modelled in terms of random effects such as φ_i, weak identification is obtained. Even the standard MNP may yield imprecise inferences about the covariance matrix Σ for small data sets. Chib *et al.* (1998) contrast the possible identifiability problems with the MNP model as against the simple MNL model (without random effects). Such identifiability problems apply to mixed MNL models also (section 6.6) and occur in classical estimation as well as Bayesian sampling-based estimation (Keane, 1992; Greene, 2000).

It may be noted that latent utilities W_{ij} under the MNL model analogous to those in (6.6) may be generated by assuming that ε_{ij} are sampled from a type I extreme value density (McFadden, 1974), as opposed to the MVN error density for the MNP model. Consider the MNL model as

$$\Pr(D_i = j) = \pi_{ij} = \lambda_{ij} / \left(1 + \sum_{k=1}^{J-1} \lambda_{ik} \right) \propto \lambda_{ij} \qquad j = 1, \ldots, J-1$$

$$\Pr(D_i = J) = \pi_{iJ} = 1 / \left(1 + \sum_{k=1}^{J-1} \lambda_{ik} \right)$$

where, for example, $\lambda_{ij} = \exp(\beta_j X_i + \gamma c_{ij})$ or $\lambda_{ij} = \beta_j X_i + \gamma c_{ij} + \eta_i Z_{ij}$ (for $j < J$) and $\lambda_{iJ} = 1$ for identifiability. Scott (2003) proposed sampling exponential variables $Z_{ij} \sim E(\lambda_{ij})$ with $Z_{ij} = \min(Z_{i1}, \ldots, Z_{iJ})$ when $D_i = j$. It is necessary to sample $Z_{iJ} \sim E(1)$ to ensure the scale is defined. If $D_i = \text{argmin}_k(Z_{ik})$ then $\Pr(D_i = j)$ is proportional to λ_{ij} and so follows the MNL model.

6.5 NON-LINEAR PREDICTOR EFFECTS

A few authors have considered how to modify MNL or MNP models to reflect non-linear impacts of predictors. For example, Abe (1999) proposes a form of conditional logistic model (see Chapter 4) with each choice occasion or consumer defining a stratum. Examples of non-linear effects include a region of price variation within which consumers are relatively insensitive, whereas once a certain threshold is exceeded sensitivity to price changes is noticed. While polynomials in, say, cost or income may be used in consumer applications to model non-linearity, a more flexible approach is provided by non-parametric regression involving spline or GAM methods. For example, the MNL model

$$\Pr(y_{ij} = 1) = \pi_{ij} = \exp(\eta_{ij}) / \sum_{k=1}^{J} \exp(\eta_{ik})$$

where

$$\eta_{ij} = \beta_j X_i + \gamma C_{ij} \quad j < J$$
$$\eta_{iJ} = \gamma C_{iJ}$$

may be replaced by one in which some predictor effects are modelled by smooth functions. So for $j < J$ one might assume that two attributes X_{i1}

and X_{i2} follow a linear effect, while the impacts of X_{i3} and C_{ij} are modelled as smooth functions. The smooth in the attribute would be specific to the choice category j, but not the smooth in C_{ij}, so that

$$\eta_{ij} = \beta_{0j} + \beta_{1j}x_{i1} + \beta_{2j}x_{i2} + S_{1j}(x_{i3}) + S_2(C_{ij})$$

As for other multiple choice models, identifiability is an issue if, for example, subjects at higher levels of X_3 are mostly confined to a subset of the range of choices $j = 1, \ldots, J$; for instance, higher income consumers may be confined to a few brands.

Example 6.4   Occupation attainment (continued)   To illustrate non-linear regression effects in the occupation choice data, let the impact of experience assume a conventional linear form but that of years of education be modelled by a penalized random effects quadratic spline (Ruppert *et al.*, 2003) with

$$S(x_{i3}) = \delta_1 x_{i3} + \delta_2 x_{i3}^2 + \sum_{k=1}^{K} \phi_k (x_{i3} - t_k)_+^2$$

where $\phi_k \sim N(0, \sigma_\phi^2)$ and $1/\sigma_\phi^2$ is assigned a Ga(0.1,0.1) prior. $K = 6$ knots are taken at $t = \{9.6, 12, 13.6, 14, 16, 16.4\}$; note that the 20th, 30th, 40th and 50th percentiles of education are all 12 years. This model yields a slight improvement in average deviance (minus twice the log likelihood), i.e. 864 versus 871 for the MNL model in Example 6.1. However, there are 26 effective parameters compared with 16 under the

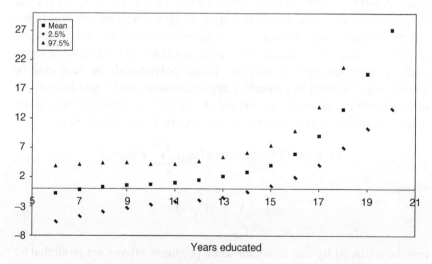

Figure 6.1   Smooth education effect for professional jobs

linear effects MNL so the DIC is slightly worse at 890. Plots of the effects of education years on allocation to professional, white-collar and blue-collar jobs suggest non-linear effects. The posterior means $S(x_{i3})$ and their 95% intervals are in Figures 6.1 to 6.3.

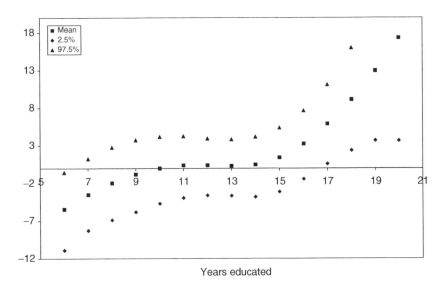

Figure 6.2   Smooth education effect for white-collar jobs

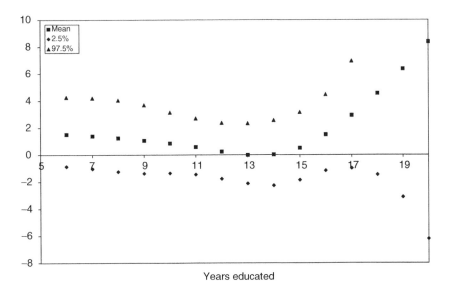

Figure 6.3   Smooth education effect for blue-collar jobs

6.6   HETEROGENEITY VIA THE MIXED LOGIT

As mentioned above, there may be heterogeneity in the impact of predictors and interdependent choices, and discrete or continuous mixture models may be applied to model such effects. The mixed MNL model is an extension of the MNL model that includes heterogeneity between subjects. Such heterogeneity may represent variation in tastes and valuations (in consumer applications) or risk aversion (in mobility applications), or ambivalence (in voting or attitudinal applications). There may also be unobserved aspects of utility for alternative choices (Glasgow, 2001).

In the mixed MNL model, heterogeneity is defined in terms of random effects for certain regression coefficients or for intercepts. Variation in regression effects between subjects may in theory apply to intercepts and all predictors, but such a model is likely to strain identifiability and instead heterogeneity is confined to a subset of predictors that may include the intercept. Consider the MNL with likelihood

$$y_i \sim M(1, \pi_i)$$

where $y_i = (y_{i1}, y_{i2}, \ldots, y_{iJ})$, $\pi_i = (\pi_{i1}, \pi_{i2}, \ldots, \pi_{iJ})$ and

$$\pi_{ij} = \exp(X_i \beta_j + C_{ij}\gamma) / \sum_{k=1}^{K} \exp(X_i \beta_k + C_{ij}\gamma)$$

Suppose now that random variability is introduced in one or more coefficients. For example, this might be in the coefficient γ_m for a predictor C_{mij}, such that for subjects i

$$U_{ij} = \eta_{ij} + \omega_{ij} = X_i \beta_j + C_{1ij}\gamma_1 + \cdots + C_{mij}\gamma_{im} + \cdots + \omega_{ij} \qquad (6.9a)$$

where the heterogeneity model itself potentially includes systematic as well as random elements

$$\gamma_{im} = \gamma_m + \eta_m H_i + \varepsilon_{im} \qquad (6.9b)$$

The H_i are known subject attributes that may be relevant to explaining heterogeneity in the coefficients γ_{im}.

For coefficients that are not choice specific ε_{im} may be a univariate density. Multivariate densities on ε might apply if, say, coefficients γ_m and γ_k on two of the predictors C were subject to random variation, or if certain β_{jm} were to vary randomly. While normal priors for variability in the regression deviations γ_{im} are possible, they may allow infeasibly large deviations from the average coefficient γ_m and other options may be substantively preferable. For example, Glasgow (2001) considers heterogeneity in the impact of union membership on voting in the 1992 US presidential election, and suggests triangular densities that are zero beyond end points $[m - a, m + a]$ and descend linearly to a peak at m.

Consider an example for coefficients that vary over choices (such as the intercept). Suppose $J = 4$ and there is heterogeneity (taken to be normal) in the intercepts; then for $j = 1, 3$

$$\eta_{ij} = \beta_{0j} + X_i\beta_j + C_{ij}\gamma + u_{ij} \qquad (6.10a)$$

where $u_i = \{u_{i1}, u_{i2}, u_{i3}\}$ is MVN with mean 0 and covariance matrix Σ, while for $j = 4$

$$\eta_{i4} = C_{i4}\gamma \qquad (6.10b)$$

Note that instead of adopting a full multivariate error model one might assume $Q < J - 1$ factors and model the correlations between the u_i in terms of factor loadings; this may be more tractable, i.e. less subject to identifiability problems (McFadden and Train, 2002). Thus for $Q = 1$, (6.10) becomes

$$\eta_{ij} = \beta_{0j} + X_i\beta_j + C_{ij}\gamma + \lambda_j u_i \quad j = 1, 2, 3$$
$$\eta_{i4} = C_{i4}\gamma$$

where u_i has a set scale (e.g. $u_i \sim N(0, 1)$).

Example 6.5 Travel mode choice This data set consists of mode choice observations ($j = 1, \ldots, 4$ and respectively air, train, bus, car) for travel between Sydney and Melbourne. For 210 subjects, there is a single chooser attribute (household income X_{i1}) and two choice attributes that are subject specific. These are C_{1ij} for generalized cost (GC) of travel mode j for subject i, derived by considering in-vehicle cost and a wage measure times time travelled; and C_{2ij} for terminal time (TT), which is zero for cars by definition (i.e. $C_{2i4} = 0$). Greene (2000) considers a utility function in which the dependence of air travel choice on household income is compared with all other modes together, so that

$$\eta_{i1} = \beta_{01} + X_{i1}\beta_{11} + C_{1i1}\gamma_1 + C_{2i1}\gamma_2$$
$$\eta_{i2} = \beta_{02} + C_{1i2}\gamma_1 + C_{2i2}\gamma_2$$
$$\eta_{i3} = \beta_{03} + C_{1i3}\gamma_1 + C_{2i3}\gamma_2$$
$$\eta_{i4} = C_{1i4}\gamma_1$$

where $\beta_{04} = 0$ for identifiability.[2] By contrast, the Chen and Kuo (2002) model for these data compares the impact of household income for

---

[2] An alternative scheme defines intercepts $\kappa_1, \kappa_2, \kappa_3, \kappa_4$ for all $J = 4$ choices but imposes the constraint $\sum \kappa_j = 0$ (i.e. three effective parameters).

air travel against the car option, train against car, and bus against car, so that

$$\eta_{i1} = \beta_{01} + X_i\beta_{11} + C_{1i1}\gamma_1 + C_{2i1}\gamma_2$$
$$\eta_{i2} = \beta_{02} + X_{i1}\beta_{12} + C_{1i2}\gamma_1 + C_{2i2}\gamma_2$$
$$\eta_{i3} = \beta_{03} + X_{i1}\beta_{13} + C_{1i3}\gamma_1 + C_{2i3}\gamma_2$$
$$\eta_{i4} = C_{1i4}\gamma_1$$

A simple conditional logit model for the latter specification, conforming to the possibly unrealistic IIA assumption, is fitted using Code A in Program 6.5. A two-chain run, using null start values and values from a trial run, shows early convergence (at around 500 iterations in a run of 2500). The DIC for this model, with a known total of eight parameters, is 395, and the regression parameter estimates show that lower income households are significantly more likely to choose train travel (mean $\beta_{11} = -0.059$ with s.d. $= 0.013$). Such households are also more likely to choose bus travel (with $\beta_{13} = -0.030$, s.d. $= 0.014$). Increased generalized cost and terminal time for a mode are associated with reduced chances of selecting that mode, with $\gamma_1 = -0.011$ (0.004) and $\gamma_2 = -0.099$ (0.011).

Next, two types of mixed logit model are considered. One considers random intercept variation with the model

$$\eta_{i1} = b_{i1} + X_{i1}\beta_{11} + C_{1i1}\gamma_1 + C_{2i1}\gamma_2$$
$$\eta_{i2} = b_{i2} + X_{i1}\beta_{12} + C_{1i2}\gamma_1 + C_{2i2}\gamma_2$$
$$\eta_{i3} = b_{i3} + X_{i1}\beta_{13} + C_{1i3}\gamma_1 + C_{2i3}\gamma_2$$
$$\eta_{i4} = C_{1i4}\gamma_1$$

where $b_i = \{b_{i1}, b_{i2}, b_{i3}\}$ is taken as MVN with means $\{\beta_{01}, \beta_{02}, \beta_{03}\}$. The matrix partition method (equation 6.8) is used to estimate the covariance among the b_{ij}, with $b_{i1} \sim N(\beta_{01}, 1)$. To improve identifiability, priors for β_{0j} and β_{1j} are based on the posterior means and standard deviations of the baseline categorical logit model, but with downweighting of precision by 100. A two-chain run of 50 000 iterations shows convergence after 25 000 iterations and similar results on the impacts of the GC and TT variables to the baseline categorical logit model. It also suggests slightly greater variability in b_{ij} for airline passengers, with $\sigma_1 = 1$ against $\sigma_2 = 0.9$ and $\sigma_3 = 0.7$. The DIC suggests that this model provides no marked gain over the simple conditional logit – the mean deviance improves (344 vs. 387) but the effective parameter count is around 53 compared with 8 under the simple model, so the DIC remains around 395. The two ω parameters in (6.8) are not well identified, suggesting that b_{i2} and b_{i3} could be taken as independent.

By contrast a mixed logit model with varying individual coefficients on the GC and TT variables, as in (6.9), shows a clear gain in fit. Independent normal priors are assumed on γ_{i1} and γ_{i2}, specifically $\gamma_{i1} \sim N(\Gamma_1, \tau_1)$ and $\gamma_{i2} \sim N(\Gamma_2, \tau_2)$. Priors for Γ_1 and Γ_2 are based on the posterior means and standard deviations of the baseline categorical logit model, but with downweighting of precision by 100; specifically

$$\Gamma_1 \sim N(-0.011, 0.002), \Gamma_2 \sim (-0.1, 0.01)$$

Gamma $G(a_j, 1)$ priors are assumed for $1/\tau_j$ with parameters a_j taken to have an approximately uniform prior, namely $a_j \sim G(1, 0.001)$.

A two-chain run of 10 000 iterations shows convergence after 5000 iterations and an enhanced central effect for both general cost and terminal time, with Γ_1 and Γ_2 having respective means -0.043 (s.d. $= 0.013$) and -0.30 (0.05). Variability in the travel time effect is greater than that in the cost effect ($\tau_1 = 0.006$ vs. $\tau_2 = 0.14$). Note that results for this model may be sensitive to the priors adopted, with less informative priors on Γ_j leading to greater departure from the homogeneous effects logistic. The DIC is considerably reduced compared with the varying intercept model, to around 230.

For the MNP applied to these data, a Cholesky decomposition prior is used with an N(0,1) prior for the elements of $h = [h_{21}, \log h_{22}, h_{31}, h_{32}, \log h_{33}]$. A two-chain run of 5000 iterations shows convergence from 2500 iterations. The DIC improves on the MNL model standing at 295 (with 18 effective parameters). As for the MNL models, lower income households are more likely to choose train travel (the mean of β_{12} is -0.035 with s.d. 0.012). However, the income effect on bus travel is no longer apparent. Again increased generalized cost and terminal time reduce chances of selecting a mode: thus $\gamma_1 = -0.0066$ (with s.d. $= 0.002$) and $\gamma_2 = -0.047$ (s.d. $= 0.008$). The largest correlation is -0.45 between air and bus, but is not significant in the sense of its 95% interval being confined to negative values. The standard deviation (square root of diagonal element of Σ) for bus is slightly lower (0.79) than for train (1.17) and air (1 by default).

6.7 AGGREGATE MULTICATEGORY DATA: THE MULTINOMIAL–DIRICHLET MODEL AND EXTENSIONS

Consider again the problem of analysing aggregated multinomial data rather than individual choice or allocation data. Often administrative data may be in this form for confidentiality reasons, while large-scale national data (e.g. distributions between income or demographic groups) can only be analysed by aggregating over individuals. Analysis techniques for multinomial discrete observations, such as electoral data (votes by party) or religious affiliation data, may overlap with those for purely compositional data (namely vectors of proportions describing the composition of the whole in terms of J components). Examples of the latter type of data include composition of river water by source (Soulsby *et al.*, 2003) and species composition data (Billheimer *et al.*, 2001).

Let f_{ij} denote the total counts of observations in units $i = 1, \ldots, I$ which may belong to one of $j = 1, \ldots, J$ categories, ranked or unranked. Examples of unranked category data include multiparty electoral data (Katz and King, 1999), while ranked data examples include job satisfaction category j by income group i (Chen *et al.*, 2000), or housing satisfaction in survey neighbourhoods (Wilson, 1989). The conjugate multinomial–Dirichlet model for such data may be inflexible in terms of modelling correlations between categories and may be modified in several ways. Possibilities include modelling the Dirichlet hyperparameters (e.g. in terms of predictors or higher stage priors), specifying mixture densities on transformations of the probabilities π_{ij}, or adopting compositional analysis techniques (e.g. converting J proportions to $J - 1$ odds ratios).

Let $N_i = \sum_j f_{ij}$ denote the total observations for unit i. A baseline model for the data $f_i = \{f_{i1}, \ldots, f_{iJ}\}$ assumes they are independent multinomial distributions of dimension J

$$f_i \sim M_J(N_i, \pi_i)$$

where the probability parameter vectors $\pi_i = (\pi_{i1}, \ldots, \pi_{iJ})'$ model the relative frequencies in each unit subject to $\pi_{iJ} = 1 - \sum_{j=1}^{J-1} \pi_{ij}$. The conjugate model involves Dirichlet priors on each of the I sets of probabilities. Thus for unit i

$$P(f_{i1}, f_{i2}, \ldots, f_{iJ} \mid \pi_i) = \frac{N_i!}{f_{i1}! f_{i2}! \ldots f_{iJ}!} \pi_{i1}^{f_{i1}} \pi_{i2}^{f_{i2}} \ldots \pi_{iJ}^{f_{iJ}}$$

$$\pi_i \sim \text{Dir}(\alpha_{i1}, \alpha_{i2}, \ldots, \alpha_{iJ})$$

where default values such as $\alpha_{ij} = 1$ or $\alpha_{ij} = 0$ for all i and j are often used. This will yield a Dirichlet posterior for π_i with means

$$P_{ij} = \text{E}(\pi_{ij} \mid f_i) = \frac{\alpha_{ij} + f_{ij}}{\sum_j (\alpha_{ij} + f_{ij})}$$

variances $V(\pi_{ij} \mid f_i) = P_{ij}(1 - P_{ij})/[1 + \sum_j(\alpha_{ij} + f_{ij})]$ and covariances $\text{Cov}(\pi_{ij}, \pi_{ik} \mid f_i) = -P_{ij}P_{ik}/[1 + \sum_j(\alpha_{ij} + f_{ij})]$.

6.8 MULTINOMIAL EXTRA VARIATION

Often extra variation occurs relative to this baseline model, owing for instance to clustering of high rates in certain observations i on some of the J options. Overdispersion may be expressed in terms of a scaled covariance matrix under a model with predictions of f_{ij} given by $\mu_{ij} = \pi_{ij}N_i$. Writing $\Pi_i = \text{diag}(\pi_{ij})$ as a $J \times J$ diagonal matrix, the covariance matrix for $f_i = (f_{i1}, f_{i2}, \ldots, f_{iJ})'$ is

$$\text{E}[(f_i - \mu_i)(f_i - \mu_i)'] = \phi\mu_i(\Pi_i - \pi_i\pi_i')$$

where $\phi = 1$ for multinomial data and $\phi > 1$ for overdispersion. The factor ϕ may be estimated by calculating the usual Pearson chi square between f_{ij} and μ_{ij} and dividing by the degrees of freedom (McCullagh and Nelder, 1989, p 175; Mebane and Sekhon, 2004). As noted above, an alternative view of the multinomial is as the joint density of $\{f_{i1}, \ldots, f_{iJ}\}$, given that the total count $\sum_j f_{ij} = N_i$ in cluster i is Poisson, with parameters $\pi_{ij} = \mu_{ij}/\mu_i$. Overdispersion will be apparent when $\text{Var}(N_i) > \text{E}(N_i)$.

For example, in research on different types of road accident at different monitoring sites (Bolduc and Bonin, 1998), $J = 3$ categories were distinguished according to day of the week (Example 6.6). There may be subsets of sites with sufficiently high weekend rates (both temporal and spatial clustering may also be involved) such as to invalidate the multinomial assumptions. In data of this type there may also be extra variation because some sites have unusually high overall totals N_i.

In cases such as the latter where N_i is not preset, one might allow variability in the overall rate μ_i (e.g. using a gamma density to represent variability in the Poisson rate), and adopt simplified models for the multinomial choice or allocation probabilities. For instance, if the allocation probabilities π_{ij} are taken as equal for all sites in the accident application above, the rate μ_{ij} for accident type j at site i is modelled as the product $\mu_i\pi_j$.

Whether the N_i are taken as given or themselves taken as stochastic, there may, however, be heterogeneity over subjects in the allocation probabilities. One way to model heterogeneity in the π_{ij} is in terms of the multiple logit link. Thus let η_{ij} be defined by

$$\pi_{ij} = \exp(\eta_{ij}) / \sum_k \exp(\eta_{ik}) \qquad (6.11)$$

One might take the η_{ij} to follow a hyperdensity that aims for 'global smoothing' towards the population-wide proportions π_j across sites. Leonard and Hsu (1994) propose an MVN model for the η_{ij} with mean $\mu = (\mu_1, \mu_2, \ldots, \mu_J)$ and covariance matrix Σ_η of dimension J. A sum to zero constraint on the μ_j is needed for identification (see Example 6.6). Heterogeneity across sites in the π_{ij} may also involve expressing the η_{ij} in regression terms, using predictors specific to subject i, choice j, or both. For example, in the accident analysis example, certain sites may be closer to weekend sports or entertainment venues, where specific dates for events during the year generate extra variation.

A compositional model appropriate for relatively large samples in each category involves taking logs of percentage ratios to a reference category (such as the Jth), so that

$$r_{ij} = \log(f_{ij}/f_{iJ})$$
$$= \log(p_{ij}/p_{iJ}) \quad j = 1, \ldots, J - 1$$

where $p_{ij} = f_{ij}/\sum_j f_{ij}$. Then the $J - 1$ dimensional data $r_i = (r_{i1}, r_{i2}, \ldots, r_{i,J-1})$ are modelled via a multivariate form such as the MVN or multivariate t (Aitchison, 1986), e.g.

$$r_i \sim N_{J-1}(\mu_i, \Sigma)$$

where

$$\mu_{ij} = \beta_j X_{ij}$$

and X_{ij} is a vector of predictors specific to clusters and choices (e.g. constituencies and parties). The original p_{ij} are obtained from

$$p_{ij} = \exp(r_{ij}) / \left[1 + \sum_{j=1}^{J-1} \exp(r_{ij}) \right]$$

Like the Leonard and Hsu model this allows a flexible covariance structure between categories, but assumptions (e.g. of multivariate normality) may not be applicable to real-life data. Katz and King (1999) discuss sources of non-normality in voting composition data and propose as a robust alternative an additive logistic Student distribution for r_{ij}. This may be attained by scale mixing as well as by formally taking

$$r_i \sim T_{J-1}(\mu_i, \Sigma, \nu)$$

6.8.1 Uncertainty in Dirichlet parameters

An alternative way to model heterogeneity in the π_{ij} is to allow for uncertainty in Dirichlet hyperparameters $\alpha = (\alpha_1, \alpha_2, \ldots, \alpha_J)$ rather than take them as known (e.g. as in an assumption $\alpha_j = 1$ for all j). This leads to the multinomial version of the beta–binomial model for binomial extra variation. Thus one may assume

$$f_i \mid \pi_i \sim M_J(N_i, \pi_i)$$
$$\pi_i \mid \alpha \sim \mathrm{Dir}(\alpha_1, \alpha_2, \ldots, \alpha_J) \qquad (6.12)$$
$$\alpha_j \sim G(a_j, b_j)$$

with $\{a_j, b_j\}$ known. A prior on $\log(\alpha_j)$ might also be used. Posterior probabilities of higher or lower than average rates $\rho_j = \alpha_j / \sum_j \alpha_j$ of each type of outcome can then be estimated by counting iterations:

$$H_{ij} = \Pr(\pi_{ij} > \rho_j) = \sum_{t=1}^{T} 1(\pi_{ij}^{(t)} > \rho_j^{(t)})/T \qquad (6.13)$$

The Dirichlet prior on π_i in (6.12) may be implemented indirectly using a set of gamma priors (see Gelman *et al.*, 1995, p 482; also Example 6.6). Thus for sampled values of $\alpha_1, \alpha_2, \ldots, \alpha_J$, sample $\psi_{ij} \sim \mathrm{Ga}(\alpha_j, 1)$ and then obtain π_{ij} as $\pi_{ij} = \psi_{ij}/\sum_j \psi_{ij}$.

Nandram (1998) proposes the form

$$\alpha_j = \rho_j \tau \qquad (6.14)$$

where $\rho = (\rho_1, \rho_2, \ldots, \rho_J)$ and $\sum \rho_j = 1$ represent the population-wide category proportions. The jth component of α_j then has prior mean π_j and variance

$$\alpha_j(\tau - \alpha_j)/[\tau^2(1 + \tau)] = \pi_j(1 - \pi_j)/(\tau + 1)$$

where $\sum \alpha_j = \tau$ is the total prior sample size. At the higher stage one may assume

$$\rho \sim \mathrm{Dirichlet}(R)$$

where the components of $R = (R_1 \ldots R_J)$ are taken to be known. The sample size parameter may take a gamma prior

$$\tau \sim \mathrm{Gamma}(t_1, t_2)$$

where t_1 and t_2 are taken as known (e.g. $t_1 = 1, t_2 = 0.001$). In the absence of subject matter or elicited knowledge regarding the R_j, a default such as $R_1 = R_2 = \cdots = R_J = 1$ may be taken.

One might, subject to identifiability, allow variation over observation units i in the Dirichlet parameters. For example, one might take

$$\pi_i \sim \text{Dir}(\alpha_{i1}, \alpha_{i2}, \ldots, \alpha_{ij})$$
$$\log(\alpha_{ij}) = \gamma_j x_{ij} + u_{ij} \qquad i = 1, \ldots, I, \; j = 1, \ldots, J-1$$

and consider multivariate unstructured priors (e.g. MVN) or structured priors for the u_{ij}. For spatial data, structured priors might allow spatial correlation parameters ρ_j for each outcome (Bolduc and Bonin, 1998). This approach resembles the Leonard–Hu model but retains the multi-nomial–Dirichlet structure. Alternatively

$$\pi_i \sim \text{Dir}(\tau \rho_{i1}, \tau \rho_{i2}, \ldots, \tau \rho_{iJ})$$

$$\rho_{ij} = \exp(\gamma_j x_{ij} + u_{ij}) / \left[1 + \sum_j^{J-1} \exp(\gamma_j x_{ij} + u_{ij}) \right] \qquad j < J$$

$$\rho_{ij} = 1 / \left[1 + \sum_j^{J-1} \exp(\gamma_j x_{ij} + u_{ij}) \right]$$

$$(6.15)$$

with τ as in (6.14).

Example 6.6   Quebec accident data   Bolduc and Bonin (1998) present accident totals for 90 intersection sites in Quebec for 1990–1993. Three types of accident are defined: accidents on Monday–Wednesday, accidents on Thursday or Friday, and weekend accidents. We first consider a fixed effects model with no pooling to global proportions or model for dispersion:

$$f_i \mid \pi_i \sim M_J(N_i, \pi_i)$$
$$\pi_i \mid \alpha_i \sim \text{Dir}(\alpha_{i1}, \alpha_{i2}, \ldots, \alpha_{iJ})$$

with $\alpha_{ij} = 1$ preset. This yields a deviance at $\bar{\theta} = (\bar{\pi})$ of 582 but has 185 parameters (DIC $= 953$), approximately equal to the number, i.e. 180, of free proportion parameters π_i. The predictions from this model are too overdispersed with $\Pr(\chi^2_{\text{new}} > \chi^2_{\text{obs}}) = 1$, where χ^2 is the usual Pearson statistic. A model as in (6.12) with three population-wide proportions to be estimated yields $\text{Dev}(\bar{\theta}) = 756$, where $\bar{\theta} = (\bar{\pi}, \bar{\alpha})$ and $\bar{\alpha} = (0.45, 0.34, 0.21)$. The parameter count is much reduced and the penalized fit improves despite the poorer fit at $\bar{\theta}$ ($p_e = 31$, DIC $= 818$). The predictive checks from this model are more satisfactory with

$\Pr(\chi^2_{\text{new}} > \chi^2_{\text{obs}}) = 0.22$, but suggest the model for heterogeneity could be improved.

By contrast the multiple logit model (6.11) with MVN effects η_{ij} generates more overdispersion than is consistent with the data, with $\Pr(\chi^2_{\text{new}} > \chi^2_{\text{obs}}) = 0.88$. For this model $p_e = 83$ and $\text{DIC} = 826$. The correlations obtained from Σ_η show that accident rates for Monday–Wednesday and for Thursday–Friday are positively correlated at 0.60 (95% interval 0.10 to 0.92).

The reparameterized Dirichlet model of (6.14) shows a closer consistency with the actual overdispersion with $\Pr(\chi^2_{\text{new}} > \chi^2_{\text{obs}}) = 0.36$. The DIC of 809 ($p_e = 31$) shows a parsimonious fit. Using the H_{ij} statistics in (6.13) one finds much higher than average weekend accident rates at sites 15 and 22 but low weekend rates at site 9. Finally a model for α_{ij}, namely

$$\alpha_{ij} = \tau \rho_{ij}$$

as in (6.15), is then applied, with a regression using intercepts only

$$\rho_{ij} = \exp(\gamma_j^c) / \sum_j^J \exp(\gamma_j^c)$$

with ρ_{ij} following the parameterization (6.2c). A two-chain run of 25 000 iterations produces a DIC of 813.5 ($p_e = 39$), a satisfactory check against the actual overdispersion with $\Pr(\chi^2_{\text{new}} > \chi^2_{\text{obs}}) = 0.38$, and a mean for τ of 77. Ideally, however, such a model would involve predictors or, for area clusters, a spatial interaction structure.

Example 6.7 Incumbency advantage Katz and King (1999) consider data from the 1997 UK election, namely proportions (for 526 English constituencies) voting Labour, Conservative or Liberal. These are converted to a 526×2 matrix of log odds ratios with the Liberal proportions as denominator, so that

$$r_{ij} = \log(p_{ij}/p_{iJ}) \qquad j = 1, \ldots, J - 1$$

where $J = 3$. The analysis investigates the benefits of incumbency on voting proportions: x_1 and x_2 are lagged values from the 1992 election of the two log odds ratios, while $x_3 = 1$, $x_4 = 1$ and $x_5 = 1$ according as the Conservative, Labour or Liberal candidate is the incumbent. The regression parameters are specific to the Conservative and Labour Parties so there are 12 regression parameters in all.

Since the log odds ratio data appear heavy tailed relative to the normal, Katz and King adopt a multivariate t distribution with unknown degrees of freedom. As a baseline, the first analysis here uses an MVN prior. To assess how far overdispersion in the data is reproduced, chi-square statistics over constituencies i and parties j are then calculated comparing (a) observed voting totals ($i = 1, \ldots, 526, j = 1, \ldots, 3$) with voting totals predicted by the model and (b) replicate voting totals (new data sampled from the model parameters) with the model predictions. A two-chain run of 1000 iterations shows early convergence. The model appears to generate predictions that are rather overdispersed relative to the observations: the posterior predictive check p value is 0.98. The incumbency effect for Conservative candidates, namely the coefficient of x_3 on r_{i1}, with mean 0.075 (95% interval -0.015 to 0.17), is weaker than that for Labour candidates. Thus the coefficient of x_4 on r_{i2} has mean 0.13 (0.03, 0.22).

A multivariate T model (MVT) model improves the predictive check slightly, but suggests scope for other modelling approaches: the p value is now just under 0.9. One feature of the MVT model is the reduced incumbency effect for Labour candidates. The mean value of ν, the degrees of freedom, i.e. 7 with 95% interval (5.4, 9.2), supports the need for a heavy-tailed model.

6.9   LATENT CLASS ANALYSIS

The other major approach to heterogeneity and overdispersed categorical data involves discrete mixture regression using the observed multinomial outcomes. This involves modifying the framework of section 5.3 to multinomial responses. Here we focus on a discrete mixture framework where the multinomial outcome is not observed. Latent class analysis is a particular form of discrete mixture model applied where a set of observed categorical or binary indicators X are taken to be imperfect measures of an underlying latent category variable G, or possibly of more than one latent categorization.

There may be a substantive basis for assuming the latent class variable is categorical, as in staging applications in psychological development studies. In medical or psychometric settings the X are typically a set of categorical symptom or diagnosis ratings. In this situation, there is often no gold standard way of classifying individuals, but instead a number of observed diagnostic measures subject to misclassification, and the goal is to identify the number of true diseased subjects. Thus in about a third of evaluation studies of medical diagnostic tests, there is no clear gold standard or reference test (Sheps and Schechter, 1984). In recent extensions of latent class models it is proposed that a continuous scale be envisaged underlying the latent category (Uebersax, 1999).

The most common model assumes that the interaction between the p manifest variables is explicable by a smaller set of latent categorical variables, such that given the level on the latent variable or variables, the manifest variables are independent (the conditional independence assumption). The most frequent applications assume either several (though fewer than p) binary latent categorical variables, or a single latent variable with several levels.

Assume p binary items are observed and that there is a single latent variable with C categories ($C \geq 2$) which in a medical application might be latent diagnosis or morbidity. The parameters of a latent class model in this setting are the allocation probabilities $\lambda_1, \ldots, \lambda_C$ of belonging to the different categories of the latent variable, and the item probabilities of a positive response δ_{ic} to the ith item ($c = 1, \ldots, C$; $i = 1, \ldots, p$) for someone allocated to the cth category of the latent variable. Suppose the latent variable has $C = 3$ categories. Then latent class analysis (LCA) under conditional independence implies the following:

$$\theta_1 + \theta_2 + \theta_3 = 1$$
$$P_i = \theta_1\delta_{i1} + \theta_2\delta_{i2} + \theta_3\delta_{i3} \qquad\qquad i = 1, \ldots, p$$
$$P_{ij} = \theta_1\delta_{i1}\delta_{j1} + \theta_2\delta_{i2}\delta_{j2} + \theta_3\delta_{i3}\delta_{j3} \qquad i, j = 1, \ldots, p$$
$$P_{ijk} = \theta_1\delta_{i1}\delta_{j1}\delta_{k1} + \theta_2\delta_{i2}\delta_{j2}\delta_{k2} + \theta_3\delta_{i3}\delta_{j3}\delta_{k3} \quad i, j, k = 1, \ldots, p$$

where P_i is the probability of a positive response to item i, P_{ij} is the joint probability of a positive response to both items i and j, etc.

Example 6.8   Antisocial behaviour   Muthén and Muthén (2000) present a latent class analysis of 17 binary indicators of antisocial behaviour (ASB) among American youth, using data from the National Longitudinal Survey of Youth and administered in 1980. Their analysis relates to 7326 subjects who provided complete response profiles on the 17 items. These items are all 'positive' indicators of ASB (whether or not the behaviour occurred in the previous year) such as fighting (1 = yes), using drugs (1 = yes) and taking an auto (1 = yes). The Muthén and Muthén analysis adopted $C = 4$ latent classes and shows a large 'low ASB' group (with probability 0.47) with low item probabilities on all items, and a small minority group (allocation probability 0.09) with high item probabilities on most crimes and ASB. Two minority groups (probabilities 0.18 and 0.28 respectively) are characterized by drug use (but no item probabilities exceeding 0.5) and by fighting and threatening behaviour.

Here a 10% subsample of the original survey respondents is taken ($n = 754$) and an unconstrained trial run of 1000 iterations (a 100 burn-in) shows a very similar pattern to that reported by Muthén and Muthén (Table 6.4).

Table 6.4 Item probabilities and allocation probabilities, trial and final estimates, antisocial behaviour data

Initial estimates	Class 1	Class 2	Class 3	Class 4
Property	0.05	0.30	0.27	0.92
Fighting	0.13	0.65	0.31	0.86
Shoplifting	0.10	0.32	0.51	0.79
Stole (under $50)	0.08	0.24	0.38	0.77
Stole (over $50)	0.01	0.08	0.05	0.47
Use of force	0.02	0.16	0.06	0.27
Threatening behaviour	0.17	0.91	0.44	0.77
Intent to injure	0.00	0.30	0.15	0.46
Took pot	0.29	0.42	0.94	0.90
Took other drug	0.07	0.03	0.56	0.65
Sold pot	0.01	0.02	0.29	0.58
Sold other drug	0.00	0.01	0.03	0.24
Confidence trick	0.14	0.37	0.35	0.61
Took auto	0.03	0.15	0.21	0.43
Broke into building	0.00	0.04	0.09	0.49
Held stolen goods	0.02	0.12	0.15	0.65
Gambling	0.01	0.02	0.03	0.25
Allocation probabilities	0.58	0.16	0.19	0.07

Final estimates (two chains)

	Class 1	Class 2	Class 3	Class 4
Property	0.05	0.30	0.25	0.92
Fighting	0.12	0.64	0.30	0.85
Shoplifting	0.10	0.30	0.49	0.80
Stole (under $50)	0.08	0.24	0.37	0.76
Stole (over $50)	0.01	0.07	0.04	0.46
Use of force	0.02	0.16	0.05	0.26
Threatening behaviour	0.16	0.90	0.43	0.77
Intent to injure	0.00	0.28	0.15	0.45
Took pot	0.29	0.41	0.93	0.90
Took other drug	0.06	0.03	0.55	0.64
Sold pot	0.00	0.02	0.27	0.57
Sold other drug	0.00	0.01	0.03	0.23
Confidence trick	0.14	0.37	0.34	0.61
Took auto	0.03	0.15	0.20	0.42
Broke into building	0.00	0.04	0.09	0.47
Held stolen goods	0.02	0.11	0.14	0.64
Gambling	0.01	0.02	0.03	0.23
Allocation probabilities	0.56	0.17	0.20	0.07

However, inferences from an LCA applied to this data set are subject to identifiability issues around consistent labelling of the latent classes through a large number of iterations. It is possible by chance that a unique labelling (with no switching) is obtained, typically in a relatively short single chain. However, in multiple chain sampling it is more than likely that different labelling schemes will emerge between one chain and another.

In the present situation the results of Muthén and Muthén (based on an EM method) suggest a clear high-probability group (group 1 in Table 6.4) and a low-probability group (group 4 in Table 6.4), and an identifiability constraint on λ_1 and λ_4 is applicable to reproduce this feature, involving constraints $\lambda_1 > \max(\lambda_2, \lambda_3)$ and $\lambda_4 < \min(\lambda_2, \lambda_3)$. However, the two other groups are similar in relative frequency and are identifiable in terms of their item probability profile, rather than the size of λ_2 versus λ_3. Table 6.4 shows that the third group has a much higher item probability on cannabis use and so an identifiability constraint that $\delta_{93} > \delta_{92}$ may be imposed.

The standard Dirichlet may be used for suitably constrained sampling of the λ_c in this problem but there are other ways. One takes a series of gamma priors, such as

$$\kappa_1 \sim G(1, 0.001)I(\kappa_h,)$$
$$\kappa_h = \max(\kappa_2, \kappa_3)$$
$$\kappa_2 \sim G(1, 0.001)I(\kappa_4, \kappa_1)$$
$$\kappa_3 \sim G(1, 0.001)I(\kappa_4, \kappa_1)$$
$$\kappa_4 \sim G(1, 0.001) \ I(, \kappa_g)$$
$$\kappa_g = \min(\kappa_2, \kappa_3)$$

and then sets $\lambda_c = \kappa_c / \sum_c \kappa_c$. Another approach (the one used here) sets normal priors on φ_c with $\varphi_1 = 0$, and V assumed known (i.e. $V = 1$),

$$\varphi_2 \sim N(0, V)I(\varphi_4, \varphi_1)$$
$$\varphi_3 \sim N(0, V)I(\varphi_4, \varphi_1)$$
$$\varphi_4 \sim N(0, V)I(, \varphi_g)$$
$$\varphi_g = \min(\varphi_2, \varphi_3)$$

and then takes $\kappa_c = \exp(\varphi_c)$, and $\lambda_c = \kappa_c / \sum_c \kappa_c$.

Unconstrained sampling is used for the second and third groups but an identifiability check, i.e. whether $\delta_{93} > \delta_{92}$, is applied. This is an alternative to a more rigorous cluster analysis procedure such as that of

Celeux *et al.* (2000), and will be suitable if the constraint holds (i.e. there is no label switching) within a long sampling run.

In a two-chain run of 5000 iterations we find the first set of initial values leads to an opposite labelling of groups 2 and 3 than the second set of initial values. After the first 50 iterations both chains are internally stable with regard to their labelling schemes (using the check whether $\delta_{93} > \delta_{92}$ just mentioned). So a consistent labelling over the two chains is achieved just by relabelling the output from one chain (i.e. relabelling group 2 to 3 and vice versa). Had there been label switching within either of the two chains then the full procedure of Celeux *et al.* would be necessary.

The final set of estimates in Table 6.4 shows a 'low ASB' group (with probability 0.56) with low item probabilities on all items, violent and drug use groups, and a 'high ASB' group with probability 0.07.

EXERCISES

1. In Example 6.1 use the data with education and experience grouped into five and seven categories (rather than in continuous form) and estimate the MNL model with data in this form using a multinomial likelihood as opposed to Poisson log-linear regression. Use the Albert (1996) method (see section 4.5) to compare this 'main effects only' log-linear model with one including interactions between ethnicity and education subject to a hierarchical mixture prior. How does the fit compare using the DIC and a predictive loss criterion based on sampling replicate counts Y_{ijkl} (i denotes ethnicity, j denotes education, k denotes experience, l denotes occupation)?

2. In Example 6.1 apply the exponential sampling model of Scott (2003) to generate latent data W_{ij} (i subjects, j occupations) underlying the MNL choice model.

3. Try fitting the analysis in Example 6.2 using a Poisson regression with dummy fixed effects parameters for schools.

4. In Example 6.2 fit the model via a multinomial likelihood but with scale mixing, namely
$$\gamma_i \sim N_J(\mu, \Sigma/\lambda_i)$$
where $\lambda_i \sim Ga(\nu/2, \nu/2)$ with ν unknown. Can outliers be identified?

5. In Example 6.4 try a cubic penalized spline for the effect of years of education. Compare its fit with that of a conventional MNL with linear regression effects for all predictors.

6. In Example 6.5 try the random intercept version of the mixed logit using a factor model ($Q = 1$). Does this improve in fit over the baseline conditional logit model?

7. In Example 6.7 try a two-group discrete mixture models

$$r_i \sim N_{J-1}(\mu_{iG_i}, \Sigma_{G_i})$$

where G_i is the latent group for constituency i and both the dispersion Σ_g and the regression model differ by group

$$\mu_{ijg} = \beta_{jg} X_{ij}$$

Assess how well this model represents overdispersion.

REFERENCES

Abe, M. (1999) A generalized additive model for discrete-choice data. *Journal of Business and Economic Statistics*, **17**, 271–284.

Aitchison, J. (1986) *The Statistical Analysis of Compositional Data*. Chapman and Hall: London.

Albert, J. (1996) Bayesian selection of log-linear models. *Canadian Journal of Statistics*, **24**, 327–347.

Albert, J. and Chib, S. (1993) Bayesian analysis of binary and polychotomous response data. *Journal of the American Statistical Association*, **88**, 669–679.

Billheimer, D., Guttorp, P. and Fagan, W. (2001) Statistical interpretation of species composition. *Journal of the American Statistical Association*, **96**, 1205–1214.

Bolduc, D. and Bonin, S. (1998) Bayesian analysis of road accidents: a general framework for the multinomial case. *Cahiers de Recherche* 9802, Université Laval – Département d'Economique.

Celeux, G., Hurn, M. and Robert, C. (2000) Computational and inferential difficulties with mixture posterior distributions. *Journal of the American Statistical Association*, **95**, 957–970.

Chen, M., Ibrahim, J. and Shao, Q. (2000) *Monte Carlo Methods in Bayesian Computation*. Springer: New York.

Chen, Z. and Kuo, L. (2002) Discrete choice models based on the scale mixtures of multivariate Normal distributions. *Sankhya*, **64B**, 192–213.

Chib, S., Greenberg, E. and Chen, Y. (1998) MCMC methods for fitting and comparing multinomial response models. *Economics Working Paper Archive, Econometrics*, No. 9802001, Washington University of St Louis.

Daganzo, C. (1979) *Multinomial Probit: The Theory and its Application to Demand Forecasting*. Academic Press: New York.

Epstein, A. and Fienberg, S. (1991) Bayesian estimation in multidimensional contingency tables. In Computer Science and Statistics: Proceedings of the Twenty-Third Symposium on the Interface, Keramidas, E (ed), Interface Foundation of North America: Fairfax, 37–47.

Gelman, A., Carlin, J., Stern, H. and Rubin, D. (1995) *Bayesian Data Analysis*. Chapman and Hall: London.

Geweke, J., Keane, M. and Runkle, D. (1994) Alternative computational approaches to inference in the multinomial probit model. *Review of Economics and Statistics*, **76**, 609–632.

Glasgow, G. (2001) Mixed logit models for multiparty elections. *Political Analysis*, **9**, 116–136.

Good, I. (1976) On the application of symmetric Dirichlet distributions and their mixtures to contingency tables. *Annals of Statistics*, **4**, 1159–1189.

Greene, W. (2000) *Econometric Analysis*, 4th edition, Prentice Hall: Englewood Cliff, NJ.

Gunel, E. and Dickey, J. (1974) Bayes factors for independence in contingency tables. *Biometrika*, **61**, 545–557.

Hausman, J. and Wise, D. (1978) A conditional probit model for qualitative choice: discrete decisions recognizing interdependence and heterogeneous preferences. *Econometrica*, **48**, 403–429.

Hensher, D. and Greene, W. (2001). The mixed logit model: the state of practice and warnings for the unwary. *Working Paper*, School of Business, The University of Sydney.

Hoffman, S. and Duncan, G. (1988) A comparison of choice-based multinomial and nested logit models: the family structure and welfare use decisions of divorced or separated women. *Journal of Human Resources*, **23**, 550–562.

Katz, J. and King, E. (1999) A statistical model for multiparty electoral data. *American Political Science Review*, **93**, 15–32.

Keane, M. (1992) A note on identification of the multinomial probit model. *Journal of Business and Economic Statistics*, **10**, 193–200.

Lahiri, K. and Gao, J. (2002) Bayesian analysis of nested logit model by Markov chain Monte Carlo. *Journal of Econometrics*, **111**, 103–133.

Lenk, P., DeSarbo, W., Green, P. and Young, M. (1996) Hierarchical Bayes conjoint analysis: recovery of partworth heterogeneity from reduced experimental designs. *Marketing Science*, **15**, 173–191.

Leonard, T. and Hsu, J. (1994) The Bayesian analysis of categorical data – a selective review. In *Aspects of Uncertainty: A Tribute to DV Lindley*, Freeman, P. and Smith, A. (eds). John Wiley & Sons: New York.

Linardakis, M. and Dellaportas, P. (2003) Assessment of Athens' metro passenger behaviour via a multi-ranked probit model. *Journal of the Royal Statistical Society Series C*, **52**, 185–200.

Lindley, D. (1964) The Bayesian analysis of contingency tables. *Annals of Mathematical Statistics*, **35**, 1622–1643.

Long, J. (1997) *Regression Models for Categorical and Limited Dependent Variables*. Sage: Thousand Oaks, CA.

McCullagh, P. and Nelder, J. (1989) *Generalized Linear Models*, 2nd edition. Chapman and Hall: London.

McCulloch, R. and Rossi, P. (1994) An exact likelihood analysis of the multinomial probit model. *Journal of Econometrics*, **64**, 207–240.

McCulloch, R. and Rossi, P. (2000) Bayesian analysis of the multinomial probit model. In *Simulation-based Inference in Econometrics*, Mariano, R., Schuermann, T. and Weeks, M. (eds). Cambridge University Press: Cambridge, 158–175.

McCulloch, R., Polson, N. and Rossi, P. (2000) A Bayesian analysis of the multinomial probit model with fully identified parameters. *Journal of Econometrics*, **99**, 173–193.

McFadden, D. (1974) Conditional logit analysis of qualitative choice behavior. In *Frontiers in Econometrics*, Zarembka, P. (ed.). Academic Press: New York, 105–142.

McFadden, D. and Train, K. (2002) Mixed MNL models for discrete response. *Journal of Applied Econometrics*, **15**, 447–470.

Mebane, W. and Sekhon, J. (2004) Robust estimation and outlier detection for overdispersed multinomial models of count data. *American Journal of Political Science*, **48**, 391–410.

Muthén, B. and Muthén, L. (2000) Integrating person-centered and variable-centered analyses: growth mixture modeling with latent trajectory classes. *Alcoholism, Clinical and Experimental Research*, **24**, 882–891.

Nandram, B. (1998) A Bayesian analysis of the three-stage hierarchical multinomial model. *Journal of Statistical Computation and Simulation*, **61**, 97–126.

Nobile, A. (1998) A hybrid Markov chain for the Bayesian analysis of the multinomial probit model. *Statistics and Computing*, **8**, 229–242.

Nobile, A., Bhat, C. and Pas, E. (1997) A random effects multinomial probit model of car ownership choice. *Case Studies in Bayesian Statistics 3*, Gatsonis, C., Hodges, J., Kass, R., McCulloch, R., Rossi, P. and Singpurwalla, N (eds). Springer: New York, 419–434.

Rossi, P. and Allenby, G. (2003) Bayesian statistics and marketing. *Marketing Science*, **22**, 304–328.

Ruppert, D., Wand, M. and Carroll, R. (2003) *Semiparametric Regression*. Cambridge University Press: Cambridge.

Scott, S. (2003) Data augmentation for the Bayesian analysis of multinomial logit models. *Proceedings of the ASA Section on Bayesian Statistical Science*. ASA: Alexandria, VA.

Sheps, S. and Schechter, M. (1984) The assessment of diagnostic tests. A survey of current medical research. *Journal of the American Medical Association*, **252**, 2418–2422.

Soulsby, C., Petry, J., Brewer, M., Dunn, S., Ott, B. and Malcolm, I. (2003) Identifying and assessing uncertainty in hydrological pathways: a novel approach to end member mixing analysis in a Scottish agricultural catchment. *Journal of Hydrology*, **274**, 109–128.

Uebersax, J. (1999) Probit latent class analysis with dichotomous or ordered category measures: conditional independence/dependence models. *Applied Psychological Measurement*, **23**, 283–297.

Wedel, M., Kamakura, W., Arora, N., Bemmaor, A., Chiang, J., Elrod, T., Johnson, R., Lenk, P., Neslin, S. and Poulsen, C. (1999) Discrete and continuous representation of heterogeneity. *Marketing Letters*, **10**, 217–230.

Wilson, J. (1989) Chi-square tests for overdispersion with multiparameter estimates. Journal of the Royal Statistical Society, Series C, **38**, 441–453.

Yu, P. (2000) Bayesian analysis of order statistics models for ranking data. *Psychometrika*, **65**, 281–299.

Zellner, A. and Rossi, P. (1984) Bayesian analysis of dichotomous quantal response models. *Journal of Econometrics*, **25**, 365–393.

CHAPTER 7

Ordinal Regression

7.1 ASPECTS AND ASSUMPTIONS OF ORDINAL DATA MODELS

Data with ordinal dependent variables occur frequently in health surveys and clinical settings (Cox, 1995), in econometric applications including studies of income inequality (Chotikapanich and Griffiths, 2000), in psychometric applications (e.g. rating scales) and in social surveys. Let y_i be an ordinal response variable for individuals $i = 1, \ldots, n$ and with levels $1, 2, \ldots, J$. Thus

$$y_i \sim \text{Categorical}(\pi_i)$$

where $\pi_i = (\pi_{i1}, \pi_{i2}, \ldots, \pi_{iJ})$ is the vector of model probabilities that subject i will be located at level j. Since the outcome is ranked one may consider cumulative probabilities conditional on regressors X_i, namely

$$\gamma_{ij}(x_i) = \Pr(y_i \leq j | X_i) = \pi_{i1}(X_i) + \pi_{i2}(X_i) + \cdots + \pi_{ij}(X_i)$$

As for other probabilities γ_{ij} may then be regressed on predictors via a link function $g = F^{-1}$ where typical forms of F include the cumulative standard normal $F = \Phi$ or the logistic cdf

$$F(u) = \exp(u)/[1 + \exp(u)]$$

A cumulative logit model with a linear predictor η_{ij} specifies

$$\text{logit}(\Pr(y_i \leq j | X_i)) = \text{logit}(\gamma_{ij}(X_i)) = \kappa_j + \beta_j X_i = \kappa_j + \eta_{ij} \qquad (7.1)$$

The logit and probit links in (7.1) are most common but another option is the complementary log–log, with

$$\log[-\log(1 - \gamma_{ij})] = \kappa_j + \beta_j x_i$$

Bayesian Models for Categorical Data   P. Congdon
© 2005 John Wiley & Sons, Ltd

A frequent simplifying assumption is that the effect of predictors X_i is constant across ordered categories $\beta_j = \beta$. If F is logistic and the predictors are respondent characteristics only, then under this assumption, the difference in cumulative logits between subjects i and k with responses both in the jth category is $C_{ij} - C_{kj}$ where

$$C_{ij} = \text{logit}(\gamma_{ij}) = \kappa_j + \beta X_i \qquad (7.2)$$

The difference $C_{ij} - C_{kj} = \beta(X_i - X_k)$ is independent of j, this being known as the proportional odds property (McCullagh, 1980). By contrast model (7.1) is a non-proportional model. The proportional odds model may be estimated by a set of $J - 1$ regressions, with $J^* = 1, 2, \ldots, J - 1$

$$D_{ij} = 1 \quad \text{if } y_i \leq J^*$$
$$D_{ij} = 0 \quad \text{if } y_i > J^*$$
$$D_{ij} \sim \text{Bern}(\gamma_{ij})$$

with $F^{-1}(\gamma_{ij}) = \kappa_j + \beta X_i$ and $\kappa_{J-1} > \kappa_{J-2} > \cdots > \kappa_1$. The most effective method involves estimating all $J - 1$ regressions in one single all-encompassing model since β is common to all models. One may apply (7.1) or (7.2) with a negative sign before the β, so that β may be interpreted in terms of odds that increase with j (Simonoff, 2003, p 436).

If it is not assumed that all the β_j are equal (e.g. only some covariates have differing regression coefficients according to j) then a partial proportional odds model is obtained (e.g. Peterson and Harrell, 1990). A drawback with allowing β_j to be level specific is that the $J - 1$ non-parallel regression lines may cross when explanatory variables are continuous; this problem does not occur for categorical explanatory variables (Gibbons and Hedeker, 2000). Further features of non-proportionality are considered by Fahrmeier and Tutz (2001, section 3.3) and Ishwaran (2001). A non-proportional or partial proportional model amounts to making the thresholds depend on explanatory variables. Thus the partial proportional model may be framed as

$$\text{logit}[\gamma_{ij}(X_i, W_i)] = \kappa_j + \gamma_j W_i + \beta X_i \qquad (7.3)$$

where the W_i are denoted threshold predictor variables and the X_i denoted (proportional) shift variables.

7.2   LATENT SCALE AND DATA AUGMENTATION

Following McCullagh (1980), the observed response y is often taken to reflect an underlying continuous random variable y^* with $J - 1$ thresholds or cut points. In medical diagnosis a latent disease severity is plausible (Cox, 1995), though in some applications a latent variable may not be appropriate. Agresti (2002) cites academic gradings (Assistant, Associate and Full Professor) as an example where a latent scale seems dubious. Suppose y_i denotes health status category as recorded by a health survey (with $J = 5$ levels, namely $1 =$ very good, $2 =$ good $, \ldots, 5 =$ bad) then a latent ill-health on a continuous scale y_i^* may be envisaged, defined by cut points

$$-\infty = \kappa_0 < \kappa_1 < \cdots < \kappa_{J-1} < \kappa_J = \infty$$

The observed discrete responses are defined according to the value of the underlying metric response, with

$$
\begin{aligned}
y_i &= 1 \quad \text{if} -\infty < y_i^* \le \kappa_1 \\
y_i &= 2 \quad \text{if } \kappa_1 < y_i^* \le \kappa_2 \\
y_i &= 3 \quad \text{if } \kappa_2 < y_i^* \le \kappa_3 \\
y_i &= 4 \quad \text{if } \kappa_3 < y_i^* \le \kappa_4 \\
y_i &= 5 \quad \text{if } \kappa_4 < y_i^* \le \infty
\end{aligned}
$$

If $J - 1$ cut points are free parameters then identifiability requires that the intercept be omitted from the regression term. However, if one of the cut points is set to a value (e.g. $\kappa_1 = 0$), or two cut points are equated, then the intercept may be included. For instance, for p predictors and $J - 1$ free cut points $\eta_{ij} = \kappa_j + \beta_1 X_{i1} + \beta_2 X_{i2} + \cdots + \beta_p X_{ip}$ excluding an intercept.

The cumulative odds under the latent response model is based on a cdf of the form $F(y^* - \eta)$ where η is a regression mean such as $\eta = \beta x$ (Hastie and Tibshirani, 1987). The model for the latent continuous scale y^* is

$$y_i^* - \eta_i + e_i \tag{7.4}$$

with the density of e_i determined by the form of F. Since

$$y_i = j \quad \text{if } \kappa_{j-1} < y^* \le \kappa_j$$

then

$$y_i = j \quad \text{if } \kappa_{j-1} - \beta x < y^* - \beta x \le \kappa_j - \beta x$$

So for a proportional cumulative odds model with $J - 1$ free cut points it is assumed that

$$y_i \sim \text{Categorical}(\pi_i)$$

where $\pi_i = (\pi_{i1}, \pi_{i2}, \ldots, \pi_{iJ})$ and

$$\pi_{i1} = \gamma_{i1}$$
$$\pi_{ij} = \gamma_{ij} - \gamma_{i,j-1} \quad j = 2, \ldots, J-1$$
$$\gamma_{ij} = \Pr(y_i \le j) = F(\kappa_j - \eta_i) \quad j = 1, \ldots, J-1$$
$$\eta_i = \beta_1 X_{i1} + \beta_2 X_{i2} + \cdots + \beta_p X_{ip}$$

The probability of the highest rank is $\pi_{iJ} = 1 - \gamma_{i,J-1}$. If attributes of the choices C_{ij} and A_j (as in Chapter 6) are added then non-proportionality is likely to exist.

Adopting a framework with an underlying scale envisaged does not necessarily require that data augmentation be applied and the y^* values be formally sampled. However, following Albert and Chib (1993) and Johnson and Albert (1999), one may estimate the proportional logit or probit model by constrained sampling of y^*. As for binary regression via data augmentation, the problem is converted to a metric linear regression with simplified Gibbs sampling and benefits in terms of residual analysis. The interval constraints for sampling a particular subject's y_i^* are based on that individual's observed category (see Example 7.1). Thus if $y_i = 1$, y^* would lie between $-\infty$ and κ_1. The cut points are sampled in a way that takes account of the sampled y^*. Thus at iteration t

$$\kappa_j^{(t)} \sim N(0, V_\kappa) I(L_j, U_j) \quad j = 1, \ldots, J-1$$

where V_κ is preset and

$$L_j = \max(y^{*(t)}, y_i = j)$$
$$U_j = \min(y^{*(t)}, y_i = j+1)$$

If $F = \Phi$ then e_i in (7.4) will be standard normal and truncated normal sampling of y^* applies. Scale mixing may be used for greater robustness allowing an implicit mixing over links. Thus

$$y^* \sim N(\beta X, \gamma_i) I(\kappa_{y_i-1}, \kappa_{y_i})$$

where $\gamma_i \sim Ga(0.5\nu, 0.5\nu)$ and ν is itself unknown. This includes an approximation to the logistic (Albert and Chib, 1993). For F logistic, direct sampling is also possible since the e_i follow a standard logistic density. An asymmetric distribution for the latent variable might be appropriate in some applications (e.g. when the response is a grouped survival time) and a complementary log–log link may be used: the e_i then follow a standard double-exponential density (extreme value) with variance $\pi^2/6$. Lang (1999) suggests a procedure for averaging over link functions in ordinal regression, specifically mixing over the left-skewed extreme value

(LSEV), the logistic and the right-skewed extreme value (RSEV); see Example 7.2.

Some drawbacks of the augmented data approach for ordinal data have been noted in recent papers (Ishwaran, 2000; Parmigiani *et al.*, 2003), for instance the addition of many extra parameters in large samples. Direct analysis of the multinomial likelihood may be applied instead but also has possible disadvantages. For example, there are $J - 1$ residuals for each multinomial observation under the direct likelihood approach in contrast to the single residual $y_i^* - \beta X_i$ from the latent variable approach (Johnson and Albert, 1999). Other types of residual may be used instead (e.g. estimates of the conditional predictive ordinate based on average inverse likelihoods). One may also sample new y_i^* values in a data augmentation (DA) approach or new y_i in a direct multinomial approach and assess the resulting concordancy between predicted categorization of subjects and their actual categories.

Note that identifiability under either approach may require informative priors, e.g. on the κ_j, especially when the numbers observed at each ordinal category may be small and the data contain relatively little inform-ation to estimate the cut points precisely. The DA approach may particu-larly require informative priors for numeric stability as it involves doubly constrained sampling: on the κ_j themselves and the y_i^* also. One method to increase stability is to increase numbers in each response category by aggregating over outcome categories with sparse response levels. How-ever, this may affect conclusions about predictors (Murad *et al.*, 2003). Since the κ_j are expected to be broadly aligned to the discrete responses, priors with a relatively low variance (e.g. 1 or 10) are justifiable. An alter-native suggested by Ishwaran (2000) is a uniform density

$$0 = \kappa_1 \leq \kappa_2 \leq \cdots \leq \kappa_{J-1} \leq U$$

where U is equal to or less than J. Sometimes the reparameterization of Fahrmeier and Tutz (2001, section 3.3) is beneficial; assuming $\kappa_1 = 0$ then this parameterization is

$$\kappa_2 = \exp(\varphi_2)$$

$$\kappa_j = \kappa_{j-1} + \sum_{m=3}^{j} \exp(\varphi_m) \qquad (7.5)$$

where the φ_m are unconstrained. In Ishwaran (2000) the partial propor-tional odds model in (7.3) is accordingly expressed as

$$\mathrm{logit}[\gamma_{ij}(X_i, W_i)] = \beta X_i + \sum_{m=1}^{j} \exp(\varphi_m + \gamma_M W_i)$$

where $\varphi_1 = -\infty$ (i.e. $\kappa_1 = 0$) for identifiability when $\eta_i = \beta X_i$ includes an intercept.

Example 7.1   Statistics grades   Johnson and Albert (1999) present an example for $N = 30$ statistics students classified into five grades (Fail $= 1$, D grade $= 2$, C grade $= 3$, B grade $= 4$, A grade $= 5$) with predictor $X = $ SATM score. We compare the fit of different links and assess residual effects under direct sampling of the latent y^*. For an ordinal probit model mildly informative priors are assumed for κ_j to assure identification. With an overall intercept β_0 retained (and assuming $\kappa_1 = 0$) there are three free cut points with assumed priors:

$$\kappa_2 \sim N(1, 10)I(0, \kappa_3)$$
$$\kappa_3 \sim N(2, 10)I(\kappa_2, \kappa_4)$$
$$\kappa_4 \sim N(3, 10)I(\kappa_3,)$$

An $N(0,10)$ prior on the effect of the SAT score is assumed for numerical reasons, though one might alternatively scale the SAT score itself (e.g. divide the score by 100).

The second half of a two-chain run of 10 000 iterations for the ordinal probit model shows a coefficient $\beta_1 = 0.025$ (posterior s.d. $= 0.006$) on the centred SAT score, close to the estimate obtained by Johnson and Albert (1999, p 143). The estimated cut points 1.55, 2.44 and 3.79 compare with their estimates of 1.29, 2.11 and 3.56. Large residuals $y^* - \beta X$ (absolutely exceeding 1.96) are obtained for subject 1 (who failed despite a SATM score of 557) with $e_1 = -2.47$, and subject 30 who got grade A despite an average SATM score, with $e_{30} = 2.16$. One may also sample new y^* (without constraint), allocate them to a predicted category based on their value relative to the cut points and compare the predicted with the actual category. This leads to a total predictive concordancy (percentage of students correctly classified) of 34.3. The lowest probability of predicted category equalling actual category is for subject 1, namely 0.024, while subject 30 has a probability of 0.05.

Logistic and double-exponential densities are adapted as being potentially more robust to sampled values in tails but the same priors as for the ordinal probit are retained. Sampling from the standard logistic yields a mean for β_1 of 0.042 but with less precision (a posterior standard deviation of 0.01). The predictive concordancy rises to 39.9%, with subject 1 again the leading outlier. Outlying subjects are downweighted and the probability of a correct match for subject 1 is now 0.002. Residuals are now compared with the standard deviation of the standard logistic, namely $\pi/\sqrt{3}$; the scaled residual for subject 1 is now -2.59. Double-exponential sampling gives a very similar performance to the other two links, with predictive concordancy of 38.3%. The residuals $y^* - \beta X$ highlight cases 1 and 30, while the probability of a predictive match highlights subjects 1, 5 and 30.

Example 7.2   Mental health status   This example considers the Lang (1999) model and modifications of it applied to data on mental health status from Agresti (2002). Health status y has levels $1 =$ well, $2 =$ mild impairment, $3 =$ moderate impairment and $4 =$ impaired. It is related to an $X_1 =$ SES (a binary measure of low socio-economic status) and $X_2 =$ LIFE (an adverse life events total including factors such as divorce, bereavement, etc.).

As noted above, the link is averaged over three options for the cumulative density, F_k, $k = 1, 2, 3$. Thus F_1 for the LSEV distribution is

$$F_1(t) = 1 - \exp[-\exp(\eta)]$$

while F_3 for the RSEV distribution is

$$F_3(t) = \exp[-\exp(-\eta)]$$

with

$$F_2(t) = \exp(t)/[1 + \exp(\eta)]$$

and cumulative probabilities $\gamma_{ij}^{(k)}$ in

$$F_k^{-1}[\gamma_{ij}^{(k)}] = \kappa_j + \beta X_i$$

are obtained according to link $k = 1, \ldots, 3$ and response category $j = 1, \ldots, J - 1$.

The link mixture is

$$F_\lambda(t) = \pi_1(\lambda)F_1(t) + \pi_2(\lambda)F_2(t) + \pi_3(\lambda)F_3(t)$$

where probabilities on F_1 and F_3 depend on a parameter $\lambda \sim N(0, \sigma_\lambda^2)$ such that

$$\pi_1(\lambda) = \exp[-\exp(3.5\lambda + 2)]$$
$$\pi_3(\lambda) = \exp[-\exp(-3.5\lambda + 2)] \qquad (7.6)$$
$$\pi_2(\lambda) = 1 - \pi_1(\lambda) - \pi_3(\lambda)$$

A negative λ means the LSEV link form is preferred, and positive λ means RSEV is preferred, while $\lambda = 0$ means $w_1(\lambda)$ and $w_3(\lambda)$ are both near zero and leads to selection of the logit link. A Dirichlet prior is another possibility for (π_1, π_2, π_3). Then the model averaged predictions are

$$\gamma_{ij}(x_i) = \sum_k \pi_k(\lambda)\gamma_{ij}^{(k)}$$

Alternative models might make the cut points and/or regression effects specific to the link (though still proportional within each link) so that

$$F_k^{-1}[\gamma_{ij}^{(k)}] = \eta_{ijk} = \kappa_{jk} + \beta_k x_i$$

or take the parameter $\varphi = 3.5\lambda$ to differ between link probabilities so that

$$\pi_1(\varphi_1) = \exp[-\exp(\varphi_1 + 2)]$$
$$\pi_3(\varphi_2) = \exp[-\exp(-\varphi_2 + 2)]$$

A further alternative model considered here averages over four possible links, namely the three considered by Lang plus the probit. A Dirichlet mixture is used

$$\pi \sim \text{Dirch}(\alpha)$$

where $\pi = (\pi_1, \pi_2, \pi_3, \pi_4)$ and $\alpha = (1,1,1,1)$. Hence the averaged link is

$$F(t) = \pi_1 F_1(t) + \pi_2 F_2(t) + \pi_3 F_3(t) + \pi_4 F_4(t)$$

with

$$F_1(t) = 1 - \exp[-\exp(\eta)]$$
$$F_2(t) = \exp(t)/[1 + \exp(\eta)]$$
$$F_3(t) = \Phi(t)$$
$$F_4(t) = \exp[-\exp(-\eta)]$$

In model A the mixture probabilities are as in (7.6), with a prior $\lambda \sim N(0,5)$, and common cut points and regression effects are assumed. Initial runs suggested that the interaction in LIFE and SES was not an important predictor. Results from the second half of a two-chain run of 20 000 iterations for model A show a credible interval for λ straddling zero, i.e. $(-3.7, 4.6)$. The posterior mean on $\pi_2(\lambda)$ of 0.49, compared with 0.30 for $\pi_1(\lambda)$ and 0.21 for $\pi_3(\lambda)$, confirms that the simple logit link is preferred, though there is clearly averaging over the three links. Both life events and low status are associated negatively with the lowest response category (being well), and so positively with impairment. The effect of life events is better defined (95% interval entirely negative, i.e. -0.48 to -0.07) while the effect of SES straddles zero. The DIC is 115.1 and a 'concordancy index' (the proportion of subjects correctly classified into one of the four grades) averages 0.315.

In a second model (model B), the cut points are allowed to differ between links, though the regression effects remain common, with

$$F_k^{-1}[\gamma_{ij}^{(k)}] = \eta_{lijk} = \kappa_{jk} + \beta X_i$$

Priors for the cut points κ_{jk} are based on the posterior means and standard deviations of κ_j from model A, with a ten-fold downweighting of precision. Results from a 20 000-iteration run suggest that the logit cut points differ from those of the skewed links, i.e. $\kappa_2 = (0.5, 2.2, 3.6)$ compared with $\kappa_1 = (-0.1, 1.9, 3.9)$ and $\kappa_3 = (-0.2, 2, 4)$. One feature of model B is a more precise effect for SES, with posterior mean -1.1 and a 95% interval $(-2.3, -0.1)$ confined to negative values. The DIC deteriorates under this model (to 118) but the concordance index is 0.317.

In model C the Dirichlet mixture on four links is considered (with common link cut points as in model A). This shows the preference for the logit with $\pi_2 = 0.32$ but shows the probit has a weight comparable with the asymmetric options ($\pi_3 = 0.21$ as against $\pi_1 = 0.26$ and $\pi_4 = 0.21$). The DIC and concordance index are similar to model A, i.e. 114.9 and 0.316.

7.3 ASSESSING MODEL ASSUMPTIONS: NON-PARAMETRIC ORDINAL REGRESSION AND ASSESSING ORDINALITY

As for GLMs involving metric, count and binary data, a non-parametric approach may be used to assess the assumed linear regression form for all predictors. Often a non-parametric approach will be combined with a review of the proportional odds assumption (Lee and Wild, 1996). Assume a latent scale model with cumulative odds

$$\gamma_{ij} = \Pr(y_i \leq j) = F(\kappa_j - \eta_{ij}) \quad j = 1, \ldots, J - 1$$

and link g, and initially assume proportional odds

$$g(\gamma_{ij}) = \kappa_j - \eta_i$$

Then the category-specific regression means η_i might be modelled semi-parametrically, e.g. by adopting a single smooth function $s_i = s(X_i)$ for one predictor. Assuming cases are ordered in terms of X_i with remaining predictors $z_{i1}, \ldots, z_{iq}$ modelled by linear effects constant over categories, then

$$\eta_i = s(X_i) + \beta_1 z_{i1} + \cdots + \beta_q z_{iq}$$

A fully non-parametric model (Hastie and Tibshirani, 1987) would take the form

$$g(\gamma_{ij}) = \kappa_j - s_1(X_{i1}) - s_2(X_{i2}) - s_3(X_{i3}) - \cdots - s_p(X_{ip})$$

Several ways are available to obtain identifiability with a smooth in one predictor. One involves setting both $\kappa_1 = 0$ and the overall intercept β_0 to zero (if the latter step is not taken the mean of the s_i will be confounded with it). The s_i series will then identify the intercept, and sampling on the s_i can be unconstrained in the semiparametric model

$$g(\gamma_{ij}) = \kappa_j - s_i - \beta_1 z_{i1} - \beta_2 z_{i2} - \cdots - \beta_q z_{iq}$$

Another way to obtain identifiability would be to recentre the s_i at each iteration (e.g. using the carnormal function in BUGS if random walk priors are used for s_i) and then an overall intercept β_0 may be included provided $\kappa_1 = 0$. If more than one predictor is modelled non-parametrically by functions $s_i^{(k)}$ for predictor k, then identification would require devices such as setting $s_1^{(k)} = 0$ for all k, or centring all the $s_i^{(k)}$ at each iteration (with an overall intercept included if $\kappa_1 = 0$).

To assess proportionality with the impact of a single predictor X_i modelled non-parametrically, one could assume smooth functions $s_{ij} = s_j(X_i)$ differing by category (or possibly subgroupings of the categories). For example,

$$g(\gamma_{ij}) = \kappa_j - s_{ij} - \beta_1 z_{i1} - \beta_2 z_{i2} - \cdots - \beta_q z_{iq}$$

where the smoothing function for subject i would be determined by that person's observed y, i.e. $s_{ij} = s_{i,y_i}$.

Example 7.3   Attitudes towards working mothers   Long (1997) presents an analysis of attitudes towards mothers working using the GHS data aggregated over two survey years (1977 and 1989), with a total of 2293 respondents. The survey asked whether a working mother can establish as warm and secure relationship with children as a non-working mother. The response scale was $1 =$ strongly disagree, $2 =$ disagree, $3 =$ agree, $4 =$ strongly agree. Apart from survey year other predictors are age in years, gender ($1 =$ male, $0 =$ female), ethnicity ($1 =$ white, $2 =$ non-white), occupational prestige and years of education.

We initially fit a proportional cumulative logit model (7.1) with linear form assumed for the impact of all predictors. A negative sign on β is used so that regressor effects relate to more favourable views on mothers working. A two-chain run of 1000 iterations, with one set of initial values based on Long (1997), converges at under 250 iterations and estimates for all predictors are significant. Positive effects (i.e. favourable to mothers working) are estimated for 1989 (vs. 1977), greater prestige and more years of education, while negative views are estimated for males, whites and older people. The DIC for this model is 6630.

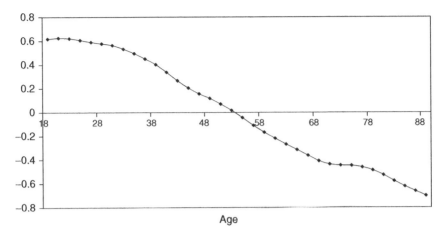

Figure 7.1 Proportional non-parametric model

For a proportional semi-parametric model (model B) with a smooth in age, a second-order random walk is assumed for $s_i = s(A_i)$ where ages are grouped into two-year bands to avoid sparsity, especially when crossed with the grades of the response in the subsequent model. So $A = 1$ for ages 18, 19, $A = 2$ for ages 20, 21, etc., up to $A = 36$ for ages 88, 89. The s_i are centred at each iteration (so that their level is identified) and so an intercept β_0 can be included, provided $\kappa_1 = 0$. The prior on the precision τ_s of the s_i is non-conjugate with $\tau_s \sim \text{Ga}(1, b)$ where b itself is assigned a $\text{Ga}(1, 0.001)$ prior. The second half of a two-chain run of 5000 iterations produces a smooth curve which suggests some non-linearity, namely a plateau effect up to around age 30 and a 'liberal views' blip for people in their seventies (Figure 7.1). The DIC improves to around 6626 under this model.

There is a slight loss in fit compared with model B (a DIC of 6630) in moving to a non-proportional semi-parametric model (model C) where the smooths in age are specific to response category. However, there is clear evidence of non-proportionality. Here full conditionals for the standard conjugate gamma priors on the three precision parameters are used, with convergence apparent from around 3000 iterations in a two-chain 5000 iteration run. For the category 'strongly disagree' the decline in approval for working mothers stops at older ages (with a plateau effect past 60). For the less extreme 'disagree' category the smooth gets close to being linear, while for the 'agree' response there is a plateau at younger ages (Figures 7.2 to 7.4).

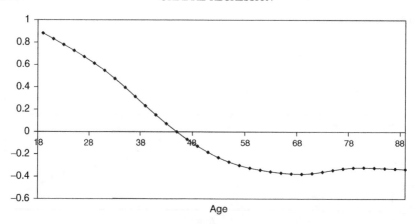

Figure 7.2   Strongly disagree by age

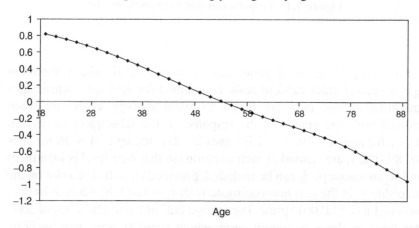

Figure 7.3   Disagree by age

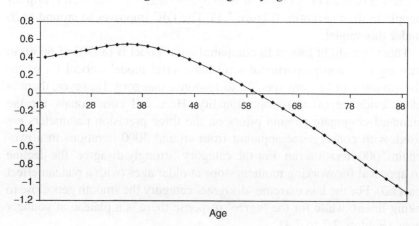

Figure 7.4   Agree by age

7.4 LOCATION-SCALE ORDINAL REGRESSION

A generalization of the ordinal regression model models both the location and scale of the underlying variable y^*, such that

$$\Pr(y_i \leq j | X_i, u_i) = F\left[\frac{\kappa_j - \beta X_i}{\exp(\delta u_i)}\right] \tag{7.7}$$

If $F = \Phi$ (the cumulative standard normal) then $y_i = j$ if $\kappa_{j-1} < y_i^* \leq \kappa_j$ where

$$y_i^* = \beta X_i + Z_i \exp(\delta u_i)$$
$$Z_i \sim N(0, 1)$$

Thus y_i^* has mean βX_i and variance $\exp(2\delta u_i)$ and, if δ is positive or negative, no longer leads to proportional odds. Tosteson and Begg (1988) consider analysis of diagnostic tests with a single predictor $X_i = u_i$ defined to be -0.5 for diseased (positive) subjects and $+0.5$ for well subjects. Then δ_1 describes how variability on the latent scale differs between well and ill subjects, while β_1 measures the predictive accuracy of the diagnostic test. For δ_1 given, more accurate tests have larger β_1.

Cox (1995) considers a further generalization, but with a logit link, such that

$$\mathrm{logit}[\Pr(y_i \leq j | X_i, u_i, A_{ij}, v_j)] = \gamma_{ij} = \left[\frac{\kappa_j - \beta X_i - \phi A_{ij}}{\exp(\delta u_i + \omega v_j)}\right] \tag{7.8}$$

where A_{ij} are attributes both of the categories and the subjects and v_j are attributes of the categories. These might be dummy variables identifying certain categories for certain classes of subjects (e.g. certain categories of subject i may be more likely to respond for more extreme categories j). If $\phi \neq 0$, $\omega \neq 0$ or $\delta \neq 0$ then a non-proportional model results (Peterson and Harrell, 1990). Recent Bayesian location-scale models have focused especially on receiver–operator curves (ROCs) involving models of this type (e.g. see Peng and Hall, 1996; Ishwaran and Gatsonis, 2000; Hellmich *et al.*, 1998).

Another application of the location-scale approach illustrates how an unknown scale may be part of a model allowing estimation of the link, rather than taking it known (cf. the discrete mixture over three or more possible links discussed above). Gelfand and Kuo (1991) consider an unknown link model in the analysis of assay data, where increasing standardized dosages $t_k, k = 1, \ldots, K$ are applied and deaths r_k at that dosage are binomial

$$r_k \sim \mathrm{Bin}(n_k, p_k)$$

By assumption, the 'potency' curve $p_k = F(t_k)$ is increasing as t_k increases (i.e. as the lethality of the dose increases or the strength of the input in general) and so the sequence of values $(p_1, \ldots, p_k)$ describes part of a cumulative density function with boundary values

$$t_0 = -\infty, \quad p_0 = F(t_0) = 0$$

and

$$t_{K+1} = \infty, \quad p_{K+1} = F(t_{K+1}) = 1$$

Suppose original doses in their original measure scale are $T_1, \ldots, T_K$ and the standardized dosages are

$$t_k = (T_k - \mu)/\sigma$$

A parametric analysis with a standard link might therefore assume

$$p_k = \Phi(t_k)$$

or

$$p_k = \exp(t_k)/[1 + \exp(t_k)]$$

However, a non-parametric approach allows uncertainty around such a central 'baseline' link such as $F_b = \Phi$ according to a prior sample size ('concentration parameter') M, with the prior on the unknown cdf $F(t)$ then being

$$F(t) \sim \text{Be}(MF_b(t), M(1 - F_b(t)))$$

Larger values of M imply greater confidence in the suitability of F_b and greater closeness of the estimated $F(t)$ to $F_b(t)$. This induces a Dirichlet prior on the differences

$$F(t_1) - F(t_0) = p_1 - p_0 = p_1$$
$$F(t_2) - F(t_1) = p_2 - p_1$$
$$F(t_3) - F(t_2) = p_3 - p_2$$
$$\vdots$$
$$F(t_K) - F(t_{K-1}) = p_K - p_{K-1}$$
$$F(t_{K+1}) - F(t_K) = p_{K+1} - p_K = 1 - p_K$$

with parameters

$$\gamma_k = M[F_b(t_k) - F_b(t_{k-1})]$$

Let $\hat{F}(t_k)$ be the mean potency curve estimate at dosage t_k. The expected potency at an intermediate dosage t^* to those at which applications are made is obtainable as

$$([\gamma_{k+1} - \gamma^*]/\gamma_{k+1})\hat{F}(t_k) + (\gamma^*/\gamma_k)\hat{F}(t_{k+1})$$

where

$$\gamma^* = M[F_b(t^*) - F_b(t_k)]$$

Possible developments on this approach include a prior on the scale σ, and possibly the origin μ, though this can be set to $\bar{t}$ to reduce parameterization. The scale may also be made a function of covariates as above. While M can be preset, a prior on M might be used but would need to be tuned to avoid low values that cause numerical problems in sampling the Dirichlet. One might also include regression effects both on the varying impact of the dosage, and on other relevant factors z_k such as measured frailty, soil fertility, etc., of the samples exposed to application level k. Thus with $\eta_k = \beta_1 T_k + \beta_2 z_k$ and $\mu = \bar{\eta}$

$$t_k = (\beta_1 T_k + \beta_2 z_k - \mu)/\sigma$$
$$p_k = F_b(t_k)$$
$$F(t) \sim Be(MF_b(t), M(1 - F_b(t)))$$

Example 7.4 Nausea and cisplatin   Cox (1995) presents data on severity of nausea in two sets of chemotherapy patients, one group ($N_1 = 161$) a control group, the other ($N_2 = 58$) in receipt of cisplatin (Table 7.1). The response is graded 0 to 5 (recoded 1 to 6 in Program 7.4), namely no nausea (0), intermediate levels of nausea (1, 2, 3, 4) and severe nausea (5).

Table 7.1   Nausea by cisplatin group

	Severity of Nausea						Total
	0	1	2	3	4	5	
No cisplatin	43	39	13	22	15	29	161
Cisplatin	7	7	3	12	15	14	58

Cox considers model (7.7) and extensions drawing partly on Peterson and Harrell to allow non-proportionality. First fitting model (7.7) with $x_i = u_i = 1$ for the cisplatin group gives $\beta_1 = 0.83$ (s.d. $= 0.24$) so that in fact cisplatin patients experienced more severe nausea (results from a two-chain run of 2500 iterations that shows convergence at under 500 iterations). The coefficient δ is marginally negative with mean -0.15 but with the 95% interval straddling zero. So there is weak evidence of less variability in y^* for the cisplatin than the non-cisplatin group. The DIC is 777, and the percentage of patients correctly classified is 19.3. A second (proportional odds) model omits u_i from (7.7). This raises β_1 to around 0.86 and has a DIC of 776.

Peterson and Harrell (1990) introduce a patient–category attribute to the model just considered, as in (7.8). Thus $A_{ij} = 1$ for cisplatin patients in the last severity category but one (note that $\gamma_{iJ} = 1$ and only categories 0, 1, 2, 3, 4 are actually modelled) and $A_{ij} = 0$ for all other patient-outcome combinations. This reduces the DIC to around 775.3 but the percentage of patients correctly classified is lower than model A, at 19.1. In contrast to the results presented by Cox (1995) the coefficient ϕ_1 straddles zero with mean 0.63 and 95% interval $(-0.18, 1.44)$.

Cox replaces A_{ij} by a dummy variable common to both patient groups, so modelling the scale, not the location, in (7.8), i.e. $v_j = 1$ if $y_i = 4$ and $v_j = 0$ otherwise. This improves the DIC to 769 under an informative $N(0,0.2)$ prior for ω_1.

Example 7.5   Cox and Snell assay data   Gelfand and Kuo (1991) consider data from Cox and Snell (1989) on deaths r_k at $K = 5$ doses $T_1 = 0$, $T_2 = 1$, $T_3 = 2$, $T_4 = 3$, $T_5 = 4$. The number of subjects at all levels is $n_k = 30$. The baseline cdf assumed is a cumulative normal Φ with mean 2 and variance 0.25. Hence the prior baseline cdf has values

$$F_b(0) = 3.17 \times 10^{-5}$$
$$F_b(1) = 0.023$$
$$F_b(2) = 0.5$$
$$F_b(3) = 0.977$$
$$F_b(4) = 0.99997$$

Gelfand and Kuo consider alternative (preset) values of M, namely $M = 1$, $M = 30$, $M = 100$ and $M = 200$. Taking $M = 30$ we sample the Dirichlet using a series of gamma densities to estimate the differences $S_k = p_k - p_{k-1}$, with $S_{K+1} = 1 - p_K$. The estimated means and standard deviations for $p_k = F(t_k)$ are then

$$F(1) = 0.061(0.035)$$
$$F(2) = 0.135(0.038)$$
$$F(3) = 0.503(0.060)$$
$$F(4) = 0.871(0.038)$$
$$F(5) = 0.921(0.035)$$

with a DIC of 29.3.

An extension is to take the scale σ as unknown – with Ga(1,0.001) prior – but M still set at 30. This improves the DIC considerably to 21.5, with $F(t)$ now estimated as 0.086 (0.040), 0.255(0.055), 0.503 (0.058), 0.752 (0.053), 0.901 (0.040) and the estimated scale in F_b standing at 1.56. One may further (model C in Program 7.5) let both the scale and Dirichlet concentration M be unknowns, with the prior on the latter assumed to be $M \sim \text{Ga}(1,0.1)$. Results may be sensitive to the priors on σ and M. In fact, this leads to a worsening of the DIC to 26.5, with M estimated at 3 and the scale now 3.3.

7.5   STRUCTURAL INTERPRETATIONS WITH AGGREGATED ORDINAL DATA

Most of the model features as considered above (e.g. use of cumulative probabilities) for individual response data also pertain to aggregated ordinal data, except that analysis typically takes place via the direct multinomial likelihood. To adopt the latent variable method where y^* is sampled would require disaggregating to the equivalent individual-level data. The discussion of aggregated ordinal data applications focuses on structural indices that are functions of the parameters (and possibly data also) in models relating to health status and to consumer choice.

Suppose aggregated counts $f_{i1}, f_{i2}, \ldots, f_{iJ}$ denote persons in ranked health categories $j = 1, \ldots, J$ for social or income groups $i(i = 1, \ldots, I)$ of size N_i. For monitoring the impact of economic and health policies it may be important to assess trends in health inequality and this might involve comparing two or more periods to assess whether health status contrasts over income groups have narrowed or widened. If the data for two periods are denoted f_{1ij} and f_{2ij} then one possible model for health status contrasts in period t is

$$f_{ti} \sim \text{Mult}(N_{ti}, \pi_{ti})$$

where $\pi_{ti} = (\pi_{ti1}, \pi_{ti2}, \ldots, \pi_{tiJ})$ is the distribution of income groups over illness categories (good health at low ranks to poor health at high ranks), with cumulative probabilities

$$\gamma_{tij} = \pi_{ti1} + \pi_{ti2} + \cdots + \pi_{tij}$$

modelled via links such as

$$\text{logit}(\gamma_{tij}) = \kappa_{tj} - \mu_{ti}$$

or

$$\Phi^{-1}(\gamma_{tij}) = \kappa_{tj} - \mu_{ti}$$

The κ_{tj} are period-specific cut points and the μ_{ti} represent the average ill-health in group i in period t. Survey data (or other sources such as the Census) may also provide information on the changing distribution of the population across income or social groups. This additional data may be modelled as

$$N_{ti} \sim \text{Mult}(N_t, p_{ti})$$

where p_{ti} are the proportions in group i in year t. The average morbidity level in period t is then $\psi_t = \sum_{i=1}^{G} \mu_{ti} p_{ti}$. Note that one might constrain the effects μ_{ti} to being monotonic as well as the κ_{tj}. Social status bands are usually also ordinal, and a gradient of health over social rankings is commonly observed.

An inequality index C_t, based on the model parameters such as κ_{jt} and μ_{it}, may be used to gauge whether inequality has changed. This would involve comparing C_t with C_{t-1} and counting the proportion of iterations where $C_t > C_{t-1}$. Take

$$r_{ti} = p_{t1} + p_{t2} + \cdots + 0.5 p_{ti}$$

as the relative rank of group i in period t (Kakwani *et al.*, 1997). One may then derive a 'concentration index' as

$$C_t = \left(\sum_{i=1}^{G} p_{ti} \mu_{ti} r_{ti} - \psi_t \right) / \psi_t \qquad (7.9)$$

This type of discrepancy is defined in Lorenz concentration curves as twice the area between the horizontal axis (health equality) and a plot of cumulative ill-health $H(s)$ against s, where s is socio-economic status or income. If $H(s)$ lies above the diagonal then inequalities in illness favour more advantaged social groups. If $H(s)$ is taken as piecewise linear then the index C_t is lower (less positive, more negative) and the more ill-health is concentrated among the disadvantaged.

As another example of structural inferences consider aggregated consumer ratings for $i = 1, \ldots, I$ products on ordinal scales with J levels f_{ij}. A multinomial distribution applies to each row of this tabulation, i.e. across the total ratings for each product. So $f_i \sim \text{Mult}(N_i, \pi_i), i = 1, \ldots, I$ where $\pi_i = (\pi_{i1}, \pi_{i2}, \ldots, \pi_{iJ})$ is the distribution over consumer ratings. Allowing for both mean and scale in the underlying ordered utilities as in section 7.4, one might specify, as in Nandram (1998, p 111),

$$\text{logit}(\gamma_{ij}) = (\kappa_j - \mu_i) / \exp(\delta_i) \quad i = 1, \ldots, I; \ j = 1, \ldots, J-1$$

with μ and δ parameters centred for identifiability, and all κ parameters free. Following Nandram (1997) one might be interested in the product having the highest mean ranking (i.e. highest consumer rating). Specifically the interest is in the posterior probabilities

$$q_{1i} = \Pr(\mu_i > \mu_k, k \neq i | y) \tag{7.10}$$

which may be estimated from the MCMC sampling chain. However, an alternative perspective on optimal ranking (with analogies to the health status application above) involves the relative distribution of the π_{ij} and associated ranks. Nandram (1997, p 404) proposes the alternative mean satisfaction measure

$$\nu_i = \sum_j j\pi_{ij}$$

The best product is then assessed by estimating

$$q_{2i} = \Pr(\nu_i > \nu_k, k \neq i | y) \tag{7.11}$$

Example 7.6   Self-assessed health application   Wagstaff and van Doorslaer (1994) consider survey data on self-assessed health in the Netherlands in the late 1980s and early 1990s, with responses on a health scale with $J = 5$ categories ($j = 1$ for health 'very good', 'good', 'fair', 'sometimes good, sometimes bad' and $j = 5$ for bad health). The data are disaggregated to six income groups; the lowest income group in 1989/ 1990 accounts for 25% of household heads and in 1991/1992 for 20%, while the highest income group accounts for 9% and 12% in the two periods. Wagstaff and van Doorslaer consider both standard normal and standard lognormal latent ill-health scales. They obtain evidence of a rise in health inequality between 1989 and 1990 and 1991 and 1992, using a concentration curve methodology as just described.

Here we consider the data for the two periods separately using a probit link. So

$$f_{ti} \sim \text{Mult}(N_{ti}, \pi_{ti})$$

$$\gamma_{tij} = \pi_{ti1} + \pi_{ti2} + \cdots + \pi_{tij}$$

$$\Phi^{-1}(\gamma_{tij}) = \kappa_{tj} - \mu_{ti}$$

The data are arranged so that the lowest category is good health and the highest is poor health; income groups are arranged from 1 (lowest) to 6 (highest) (see Table 7.2 for the 1989–1990 data).

κ_{t1} is set at zero so all health by income parameters are free; they are assigned N(0,100) priors. Two-chain runs of 20 000 iterations converge after 1000 iterations. The concentration index in (7.9) for the second period is found to be lower than for the first, i.e. −0.195 vs. −0.156

Table 7.2   1989–1990 health by income data

Income	Health				
	V. good	Good	Fair	Mixed	Bad
1 (lowest)	455	1140	420	186	98
2	388	794	225	109	45
3	418	872	202	88	35
4	424	850	158	55	19
5	420	914	110	51	17
6 (highest)	275	490	65	8	8

Table 7.3   Health in the Netherlands

	1989–1990			1991–1992		
	Mean	2.5%	97.5%	Mean	2.5%	97.5%
Concentration index	−0.156	−0.176	−0.137	−0.195	−0.216	−0.174
Mean ill-health by income group						
μ_1	0.935	0.887	0.983	1.032	0.978	1.085
μ_2	0.752	0.694	0.809	0.783	0.723	0.844
μ_3	0.663	0.608	0.720	0.669	0.613	0.724
μ_4	0.539	0.481	0.595	0.554	0.499	0.611
μ_5	0.494	0.438	0.551	0.462	0.411	0.514
μ_6	0.365	0.290	0.443	0.334	0.268	0.401

(Table 7.3). The mean health parameters by income band μ_{ti} reflect the worse health in lower income bands. The gap in mean health in 1991–1992 between highest and lowest income groups is wider than in 1989–1990. A suggested exercise is to code the analysis (i.e. pooling the likelihood over the two periods) such that one can assess the probability that $C_2 < C_1$.

Example 7.7   Sensory evaluation, entrées in army meals   Nandram (1998) presents data on 12 entrées in a military ration. The entrées were each rated for taste by 36 panelists (a different panel for each entrée) on a nine-point hedonic scale. The intention is to assess the best rated entrée. The frequencies f_{ij}, $i = 1,\ldots,12$ and $j = 1,\ldots,9$, are analysed by the location-scale cumulative logit model

$$\text{logit}(\gamma_{ij}) = (\kappa_j - \mu_i)/\exp(\delta_i)$$

A two-chain run of 2500 iterations shows early convergence. The posterior probabilities (7.10) show entrée 9 (chicken) as best with $q_{1,9} = 0.47$, followed by the 11th entrée (ham slices) with $q_{1,11} = 0.20$, in turn closely followed by the first entrée (pork sausage) with $q_{1,1} = 0.19$. Under (7.11), the respective probabilities for these entrées are $q_{2,9} = 0.28$, $q_{2,11} = 0.47$ and $q_{2,1} = 0.10$.

7.6 LOG-LINEAR MODELS FOR CONTINGENCY TABLES WITH ORDERED CATEGORIES

In contingency tables one or more of the dimensions may be ordinally ranked. To take this ordering into account one may adopt random walk priors for the main effects or interactions. For example, for mortality or disease data recorded by discrete age groups and period, another dimension (the cohort, a form of 'interaction' between age and period) can be derived leading to age–period–cohort models. Each of these dimensions is ordered and random walk priors of first or second order can be adopted to reflect the likely auto-correlation between successive effects; an order constraint may be applied, e.g. to age effects, to reflect their ordinality (Besag *et al.*, 1995).

Alternatively in the row–column (RC) model for a two-way table y_{ij} (with $i = 1, \ldots, I$ and $j = 1, \ldots, J$ and one or both categorizations ordered) the priors for the main effects are taken as unordered fixed effects, but the prior for the interactions incorporates the ordering of the categories (Goodman, 1979). The usual log-linear or logit linear interaction terms u_{12ij} (involving $IJ - I - J + 1$ unknowns) are replaced by a less heavily parameterized multiplicative structure involving row effects, column effects, or both. The RC model leads to simpler tests of independence than in the usual log-linear model for unordered categories.

A simple model with only one unknown parameter when both factors are ordered is called the linear by linear association model (or constant association model). Thus with Poisson sampling $y_{ij} \sim Po(\mu_{ij})$ the means are modelled as

$$\log(\mu_{ij}) = u_0 + u_{1i} + u_{2j} + \phi \rho_i \chi_j$$

where ρ_i and χ_j are known scores, e.g. $\rho_i = i$ and $\chi_j = j$. Alternatively, for a two-way table with multinomial sampling, let π_{ij} be the probabilities of response in cell (i, j) with $\Sigma_{ij}\pi_{ij} = 1$. So with $\pi = (\pi_{11}, \pi_{12}, \ldots, \pi_{IJ})$, $y \sim \text{Mult}(y \ldots, \pi)$ then

$$\pi_{ij} = \exp(\mu_{ij})/[1 + \exp(\mu_{ij})], \quad j = 1, \ldots, J - 1$$

with μ_{ij} as above.

If the 95% interval for ϕ straddles zero then independence is supported, while a 95% interval consistent with $\phi > 0$ implies that higher totals y_{ij} occur when κ_i and λ_j increase or decrease together. $\phi < 0$ means higher counts y_{ij} occur when κ_i increases but λ_j decreases or vice versa. Under this model the log odds for any 2×2 subtable is obtained as

$$\log([\pi_{ij}\pi_{km}]/[\pi_{im}\pi_{kj}]) = \phi(\rho_i - \rho_k)(\chi_j - \chi_m)$$

so in this sense association is constant across the whole table.

If rows are not ordered but columns are, one might model interactions u_{12ij} with assigned column scores and unknown row scores, not subject to an order constraint, so that

$$u_{12ij} = j\rho_i$$
$$\log(\mu_{ij}) = u_0 + u_{1i} + u_{2j} + j\rho_i$$

with row scores subject to $\sum \rho_i = 0$ for indentifiability (this is a row effects model). A column effects model has $u_{12ij} = i\chi_j$.

When both factors are ordered, both scores ρ_i and χ_j may be taken as unknown (this is the row $\times$ column or RC model) so that

$$u_{12ij} = \rho_i\chi_j$$

The scaling and location of both ρ_i and χ_j is arbitrary and for identifiability a zero-sum, unit-length constraint may be imposed, e.g. $\sum \chi_j = 0$, $\sum \chi_j^2 = 1$ and $\sum \rho_i = 0$, $\sum \rho_j^2 = 1$. Order constraints may also be applied to both sets of scores $\rho_1 \leq \rho_2 \leq \cdots \leq \rho_I$ and $\chi_1 \leq \chi_2 \leq \cdots \leq \chi_J$ (Ritov and Gilula, 1991). Additionally an association parameter may be included, namely

$$\log(\mu_{ij}) = u_0 + u_{1i} + u_{2j} + \phi\rho_i\chi_j$$

There is also a row + column (R + C) model defined by

$$\log(\mu_{ij}) = u_0 + u_{1i} + u_{2j} + s_{2j}\rho_i + s_{1i}\chi_j$$

where ρ_i and χ_j are known scores (e.g. $\rho_i = i$, $\chi_j = j$) and the scores s_{1i} and s_{2j} are constrained to sum to zero and unit length but not constrained to be monotonic.

Albert (1997) considers how departures from the constant association model can be accommodated when they are confined to a few cells. Thus the departures are limited to a small fraction of the total cells. Specifically, for a two-way table with IJ total cells, the number of outliers might be a priori set at $k = qIJ$ where q (the fraction of outliers) is small (under 15%). The constant association RC model is therefore extended to the

RC + INT model

$$\log(\mu_{ij}) = u_0 + u_{1i} + u_{2j} + \phi \rho_i \chi_j + e_{ij}$$

where the e_{ij} measure departures from the constant association model. Because only a small proportion of the e_{ij} are expected to be non-zero, binary indicators δ_{ij} are used to choose between an outlier prior (non-zero interaction prior)

$$e_{ij} \sim N(0, P_N^{-1})$$

and a zero-interaction prior, $e_{ij} = 0$. In practice the zero interaction prior is $e_{ij} \sim N(0, P_Z^{-1})$ where P_Z is large. Typical values for P_N and P_Z might be 0.001 and 1000 respectively. The prior on the δ_{ij} is $\delta_{ij} \sim \text{Bern}(q)$ where q is the proportion of outliers assumed known as part of the prior, so q might be varied to assess sensitivity (e.g. $q = 0.05$ vs. $q = 0.01$). The prior probabilities of $0, 1, 2, \ldots$ outliers are then given by a binomial with probability of success q and IJ as the binomial denominator.

Example 7.8   Mental health status, constant association model with outliers   Albert (1997) analyses data on mental health status by parental socio-economic status (Table 7.4) and originally considered by Goodman (1979); the health state has $I = 4$ ordered categories while social status has $J = 6$.

Table 7.4   Mental health by social status

Mental health	Parental status					
	A	B	C	D	E	F
Well	64	57	57	72	36	21
Mild symptoms	94	94	105	141	97	71
Moderate symptoms	58	54	65	77	54	54
Impaired	46	40	60	94	78	71

The outlier model just outlined is applied with $k = 3$ as the expected outlier total so that $q = 3/24 = 0.125$. So model 1 is the constant association RC model

$$\log(\mu_{ij}) = u_0 + u_{1i} + u_{2j} + \phi \rho_i \chi_j$$

with $\rho_i = i$ and $\chi_j = j$ taken as known. The alternative general dependence model (model 2) is

$$\log(\mu_{ij}) = u_0 + u_{1i} + u_{2j} + \phi \rho_i \chi_j + e_{ij}$$

where one or more e_{ij} are non-zero. The prior probability of three outliers is 0.24 by the binomial formula and that of zero outliers (i.e. model 1 applies) is $(0.875)^{24} = 0.04$. Setting $P_N = 0.001$ and $P_Z = 10\,000$ a posterior probability of zero outliers of 0.972 is obtained, so the Bayes factor B_{12} supporting model 1 is 24, providing strong support for the constant association model without further interaction parameters. The parameter ϕ indicates positive association between the factors with a 95% interval (0.06, 0.12). The DIC for this approach (model A which averages over models 1 and 2) is 175 ($d_e = 11$).

Model B has an association parameter and unknown row and column scores; the latter are subject to standardization for identifiability and also constrained to be monotonic. Thus

$$\log(\mu_{ij}) = u_0 + u_{1i} + u_{2j} + \phi\rho_i\chi_j$$

with $\rho_1 \leq \rho_2 \leq \cdots \leq \rho_I$ and $\chi_1 \leq \chi_2 \leq \cdots \leq \chi_J$. This model also shows a positive association with a 95% interval on ϕ of (0.16, 0.31). However the DIC is now 204.5 with $d_e = 17$. So despite assuming known row and column scores, model A is preferred.

7.7 MULTIVARIATE ORDERED OUTCOMES

Suppose M outcomes are obtained for $i = 1, \ldots, n$ individuals and containing $J_1, J_2, \ldots, J_M$ ordered categories respectively. For simplicity, take $J = J_1 = J_2 = \cdots = J_M$ perhaps by expanding or conflating scales to achieve this simplification. Then the observed ranks are denoted y_{im}, $i = 1, \ldots, N$, $m = 1, \ldots, M$. As above, one may introduce latent variables y_{im}^* and $J - 1$ cut points κ_{mj} on their range such that

$$y_{im} = j \quad \text{if } \kappa_{m,j-1} \leq y_{im}^* < \kappa_{mj}$$
$$-\infty < \kappa_{m1} \leq \cdots \leq \kappa_{m,J-1} < \infty$$

For example, the analogue of the univariate ordinal probit ($F = \Phi$) would mean the y_{im}^* being multivariate normal with mean $\{B_1X_i, B_2X_i, \ldots, B_MX_i)$, and with a dispersion matrix being a correlation matrix $R = [\rho_{lm}]$ of order M. As above, the constraint $\kappa_{m1} = 0$ may be used, leaving $J - 2$ free cut points and the intercept β_{m0} as a free parameter, so that the regression parameter vectors are $B_m = (\beta_{m0}, \ldots, \beta_{mp})$.

However, a further option has the advantage that the dispersion matrix is replaced by an unrestricted dispersion matrix. Instead of modelling the κ_{mj} directly, Chen et al. (2000) consider transformed cut points χ_{mj} confined to the interval $(0,1)$, such that $\chi_{m,J-1} = 1$ in addition to $\chi_{m1} = 0$.

Then

$$y_{im} = j \quad \text{if } \chi_{m,j-1} \leq w^*_{im} < \chi_{mj} \qquad (7.12)$$

where the redefined latent variables w^*_{im} are (for instance) taken as multi-variate normal with mean for the mth outcome $C_m X_i$ and an unrestricted dispersion matrix Ω, such that the (l, m)th element of Ω is given by $\rho_{lm}\sqrt{\Omega_{ll}}\sqrt{\Omega_{mm}}$. The original regression parameters B_{mj} for covariates $j = 0, \ldots, p$ are obtained as

$$B_{mj} = C_{mp} / \sqrt{\Omega_{mm}}$$

and the original cut points as

$$\kappa_{mj} = \chi_{mj} / \sqrt{\Omega_{mm}}$$

including $\kappa_{mJ} = 1/\sqrt{\Omega_{mm}}$. So in the redefined analysis there are fewer cut point parameters, namely $M(J - 3)$, but an unrestricted dispersion matrix. Other links can be obtained by scale mixing. For example, O'Brien and Dunson (2004) approximate multivariate logistic sampling for multiple ordinal variables. Thus truncated sampling of y^* takes place from a multivariate normal with observation-specific scale matrix R/ξ_i or Ω/ξ_i where $\xi_i \sim \text{Ga}(0.5\nu, 0.5\nu)$ and ν may be unknown.

As for multivariate binary and count responses, an alternative approach to modelling correlation between several ordinal variables is to introduce $Q < M$ factor variables. For independent factors and (standardized) load-ings $\lambda_{mq}(m = 1, \ldots, M, q = 1, \ldots, Q)$ of variable m on factor q, the cor-relations ρ_{lm} between the original ordinal variables are represented as $\rho_{lm} = \sum_{q=1}^{Q} \lambda_{lq}\lambda_{mq}$. Let γ_{mij} be the cumulative probability of a response j or lower for response m and respondent i so that

$$\gamma_{mij} = \pi_{mi1} + \pi_{mi2} + \cdots + \pi_{mij}$$

and with logit, probit or other links g and a proportional regression effect

$$g(\gamma_{mij}) = \kappa_{mj} - \eta_{mi}$$

Under a factor approach, the regression function will combine the impact of known predictors x_i and one or more factors so that

$$\eta_{mi} = B_m X_i + \lambda_{m1} F_{i1} + \lambda_{m2} F_{i2} + \cdots + \lambda_{mQ} F_{iQ}$$

For the mth variable with J_m categories, an intercept will not be included in B_m if there are $J_m - 1$ free cut points but can be included (since it is then identified) when there are $J_m - 2$ unknown cut points. As is usual in factor analysis applications, the y_{im} are taken to be independent when the values of the latent variables are known,

$$f(y_i|F_i) = \prod_{m=1}^{M} f(y_{im}|F_i)$$

where $F_i = (F_{i1}, \ldots, F_{iQ})$, $y_i = (y_{i1}, \ldots, y_{iM})$.

Another possible approach when multivariate items are all taken to reflect the same underlying concept is to regard the observations y_{im} as repetitions on a single scale, much as in item analysis for binary responses (Gibbons and Hedeker, 2000). One may then define a subject effect u_i and assess intrasubject correlation by comparing σ_u^2 to $\sigma_u^2 + \sigma_e^2$, where σ_e^2 is defined by the link used. Assume all M variables have J ordered categories. Consider a logit link and single predictor X_i, and define the dummy variables $D_{imj} = 1$ if $Y_{im} \le j$, and $D_{imj} = 0$ otherwise. When regarded as repetitions on a single scale the data can be modelled as

$$D_{imj} \sim \text{Bern}(\gamma_{imj}) \quad m = 1, \ldots, M; j = 1, \ldots, J-1$$

and assuming proportional odds

$$\text{logit}(\gamma_{imj}) = \alpha_j + \beta_{0m} + \beta_{1m}X_i + u_i \tag{7.13}$$

where $\alpha_{J-1} > \alpha_{J-2} > \cdots > \alpha_1$ and a corner constraint is applied to item intercepts β_{0m} (e.g. $\beta_{01} = 0$) if all $J-1$ cut points are free parameters. The comparison with item analysis becomes closer if the variance of the u_i is preset, e.g. $u_i \sim N(0, 1)$, and loadings λ_j are introduced such that

$$\text{logit}(\gamma_{imj}) = \alpha_j + \beta_{0m} + \beta_{1m}X_i + \lambda_j u_i$$

Example 7.9 London suicides To illustrate sampling y^* when there are $M = 2$ ordinal responses consider suicide SMRs in 32 London boroughs for 1989–1993 in terms of quartile ranks, $J_1 = J_2 = 4$ (e.g. male suicide SMRs under 80.7 have rank 1, those between 80.7 and 96 have rank 2, etc.). There is a single predictor, a deprivation score. The cut point parameterization in (7.12) is used, and the bivariate normal therefore has variances $\{\sigma_1^2, \sigma_2^2\}$ which are not fixed. The bivariate normal is obtained conditionally; that is, y_{i1}^* is sampled according to cut points χ_{1j} (only one of which is free) from

$$\eta_{i1} = \gamma_{11} + \gamma_{12}X_i$$

giving a residual $r_{i1} = y_{i1}^* - \eta_{i1}$ and then y_{i2}^* is sampled from

$$\eta_{i2} = \gamma_{21} + \gamma_{22}X_i + \delta r_{i1}$$

Sampling of the cut points requires specifying those observations with relevant y values that determine the constraints used in their sampling.

Following Spiegelhalter and Marshall (1999) a relatively diffuse prior is adopted on $\delta = \rho\sigma_2/\sigma_1$. The correlation ρ between the scores y_{i1}^* and y_{i2}^* may be estimated using the posterior means of δ, ϕ_1 and ϕ_2. Here we find a stronger effect of deprivation on male suicide $\gamma_{12} = 0.46(0.12)$ than on female suicide $\gamma_{22} = 0.39(0.14)$ and a correlation of 0.90 between the two latent scores. This conditional specification of the bivariate (or higher order) multivariate normal may have advantages in scale mixing as it allows different implicit links for each response. Thus scale multipliers ξ_{i1}, ξ_{i2}, etc., may be sampled from gamma densities with different unknown degrees of freedom.

Example 7.10   Social attitudes   As an example of the factor analytic approach consider three ordinal attitude variables from the 2000 General Social Survey. The analysis is confined to 1665 subjects with complete responses to all items:

1. Sex relations before marriage: always wrong (1), almost always wrong (2), wrong only sometimes (3) or not wrong at all (4).
2. Divorce laws: should divorce in this country be easier or more difficult to obtain than it is now? Easier (1), Stay Same (2), More Difficult (3).
3. It is sometimes necessary to discipline a child with a good, hard spanking: Strongly Agree (1), Agree (2), Disagree (3), Strongly Disagree (4).

A single factor $(Q = 1)$ is taken to underlie associations between these items and represent the contrast between traditional and libertarian social attitudes. The known predictor age (in standard units, denoted x_i) is also used. The first cut point for each variable is taken as free, so the regression means $(m = 1, 2, 3)$ have the form

$$\eta_{mi} = \beta_m x_i + \lambda_m F_i$$

The loading λ_1 of the first question on the factor is constrained to be positive and since responses to the first question are increasing in libertarianism F_i will be a positive measure of libertarianism. Note that responses to the second question are ascending in traditionalism so λ_2 will tend to be negative.

    A two-chain run of 2000 iterations shows early convergence with posterior means from iteration 500 to 2000 on the scaled loadings being $\lambda = \{0.92, -0.56, 0.39\}$. Traditional attitudes tend to increase with age, especially for the first two questions, so β_1 is negative and β_2 is positive.

The highest absolute correlations $\rho_{m_1 m_2}$ are between variables 1 and 2, namely -0.51, as against 0.35 between the first and third questions, and -0.21 between questions 2 and 3. Further options might be considered for these data, such as alternative links and possible scope for a second factor since the third variable does not seem so well explained.

Example 7.11   NORC sexual attitudes data   Agresti and Lang (1993) and Gibbons and Hedeker (2000) both consider data from the earlier 1989 General Social Survey on sexual attitudes. Specifically they consider $M = 3$ ordinal responses of 475 subjects (Table 7.5) to three items: (a) sex relations among early teens (ages 14–16); (b) premarital sex relations; (c) extramarital sex relations. The $J = 4$ responses are as follows: $1 =$ always wrong, $2 =$ almost always wrong, $3 =$ wrong only sometimes, $4 =$ not wrong.

Table 7.5   Sexual attitudes

Teen sex	Premarital sex	Extramarital sex				
		1	2	3	4	Total
1	1	140	1	0	0	141
	2	30	3	1	0	34
	3	66	4	2	0	72
	4	83	15	10	1	109
2	1	3	1	0	0	4
	2	3	1	1	0	5
	3	15	8	0	0	23
	4	23	8	7	0	38
3	1	1	0	0	0	1
	2	2	0	0	0	0
	3	3	2	3	1	9
	4	13	4	6	0	23
4	1	0	0	0	0	0
	2	0	0	0	0	0
	3	0	0	1	0	1
	4	7	2	2	4	15

First of all consider the proportional model as in (7.13)

$$D_{imj} \sim \text{Bern}(\gamma_{imj}) \quad m = 1, \ldots, 3; j = 1, \ldots, 3$$
$$\text{logit}(\gamma_{imj}) = \kappa_j + \beta_{0m} + u_i$$

With $J - 1$ cut points κ_j as free parameters, and the first item intercept β_{01} (for teen sex) set equal to zero for identifiability, the remaining β_{0m}

are then contrast parameters measuring whether premarital and extra-marital sex are seen as more or less wrong (in terms of opinions in higher categories) than teen sex. Convergence in a two-chain run of 2000 iterations is obtained at around 250 iterations. The estimates β_{02} and β_{03} of -4.8 and 0.7 show premarital sex is seen as less wrong than teen sex, but extramarital sex is seen as more wrong. The estimated variance of the subject effects u_i is 10.8 and the intrasubject correlation

$$\sigma_u^2/(\sigma_u^2 + \sigma_e^2)$$

averages 0.76; $\sigma_e^2 = \pi^2/3$ is the variance of the standard logistic continuous variable taken to underlie the observed ordinal response. The DIC is 2062 with 259 effective parameters.

A non-proportional model takes the item intercepts to vary according to the level of the underlying scale, namely

$$\text{logit}(\gamma_{imj}) = \alpha_j + \beta_{0mj} + u_i$$

This leads to an apparent loss of fit with the DIC rising to 2066. The proportional odds model seems most at doubt for the comparison of extramarital and teen sex, with the coefficients β_{031}, β_{032} and β_{033} being 0.75 (0.31, 1.18), 0.44 (-0.14, 1.04) and 1.70 (0.52, 3.03). However, these estimates are relatively imprecise being based on low frequencies. One may also estimate the within-item contrasts (e.g. $\beta_{033} - \beta_{032}$) as in Agresti and Lang (1993).

EXERCISES

1. In Example 7.1 try a scale mixture with y^* sampled from truncated normals with unknown degrees of freedom parameter. Does the probit or logit link get more support?

2. In Example 7.2 try a model with cut points common between the skewed left and skewed right extreme value links but different cut points for the logit link (i.e. model B simplified to have six rather than nine cut points).

3. In Example 7.3 try a GAM for the impact of age using 15 five-year age bands (under 20, 20–24, 25–29, 30–34, ..., 80–84, 85–89) rather than the 72 individual ages between 18 and 89. Assess the fit for proportional and non-proportional age GAMs. Also assess whether proportional and non-proportional GAMs for the impact of prestige improve the model (with age also still a smooth function).

4. In Example 7.5 try $N(2, 1)$ as a baseline density F_b and experiment with unknown M using a discrete prior over selected values $1, 2, 3, 4, \ldots, 20$.

5. In Example 7.6 recode the program to pool the likelihood over the two periods so as to assess the probability that $C_2 < C_1$ (i.e. that health inequality increased).

6. In Example 7.6 compare the fit of logit, probit and complementary log–log models and apply a procedure to mix over the links.

7. In Example 7.7 apply the cumulative complementary log–log link rather than the cumulative logit link and assess any changes in fit or in inferences about the best entrée.

8. In Example 7.8 (mental health by social status) obtain the Bayes factor in support of the independence model

$$\log(\mu_{ij}) = u_0 + u_{1i} + u_{2j}$$

and the alternative interaction model is

$$\log(\mu_{ij}) = u_0 + u_{1i} + u_{2j} + e_{ij}$$

when $k = 2$ interactions e_{ij} are a priori expected to be non-zero.

9. In Example 7.8 try the $R + C$ model and compare its fit to the two RC models already discussed.

10. In Example 7.9 try scale mixing with $\nu_1 = \nu_2 = 4$ as preset parameters, and assess fit against the MVN using the predictive concordancy approach used in Example 7.1.

11. In Example 7.11 obtain the posterior probability that β_{032} is less than β_{033} in the non-proportional model.

REFERENCES

Agresti, A. (2002) *Categorical Data Analysis*, 2nd edition. John Wiley & Sons: New York.

Agresti, A. and Lang, J. (1993) Quasi-symmetric latent class models, with application to rater agreement. *Biometrics*, **49**, 131–139.

Albert, J. (1997) Bayesian testing and estimation of association in a two-way contingency table. *Journal of the American Statistical Association*, **92**, 685–693.

Albert, J. and Chib, S. (1993) Bayesian analysis of binary and polychotomous response data. *Journal of the American Statistical Association*, **88**, 669–679.

Besag, J., Green, P., Higdon, D. and Mengersen, K. (1995) Bayesian computation and stochastic systems. *Statistical Science*, **10**, 1–41.

Chen, M., Shao, Q. and Ibrahim, J. (2000) *Monte Carlo Methods in Bayesian Computation*. Springer: New York.

Chotikapanich, D. and Griffiths, W. (2000) Posterior distributions for the Gini coefficient using grouped data. *Australian and New Zealand Journal of Statistics*, **42**, 383–392.

Cox, C. (1995) Location-scale cumulative odds models for ordinal data: a generalized nonlinear model approach. *Statistics in Medicine*, **14**, 1191–1203.

Cox, D. and Snell, E. (1989) *Analysis of Binary Data*. Chapman and Hall: London.

Fahrmeier, L. and Tutz, G. (2001) *Multivariate statistical modelling based on generalized linear models*. Springer: Berlin.

Gelfand, A. and Kuo, L. (1991) Nonparametric Bayesian bioassay including ordered polytomous response. *Biometrika*, **78**, 657–666.

Gibbons, R. and Hedeker, D. (2000) Applications of mixed-effect models in biostatistics. *Sankyha*, **62B**, 70–103.

Goodman, L. (1979) Simple models for the analysis of association in cross-classifications having ordered categories. *Journal of the American Statistical Association*, **74**, 537–552.

Hastie, T. and Tibshirani, R. (1987) Generalized additive models: some applications. *Journal of the American Statistical Association*, **82**, 371–386.

Hellmich, M., Abrams, K., Jones, D. and Lambert, P. (1998) A Bayesian approach to a general regression model for ROC curves. *Medical Decision Making*, **18**, 436–443.

Ishwaran, H. (2000) Univariate and multirater ordinal cumulative link regression with covariate specific cutpoints. *Canadian Journal of Statistics*, **28**, 715–730.

Ishwaran, H. and Gatsonis, C. (2000) A general class of hierarchical ordinal regression models with applications to correlated ROC analysis. *Canadian Journal of Statistics*, **28**, 731–750.

Johnson, V. and Albert, J. (1999) *Ordinal Data Modelling*, Springer: New York.

Kakwani, N., Wagstaff, A. and van Doorslaer, E. (1997) Socioeconomic inequalities in health: measurement, computation and statistical inference. *Journal of Econometrics*, **77**, 87–103.

Lang, J. (1999) Bayesian ordinal and binary regression models with a parametric family of mixture links. *Computational Statistics and Data Analysis*, **31**, 59–87.

Lee, T. and Wild, C. (1996) Vector general additive models. *Journal of the Royal Statistical Society, Series B*, **58**, 481–493.

Long, J. (1997) *Regression Models for Categorical and Limited Dependent Variables*. Sage: Thousand Oaks, CA.

McCullagh, P. (1980) Regression models for ordinal data. *Journal of the Royal Statistical Society, Series B*, **42**, 109–142.

Murad, H., Fleischman, A., Sadetzki, S., Geyer, O. and Freedman, L. (2003) Small samples and ordered logistic regression: does it help to collapse categories of outcome? *American Statistician*, **57**, 1–6.

Nandram, B. (1997) Bayesian inference for the best ordinal multinomial population in a taste test. In *Case Studies in Bayesian Statistics, 3*, Gatsonis, C., Hodges, J., Kass, R. and Singpurwall, N. (eds). Springer: New York.

Nandram, B. (1998) A Bayesian analysis of the three stage hierarchical multinomial model. *Journal of Statistical Computing and Simulation*, **61**, 97–126.

O'Brien, S. and Dunson, D. (2004) Bayesian multivariate logistic regression. *Biometrics*, **60**, 739–746.

Parmigiani, G., Ashih, H., Samsa, G., Duncan, P., Lai, S. and Matchar, D. (2003) Cross-calibration of stroke disability measures: Bayesian analysis of longitudinal ordinal categorical data using negative dependence. *Journal of the American Statistical Association*, **98**, 273–281.

Peng, F. and Hall, W. (1996) Bayesian analysis of ROC curves using Markov-chain Monte Carlo methods. *Medical Decision Making*, **16**, 404–411.

Peterson, B. and Harrell, F. (1990) Partial proportional odds models for ordinal response variables. *Applied Statistics*, **39**, 205–217.

Ritov, Y. and Gilula, Z. (1991) The order-restricted RC model for ordered contingency tables: estimation and testing for fit. *Annals of Statistics*, **19**, 2090–2101.

Simonoff, J. (2003) *Analyzing Categorical Data*. Springer: New York.

Spiegelhalter, D. and Marshall, E. (1999) Inference-robust institutional comparisons: a case study of school examination results. In *Bayesian Statistics 6*, Bernardo, J., Berger, J., Dawid, A. and Smith, A. (eds). Oxford University Press: Oxford, 613–630.

Tosteson, A. and Begg, C. (1988) A general regression methodology for ROC curve estimation. *Medical Decision Making*, **8**, 204–215.

Wagstaff, A. and van Doorslaer, E. (1994) Measurement of health inequalities in the presence of multiple-category morbidity indicators. *Health Economics*, **3**, 281–291.

CHAPTER 8

Discrete Spatial Data

8.1 INTRODUCTION

Discrete data are frequent in ecological studies of social behaviour or disease outcomes and it is here that many developments of recent Bayesian methodology have occurred – see Wakefield *et al.* (2000), Bailey (2001) and Pascutto *et al.* (2000) for reviews of spatial epidemiological applications. The same issues occur for spatially configured count or rate data as considered in Chapters 5 and 6: discrete data for areas often show greater variability than under Poisson, multinomial or binomial sampling. Event totals within districts cumulate over individual events that may be clustered (e.g. within households). Such data are typically also accumulated within discrete administrative areas, so that confounded with variability between areas due to real substantive influences, there is also that due to arbitrary boundaries.

Hence, with mortality or morbidity data overdispersion in the observed event counts may be to some degree spatially unstructured, but may also reflect spatial correlation in underlying relative risks – that in turn may reflect omitted risk factors (Pascutto *et al.*, 2000; Besag *et al.*, 1991) straddling arbitrary boundaries. However, the possibly multiple form of random effects in spatial data raises distinct identifiability and parameterization issues. For example, certain models may only be identifiable by restricting certain random parameters within the set, or by enforcing identifiability during the MCMC run (Carlin and Pérez, 2000). In this chapter the focus is on discrete area analysis (e.g. using health counts for administrative areas). Recent progress in Bayesian geostatistics (i.e. in continuous space) has been substantial, exemplified in the book by Banerjee *et al.* (2004), and papers by Diggle *et al.* (1998) and Diggle *et al.* (2003).

Bayesian Models for Categorical Data   P. Congdon
© 2005 John Wiley & Sons, Ltd

8.2   UNIVARIATE RESPONSES: THE MIXED ICAR MODEL AND EXTENSIONS

Let y_i denote death totals in area i, with expected deaths E_i or populations at risk T_i. Extra variability is usually modelled via a log-linear or logit-linear regression as in (5.5). For example,

$$y_i \sim \text{Poi}(E_i \rho_i)$$

where ρ_i is the relative mortality risk. The mixed model of Besag *et al.* (1991) has been seminal in terms of wider developments such as spatio-temporal models and multivariate spatial models. The mixed model proposes

$$\log(\rho_i) = \beta X_i + u_i + s_i \qquad (8.1)$$

where the spatially unstructured error u_i is typically taken as normal with constant variance

$$u_i \sim \text{N}(0, \tau_u^2)$$

One possible joint density for the spatial effects $s = (s_1, \ldots, s_n)$ is in terms of pairwise differences in errors, and a variance term φ_s^2

$$P(s) \propto \exp\left[-0.5 \varphi_s^{-2} \sum_{i<j} (s_i - s_j)^2 \right]$$

An equivalent conditional specification is obtained using the properties of the multivariate normal (Besag and Kooperberg, 1995). Thus, the preceding joint density implies a conditional prior for area i, conditioning on the effects s_j in remaining areas $j \neq i$ that are denoted $s_{[i]}$, namely

$$P(s_i | s_{[i]}, \tau_s^2) \sim \text{N}(\omega_i, V_i) \qquad (8.2)$$

where ω_i is a weighted average

$$\omega_i = \sum_j c_{ij} s_j / \sum_j c_{ij} = \sum_j w_{ij} s_j$$

The conditional variances $V_i = \sigma_s^2 / \sum_j c_{ij}$ in (8.2) depend on the interaction structure represented by c_{ij}. Typical forms for c_{ij} are (a) binary adjacency: $c_{ij} = 1$ if areas i and j are neighbours, $c_{ij} = 0$ otherwise (and $c_{ii} = 0$); and (b) distance decay: $c_{ij} = \exp(-\gamma d_{ij})$ where $\gamma > 0$ and d_{ij} are distances between the area centres (and $c_{ii} = 0$). From c_{ij} the spatial weights w_{ij} are obtained as row standardized interactions $w_{ij} = c_{ij} / \sum_j c_{ij}$. The more neighbours there are (definition a), or the closer they are (definition b), the more precisely is s_i defined (in terms of lower V_i).

Events y_i may be modelled via a logistic linear regression when populations at risk T_i are small, not just the event totals y_i (e.g. MacNab, 2003, p 306), with $y_i \sim \text{Bin}(T_i, \mu_i)$ and the mixed model is accordingly

$$\text{logit}(\mu_i) = \beta X_i + u_i + s_i$$

Aggregate multinomial data $y_i = \{y_{i1}, \ldots, y_{iJ}\}$ with J categories (e.g. deaths by area due to different causes, or votes by constituency to different parties, or households by census area with different religious affiliations) may also be included in this framework. Thus one might specify

$$y_i \sim M_J(y_{i+}, \pi_i)$$
$$\pi_{ij} = \exp(\eta_{ij}) / \sum_k \exp(\eta_{ik})$$
$$\eta_{ij} = \beta_j X_i + u_{ij} + s_{ij}$$

with a zero-mean multivariate normal model for u_{ij}, an identifiability contraint on the β_j such as $\beta_1 = 0$ and the s_{ij} being spatial errors constrained to sum to zero over areas and over categories j.

For some purposes (e.g. defining priors for spatio-temporal effects) it is useful to note that the joint prior for $s = (s_1, \ldots, s_n)$ can be expressed as a multivariate normal of dimension n with mean 0 and precision matrix $\tau_s K$, where $\tau_s = 1/\sigma_s^2$ and the structure matrix K is defined as $K_{ij} = -c_{ij} (i \neq j)$, $K_{ii} = \sum_j c_{ij}$; see Mollié (1996), Lagazio et al. (2001) and Clayton (1996).

The most common assumption in disease mapping is to define c_{ij} by binary adjacency, so that

$$\omega_i = \bar{s}_i \tag{8.3a}$$
$$V_i = \sigma_s^2 / N_i \tag{8.3b}$$

where $\bar{s}_i$ is the average of s_j over the N_i neighbours of area i. When interaction is specified by adjacency the structure matrix K is defined as

$$
\begin{aligned}
K_{ij} &= -1 \quad &&\text{if areas } i \text{ and } j \text{ are neighbours} \\
&= 0 \quad &&\text{for non-adjacent areas} \\
&= N_i \quad &&\text{when } i = j
\end{aligned}
\tag{8.4}
$$

This prior is termed the intrinsic conditional autoregression or ICAR prior and poses identification issues since it specifies only differences in log relative risks, and not the overall level of relative risk. To achieve identification of the s_i one may centre them at each iteration in an MCMC run (Gelfand and Sahu, 1996). Other ways of achieving identification are to set a single s_i to a fixed value (e.g. $s_1 = 0$) or to omit the intercept in the log-linear or log logistic regression.

There are also identification issues regarding the separate random effects in (8.1). Only the sum of u_i and s_i is identified and the partitioning of the variance between unstructured and spatial variation may be affected by prior specifications. The prior on the variances σ_s^2 and σ_u^2 or their inverses τ_s and τ_u is especially relevant, e.g. in terms of the appropriate balance between the two types of variation. Mollié (1996, p 372) suggests using the inverse of the observed variance of the crude relative risks $\Pi_r = 1/\mathrm{Var}(r)$ where $r_i = y_i/E_i$. Thus $\tau_u \sim \mathrm{Ga}(c\Pi_r, c)$ and $\tau_s \sim \mathrm{Ga}(c\Pi_r/\bar{N}, c)$ where $\bar{N}$ is the average number of neighbours and c is a small constant (e.g. $c = 0.01$) that downweights the prior (data-based) information. Bernardinelli *et al.* (1995) suggest the marginal standard deviation φ_s of the spatial effects be proportional to the conditional standard deviation σ_s. One may estimate φ_s empirically during an MCMC run as the standard deviation, $\mathrm{sd}(s)$, of the sampled $s_i^{(t)}$ and then it is recommended that

$$\mathrm{sd}(s) \approx \sigma_s/(N_i^{0.5}C) \quad \text{with } C = 0.7$$

A neutral prior (with regard to the balance between independence and spatial clustering) might then set

$$\sigma_u^2 = \sigma_s^2/(\bar{N}C^2) \tag{8.5}$$

(see Carlin and Pérez, 2000). If preset values $\{a_s, b_s\}$ are used for the spatial precision $\tau_s \sim \mathrm{Ga}(a_s, b_s)$ then default choices such as $a_s = b_s = 0.001$ have been shown to put little mass on small variances, in contrast to choices such as $a = 0.5$, $b = 0.0005$, as reecommended by Kelsall and Wakefield (1999).

Some studies (e.g. Gelfand *et al.*, 1998) report that both unstructured and structured errors, namely u_i and s_i, may not both be necessary in many applications, and one may consider a reduced model retaining s_i but without u_i. This defines a pure spatial ICAR process

$$g(\mu_i) = \beta X_i + s_i$$

This simplification eliminates one form of identifiability issue but it is still necessary to adopt devices such as centring the s_i at each iteration.

To obtain propriety and so avoid such devices, a spatial correlation parameter ρ may be added (Sun *et al.*, 2000), leading to the ICAR(ρ) model, whereas (8.2) is the ICAR(1) model where $\rho = 1$ by default. Thus

$$P(s_i|s_{[i]}, \rho, \tau_s^2) \sim \mathrm{N}(\rho\omega_i, V_i) \tag{8.6}$$

where ω_i is as above. So under binary adjacency, $\omega_i = \bar{s}_i$ in (8.6). $\rho \in [0, 1]$ measures spatial association since the covariance between s_i

and s_j is obtained as

$$\tau_s^2 \rho c_{ij} / \left(\sum_k c_{ik} \sum_k c_{jk} + \rho^2 c_{ij}^2 \right)$$

Another ICAR prior proposed as avoiding identifiability problems with the mixed model (8.1) is suggested by Leroux et al. (1999) and applied by MacNab (2003). This has joint form

$$P(s|\kappa^2, \lambda) \sim N_n(0, \kappa^2 R^{-1})$$

where the precision matrix is $R = \lambda K + (1 - \lambda)I$, with K an $n \times n$ structure matrix as in (8.4). The corresponding conditional form in the case of binary adjacency is

$$P(s_i|s_{[i]}, \kappa^2, \lambda) \sim N\left(\lambda \sum_{j \sim i} s_j / D_i, \kappa^2 / D_i \right) \qquad (8.7)$$

where $j \sim i$ denotes that j is a neighbour of area i, and $D_i = 1 - \lambda + \lambda N_i$. This reduces to an ICAR(1) prior when $\lambda = 1$ and to unstructured variation when $\lambda = 0$, so can be viewed as allowing for both types of variation, spatially structured or unstructured.

Another spatial prior is based on the 'multiple members' model of Langford et al. (1999) and Browne et al. (2001). Thus underlying the spatial errors s_i are unstructured errors v, distinct from u in (8.1), namely

$$s_i = \sum_{j=1}^{n} w_{ij} v_j \qquad (8.8a)$$

$$v_i \sim N(0, \sigma_v^2) \qquad (8.8b)$$

where the w_{ij} are row standardized spatial interactions. For binary adjacency, $s_i = N_i^{-1} \sum_{j \sim i} v_j$. This framework has the benefit that u_i and v_i may be correlated and also provides one way of modelling multivariate spatial effects and spatially non-constant predictor effects (section 8.5). One may also extend it to allow a compromise prior as in (8.7). Thus for underlying unstructured random effects v_{i1} and v_{i2}, define

$$g(\mu_i) = \beta X_i + s_i$$

where

$$s_i = \phi_i N_i^{-1} \sum_{j \sim i} v_{j1} + (1 - \phi_i) v_{i2}$$

and ϕ_i is a beta mixture, $\phi_i \sim \text{Be}(k, k)$, with k known.

Example 8.1   Glasgow cancer deaths   This example uses data from Langford et al. (1999) on cancer deaths in 143 Glasgow postcode sectors, with a deprivation score (Carstairs index) as the single predictor X_i. Among other features of these data, overdispersion ($s_y^2 = 160$ compared with $\bar{y} = 22.3$) may be noted. Because of this, as well as likely spatial structuring of mortality, some sort of strategy involving pure hetero-geneity and spatial correlation is suggested, as in the mixed model of (8.1). To assess outliers (or areas not well fitted by the model) the usual CPO diagnostics are obtained as well as the mixed predictive p values mentioned in Chapter 2 (Marshall and Spiegelhalter, 2003). Thus y_i is compared with the replicate data z_i to gauge if it is extreme in terms of this predictive density, via p values:

$$P_c = \Pr(z_i < y_i) + 0.5\Pr(z_i = y_i)$$

To identify the mixed ICAR(1) model of Besag et al. (1991), with s_i as in (8.2), one strategy is to omit the intercept, so that

$$\log(\rho_i) = \log(E_i) + \beta X_i + u_i + s_i$$

with predicted deaths in area i being $\hat{Y}_i = E_i\bar{\rho}_i$ where $\bar{\rho}_i$ is the posterior mean.

The first analysis sets $Ga(1, 0.01)$ priors on the precisions $\tau_u = 1/\sigma_u^2$ and $\tau_s = 1/\sigma_s^2$. To assess the relative importance of clustering vs. unstruc-tured heterogeneity one can monitor the ratio

$$R_s = sd(s)/[sd(s) + sd(u)]$$

where $sd(s)$ is based on taking the standard deviation of the sampled s_i at any iteration. This shows the two components to be roughly equal in size, with $R_s = 0.47$. The mixed model seems to provide an adequate account of the overdispersion, with the predictive check (comparing the ratios $s_z^2/\bar{z}$ for data z replicated from the model with the observed ratio) stand-ing at 0.77.

For estimates of the effective parameter count one may use the DIC method or the chi-square approximation noted by Gelman et al. (2003, p 182). The latter tends to show higher effective parameters in complex random effects models and accordingly $d_e^* = 51.5$ under the chi-square method as against $d_e = 26$ under the DIC method (DIC $= 823.4$). Finally, the regressor effect has mean $\bar{\beta} = 0.025$ with a 95% interval (0.016, 0.034).

Instead one may use the principle in (8.5) to set a higher mean on the prior precision τ_u than on τ_s. Thus with $\tau_s \sim Ga(1, 0.01)$ we take $\tau_u \sim Ga(h_u, 0.01)$ where $h_u = 0.7^2(5.44)$ with 5.44 being the average $\bar{N}$ of neighbours. This reduces d_e^* to 44.8 and d_e to 21.6 (DIC $= 822.4$) and R_s now stands at 0.49. Relatively poor fits under this model are obtained for case 18 ($Y_{18} = 13$, $\hat{Y}_{18} = 24.4$, log CPO $= -6.5$) and case 49 ($Y_{49} = 43$, $\hat{Y}_{18} = 30.6$, log CPO $= -6$) and three more areas (47, 82, 88) have CPOs under 1% of the highest CPO. Mixed predictive checks for this model (Marshall and Spiegelhalter, 2003) show four areas (42, 49, 82, 88) with P_c values over 0.95 and five areas (18, 38, 47, 103, 109) have P_c values under 0.05. The most extreme points are 18 ($P_c = 0.002$) and 49 ($P_c = 0.991$).

Model B applies the mixture model (in version 1.3) but with s_i proper according to (8.6). This gives a median ρ of 0.62 and DIC via (2.17b) of 823.5 with $d_e = 24.7$. Instead of a mixed model (8.1) one may instead apply a model which incorporates both clustering and heterogeneity as extremes. Thus the Leroux *et al.* (1999) model in (8.7) is applied with Ga(1,0.01) prior on κ^2. We find $\lambda = 0.85$ with a relatively wide 95% interval (0.55, 0.995). This leads to a worse DIC (837.6) with higher effective parameters ($d_e = 53$, $d_e^* = 63$) perhaps partly because the conditional prior for s_i involves the ratio $\lambda/(1 - \lambda + \lambda N_i)$ and hence a loss of sampling precision. The impact of the predictor is slightly reduced with $\bar{\beta} = 0.022$.

8.3 SPATIAL ROBUSTNESS

While spatial patterns are often continuous (adjacent areas have similar patterns) there are circumstances leading to discontinuities in behavioural or disease patterns over space (e.g. a spatial error model will not be appropriate for deprived high-mortality areas surrounded by affluent lower mortality areas). Greater robustness in the spatial error in (8.1) may be achieved by a scale mixture of normal densities or by heavier tailed alternatives such as the double exponential DE or Laplace prior (Besag, 1989)

$$P(s|\xi) \propto \xi^{0.5} \exp\left(-0.5\xi \sum_{i<j} |s_i - s_j|^2\right) \tag{8.9}$$

where smaller values of the scaling parameter ξ imply lower spatial variability. With binary spatial interaction, the conditional prior is double exponential with mean $\omega_i = \bar{s}_i$ and variance $V_i = \sigma_s^2/N_i$. One may premultiply the average of neighbouring values under this prior as in (8.6) giving a DE(ρ) prior.

Non-parametric mixtures for modelling discontinuities have also been considered (e.g. Militino *et al.*, 2001; Knorr-Held and Rasser, 2000;

Fernandez and Green, 2002). Lawson and Clark (2002) propose a mixture of the CAR and heavy-tailed densities for different spatial errors, such as

$$\log(\rho_i) = \beta X_i + u_i + \phi_i s_{i1} + (1 - \phi_i) s_{i2} \qquad (8.10)$$

where s_{i1} is ICAR(1) or ICAR(ρ), s_{i2} might be double exponential and ϕ_i is a beta mixture, $\phi_i \sim \text{Beta}(k_1, k_2)$ where k_1 and k_2 are preset for identifiability. The same idea may be applied to the mixture model of (8.1) so that greater emphasis is put on the unstructured error in areas that show discontinuities with respect to neighbouring areas:

$$\log(\rho_i) = \beta X_i + \phi_i u_i + (1 - \phi_i) s_i \qquad (8.11)$$

One might make the unstructured error follow a Student t model here to allow for outliers (Pascutto *et al.*, 2000).

Discrete mixture approaches may also be used for robust spatial data analysis. For instance, one may adapt the prior of (8.8) to be defined by an underlying non-parametric mixture (Congdon, 2004). Thus the underlying v_i may be drawn from a discrete mixture with a known number J of components or, as in a DPP model, with an unknown and stochastically varying total of components. Under the latter option one might define categorical indicators for area i

$$D_i \sim \text{Categorical}(\pi)$$

where π is of length M (M less than or equal to n), and updated using an appropriate prior such as the stick breaking prior with concentration parameter κ_s. Associated with each cluster k is a value $V_k (k = 1, \ldots, M)$ drawn from the baseline prior G_0 (that might be normal or Student t). Then if at a particular iteration $D_j^{(t)} = k$, for areas $j \neq i$, one obtains

$$s_i^{(t)} = \sum_{j=1}^{n} w_{ij} V_{D_j^{(t)}} \qquad (8.12)$$

One might also model the unstructured error in (8.1) in terms of a discrete mixture rather than a parametric random effect, e.g. via a standard mixture of normals (see Clayton and Kaldor, 1987) or

$$u_i \sim \text{DPP}(\kappa_u, G_0)$$

where G_0 might be a normal density.

Fernandez and Green (2002) propose modelling spatial interaction effects via the weights in a discrete mixture. Thus

$$y_i \sim \sum_{j=1}^{J} \pi_{ij} \text{Po}(E_i \mu_{ij})$$

where μ_{ij} is a regression mean differing in intercepts or other regression effects. Thus suppose a fixed number of categories M. Then generate J sets of n underlying spatially correlated effects q_{ij} from a proper density such as (8.6) or (8.7) (as opposed to the CAR(1) prior). The q_{ij} are converted to area-specific mixture weights π_{ij} via

$$\pi_{ij} = \exp(q_{ij}/\phi) / \left[1 + \sum_{k=1}^{J} \exp(q_{ik}/\phi) \right]$$

where ϕ is a positive tuning parameter. As ϕ tends to infinity the π_{ij} tend to $1/J$ without any spatial patterning, whereas small values of ϕ act to reduce overshrinkage.

Example 8.2   Robust models for Glasgow data   The analysis in Example 8.1 revealed some poorly fit cases, raising the possibility that robust models with special mixture or heavy-tailed density features might improve fit. Consider, first of all, a beta mixture on ICAR and DE spatial priors together with an unstructured effect. Thus

$$\log(\rho_i) = \beta X_i + u_i + \phi_i s_{i1} + (1 - \phi_i) s_{i2}$$
$$s_{i1} \sim \text{ICAR}(\rho_1)$$
$$s_{i2} \sim \text{DE}(\rho_2)$$

This raises the DIC to 861 (i.e. fit deteriorates) and has $d_e = 82$. The fit to cases 18 and 49 does improve slightly, however, and their CPOs relative to the maximum CPO rise above 0.01.

If the unstructured error is excluded, then d_e falls considerably (to around 12) and the DIC stands at 822.5. However, the CPOs for cases 18 and 49 worsen considerably (log CPOs of -6.5 and -6.25 respectively against a maximum of -0.1). This suggests a benefit from including an unstructured error. Hence model C follows (8.11). This leads to a very similar fit to the mixed spatial error model (DIC $= 822.1$, $d_e = 17$) without any marked improvement or worsening in the fit to cases 18 and 49.

8.4   MULTIVARIATE SPATIAL PRIORS

Suppose multivariate observations of dimension M are observed for each area. For example, suppose the responses for n areas consist of multivariate counts

$$\{y_{11}, y_{12}, \ldots, y_{1M}, y_{21}, y_{22}, \ldots, y_{2M}, \ldots, y_{n1}, y_{n2}, \ldots, y_{nM}\}$$

which might for instance be data on numbers ill and numbers dying from a particular cause. Let $y_i = (y_{i1}, \ldots, y_{iM})'$. One might propose models for the means of y_{im} as follows:

$$g(\mu_{im}) = X_i \beta_m + u_{im} + s_{im} \qquad (8.13)$$

While independent priors for each s_{im} are feasible, one might suppose intercorrelation often exists between s_{ik} and s_{im} $(k \neq m)$.

One prior allowing for multivariate interdependence for $S_i = (s_{i1}, s_{i2}, \ldots, s_{iM})'$ is proposed by Gelfand and Vounatsou (2003) and generalizes the ICAR(ρ) model of Sun et al. (1999). This provides the MCAR (ρ, Φ) prior where ρ is a scalar, with

$$P(S_i | S_{[i]}, \rho, \Phi) = N_M \left(\rho \sum_j W_{ij} S_j, \Phi / \sum_j c_{ij} \right) \qquad (8.14)$$

where Φ is an $M \times M$ covariance matrix and $W_{ij} = w_{ij} I_{M \times M}$. If ρ is set to one then identifiability requires devices such as centring each of the M sets of effects at each iteration. Gamerman et al. (2002) also discuss the MCAR(1,Φ) prior in connection with spatially varying predictor effects (section 8.5 below).

A generalization of (8.14) to a vector $\rho = (\rho_1, \ldots, \rho_M)$ is complicated by the symmetry requirement for the covariance matrix in the corresponding joint prior, though Gelfand and Vounatsou (2003) suggest a method of circumventing this. If correlations between the unstructured effects in (8.13) are required, then $U_i = (u_{i1}, u_{i2}, \ldots, u_{iM})$ may be taken as multivariate normal of order M.

An alternative way to allow for correlation between spatial errors is via a spatial common factor model (Congdon, 2004; 2002; Wang and Wall, 2003). Thus a model with one spatial factor might be

$$g(\mu_{i1}) = X_i \beta_1 + u_{i1} + \lambda_1 s_i$$
$$g(\mu_{i2}) = X_i \beta_2 + u_{i2} + \lambda_2 s_i$$
$$\vdots \qquad (8.15)$$
$$g(\mu_{iM}) = X_i \beta_M + u_{iM} + \lambda_M s_i$$

where for identification one of the λ_m is preset (e.g. $\lambda_1 = 1$) or the variance of s_i is fixed. Additionally if a CAR(1) prior is used for s_i then sampled values $s_i^{(t)}$ must be centred at each iteration. A constraint such as $\lambda_1 = 1$ assists in the avoidance of label switching, which affects factor analysis as well as discrete mixtures (Congdon, 2003a).

Example 8.3 Joint life table   This example considers an analysis by
Congdon (2002) concerning mortality and health data drawn from the
1991 Census and from vital statistics returns for 44 small areas in NE
London. Health data relate to the year 1991 and deaths to the five-year
periods 1988–1992 and 1993–1997. The data are by age x (19 five-year
bands up to 90+), gender s, area i and additionally by period t for deaths
only. Populations are denoted P_{tixs}. Well populations are defined as total
populations minus the long-term ill.

A different model is assumed for ages up to 65 and ages over 65. Specifically, a shared ICAR(1) spatial effect φ_i for ages under 65 reflects the
expectation that good health and mortality will be negatively correlated.
It is at these ages that deprivation has the strongest effect on mortality and
that a latent spatial effect (proxying deprivation) is arguably more relevant. So φ_i has a preset loading of -1 in the health model and a free
loading ω in the deaths model.

For five-year age groups $x = 1, \ldots, 13$ (where $x = 13$ for ages 60–64),
deaths $Y_{tixs} \sim \text{Poi}(P_{tixs}\mu_{tixs})$ and well populations $H_{ixs} \sim \text{Bin}(P_{ixs}, \pi_{ixs})$,
the model means are

$$\log(\mu_{tixs}) = c_{11} + \gamma_{1t} + \delta^m_{x,s} + u_{mi} + \omega\varphi_i$$

$$\text{logit}(\pi_{i,x,s}) = c_{21} + \delta^h_{x,s} + u_{hi} - \varphi_i.$$

For higher ages ($x = 14$ to $x = 19$) the model is

$$\log(\mu_{tixs}) = c_{12} + \gamma_{2t} + \delta^m_{x,s} + u_{mi}$$

$$\text{logit}(\pi_{i,x,s}) = c_{22} + \delta^h_{x,s} + u_{hi}.$$

Of interest as model outputs (structural parameters) are total and healthy
life expectancies and the difference between the two which defines the
'disease burden'.

From a two-chain run of 3000 iterations (a 1000 burn-in), Table 8.1
shows the high association (correlation 0.93) between posterior means of
the spatial factor $100 \exp(\bar{\varphi}_i)$ and actual government index of multiple
deprivation (IMD) scores. The correlation with the Jarman deprivation
score (also used in health need applications) is 0.89. The loading ω has
mean 0.65 with a 95% interval (0.55,0.76), so both morbidity and mortality are reflected in the factor score. This example therefore illustrates
how spatial effects may proxy real substantive influences (cf. Pascutto
et al., 2000).

Table 8.1 Spatial factor and deprivation scores

Area number	Area (ward) name	Jarman UPA deprivation score	Life table common spatial effect (exponentiated and ×100)	Index of multiple deprivation score
1	Abbey	39.8	149.2	44.91
2	Alibon	17.5	136.6	42.91
3	Cambell	14.5	119.2	38.84
4	Chadwell Heath	6.6	91.2	21.48
5	Eastbrook	4.9	97.7	26.07
6	Eastbury	18.0	119.8	36.47
7	Fanshawe	19.2	129.5	45.34
8	Gascoigne	55.1	170.4	50.29
9	Goresbrook	20.9	131.8	38.96
10	Heath	27.1	119.1	40.93
11	Longbridge	−1.3	78.6	17.30
12	Manor	18.5	132.8	43.00
13	Marks Gate	28.9	115.6	42.47
14	Parsloes	14.5	145.9	39.62
15	River	16.4	114.6	33.33
16	Thames	34.3	121.1	40.94
17	Triptons	16.7	129.8	39.33
18	Valence	20.5	143.2	37.92
19	Village	22.4	103.9	41.70
20	Airfield	6.3	94.1	17.81
21	Ardleigh Green	−14.2	82.8	6.86
22	Brooklands	7.5	79.7	18.51
23	Chase Cross	8.3	103.9	27.43
24	Collier Row	−7.8	95.4	18.31
25	Cranham East	−12.9	67.4	10.03
26	Cranham West	−25.8	67.6	3.69
27	Elm Park	−9.7	82.0	10.47
28	Emerson Park	−20.1	69.0	7.05
29	Gidea Park	−20.3	58.9	7.02
30	Gooshays	18.5	127.2	36.02
31	Hacton	−8.8	82.1	11.82
32	Harold Wood	−9.9	83.0	10.16
33	Heath Park	−5.0	73.4	10.92
34	Heaton	14.6	142.1	35.04
35	Hilldene	22.5	146.3	42.33
36	Hylands	−4.1	73.5	11.77
37	Mawney	−2.2	87.6	16.57
38	Oldchurch	10.9	115.9	22.61
39	Rainham	−7.3	86.0	17.25
40	Rise Park	−13.6	73.2	10.49
41	St. Andrew's	−6.0	69.5	12.06
42	St. Edward's	3.4	90.9	16.55
43	South Hornchurch	−0.3	100.7	26.32
44	Upminster	−17.2	64.0	3.27

8.5 VARYING PREDICTOR EFFECT MODELS

Developments in spatial epidemiology in the 1990s focused on variations
between small areas in incidence or mortality and interpreted such variations as due to unobserved factors which vary smoothly over space. A
spatially correlated 'error' term was used to describe such effects. By
contrast, the impact of known predictors was generally assumed homogeneous, i.e. not spatially varying. In many applications in both spatial
econometrics and disease mapping it is likely that regression effects
are not in fact constant over a region, and spatially varying regression
coefficients are defined as one form of spatial heterogeneity by Anselin
(2001).

In geographical applications the spatial expansion model and Bayesian
geographically weighted regression (LeSage, 2004) allow for regression
heterogeneity, though have focused on metric responses. By contrast,
developments in the Bayesian epidemiology perspective include
Gamerman *et al.* (2003) and Congdon (2003b), while Congdon (1997)
describes a spatially varying regression coefficient model for a univariate
Poisson outcome.

One might adopt the MCAR models of section 8.4 to model spatial
regression heterogeneity. Assume all p predictors can have spatially non-
constant effects so that for y_i in the exponential family (e.g. Poisson,
binomial) with mean μ_i, one has

$$g(\mu_i) = \beta_0 + X_i\beta_i \tag{8.16}$$

where $X_i = (x_{i1}, x_{i2}, \ldots, x_{ip})$ and $\beta_i = (\beta_{i1}, \beta_{i2}, \ldots, \beta_{ip})'$ is $p \times 1$. The
relevant pairwise difference MCAR(1) joint prior has the form

$$P(\beta|\vartheta) \propto |\vartheta|^{-0.5n} \exp\left[-0.5 \sum_{i \neq j}^{n} c_{ij}(\beta_i - \beta_j)'\vartheta^{-1}(\beta_i - \beta_j)\right] \tag{8.17a}$$

where $\beta = (\beta_1, \ldots, \beta_n)$ and ϑ is a dispersion matrix. The conditional
form under binary adjacency is

$$\beta_i | \textstyle\sum_{\beta}, \beta_{[i]} \sim N_p\left(\bar{\beta}_i, \textstyle\sum_{\beta}/N_i\right) \tag{8.17b}$$

where $\sum_{\beta}$ is the conditional dispersion matrix, and $(\bar{\beta}_i = \bar{\beta}_{i1}, \ldots,$
$\bar{\beta}_{ip})$ where $\bar{\beta}_{ik}$ is the average of β_{jk} over the $j = 1, \ldots, N_i$ neighbours
of area i. The overall average effects B_k of predictor k are obtained as the
average of $\bar{\beta}_{ik}$.

A mixed prior analogous to (8.1) for the β_i in (8.16), as in

$$\beta_i = \beta_i^{\text{s}} + \beta_i^{\text{u}}$$

where β_i^{s} follows (8.17) and $\beta_i^{\text{u}} \sim N_p(0, \sum_u)$ is spatially unstructured, is likely to be difficult to identify. However, robust versions of (8.17) such as heavier tailed densities may be used if there is irregularity in the spatially varying regressor effects.

Another approach extends the spatial intercepts method of Leyland et al. (2000), Browne et al. (2001) and Langford et al. (1999) to obtain spatially varying regression coefficients for univariate or multivariate outcomes. For example, consider an M-variate count response $(y_{i1},\ y_{i2}, \ldots,\ y_{iM})$ with means $(E_{i1}\mu_{i1}, E_{i2}\mu_{i2}, \ldots, E_{iM}\mu_{iM})$ and p predictors. Then let

$$\log(\mu_{i1}) = \kappa_1 + \beta_{i11}x_{i1} + \beta_{i21}x_{i2} + \beta_{i31}x_{i3} + \cdots + \beta_{ip1}x_{ip} + u_{i1} + s_{i1}$$
$$\log(\mu_{i2}) = \kappa_2 + \beta_{i12}x_{i1} + \beta_{i22}x_{i2} + \beta_{i32}x_{i3} + \cdots + \beta_{ip2}x_{ip} + u_{i2} + s_{i2}$$

$$\vdots$$

$$\log(\mu_{iM}) = \kappa_M + \beta_{i1M}x_{i1} + \beta_{i2M}x_{i2} + \beta_{i3M}x_{i3} + \cdots + \beta_{ipM}x_{ip} + u_{iM} + s_{iM}$$

$$(8.18a)$$

It is possible to adopt separate priors for each of $(p+2)M$ possible effects. However, to allow interdependence between all effects including the unstructured error, one may assume underlying unstructured errors v_i of dimension $(p+2)M$ or possibly lower.

Consider an example with $M = 2$ and $p = 3$; then $v_i = (v_{i1}, v_{i2}, \ldots, v_{i,10})$ is of dimension 10 and linked to the spatial regression coefficients and varying intercepts as follows:

$$\beta_{i11} = \sum_{j=1}^{n} w_{ij}v_{j1} \qquad \beta_{i12} = \sum_{j=1}^{n} w_{ij}v_{j6}$$

$$\beta_{i21} = \sum_{j=1}^{n} w_{ij}v_{j2} \qquad \beta_{i22} = \sum_{j=1}^{n} w_{ij}v_{j7}$$

$$\beta_{i31} = \sum_{j=1}^{n} w_{ij}v_{j3} \qquad \beta_{i32} = \sum_{j=1}^{n} w_{ij}v_{j8} \qquad (8.18b)$$

$$s_{i1} = \sum_{j=1}^{n} w_{ij}v_{j4} \qquad s_{i2} = \sum_{j=1}^{n} w_{ij}v_{j9}$$

$$u_{i1} = v_{j5} \qquad u_{i2} = v_{j10}$$

Then one might take the $v_i = (v_{i1}, v_{i2}, \ldots, v_{i,10})$ to be parametric random effects, e.g. multivariate normal of order 10 with dispersion matrix $\sum_v$ and mean $\omega_v = (B_{11}, B_{21}, B_{31}, 0, 0, B_{12}, B_{22}, B_{32}, 0, 0)$. Thus the means of the effects $\{v_{i4}, v_{i5}, v_{i9}, v_{i10}\}$ underlying the spatial intercepts are zero, but the remainder are outcome-specific average effects $B_{km}(k = 1, \ldots, 3; m = 1, 2)$ of the predictors X_1, X_2 and X_3.

It may be possible to omit intercept variation (if deviations are not spatially correlated) so that (8.18a) reduces to

$$\log(\mu_{i1}) = \kappa_1 + \beta_{i11}x_{i1} + \beta_{i21}x_{i2} + \beta_{i31}x_{i3} + \cdots + \beta_{ip1}x_{ip}$$

$$\log(\mu_{i2}) = \kappa_2 + \beta_{i12}x_{i1} + \beta_{i22}x_{i2} + \beta_{i32}x_{i3} + \cdots + \beta_{ip2}x_{ip}$$

$$\vdots \qquad\qquad\qquad\qquad\qquad\qquad\qquad\qquad\qquad (8.18c)$$

$$\log(\mu_{iM}) = \kappa_M + \beta_{i1M}x_{i1} + \beta_{i2M}x_{i2} + \beta_{i3M}x_{i3} + \cdots + \beta_{ipM}x_{ip}$$

In this case the MVN or MVT prior would be of dimension Mp, and in the above example

$$\beta_{i11} = \sum_{j=1}^{n} w_{ij}v_{j1} \qquad \beta_{i12} = \sum_{j=1}^{n} w_{ij}v_{j4}$$

$$\beta_{i21} = \sum_{j=1}^{n} w_{ij}v_{j2} \qquad \beta_{i22} = \sum_{j=1}^{n} w_{ij}v_{j5}, \qquad (8.18d)$$

$$\beta_{i31} = \sum_{j=1}^{n} w_{ij}v_{j3} \qquad \beta_{i32} = \sum_{j=1}^{n} w_{ij}v_{j6}$$

While the covariance matrix Φ_v in such models may be taken as unstructured, reduced parameterizations are possible, such as those allowing correlations between parallel effects (e.g. correlations between v_{i1} and v_{i6}, and hence between β_{i11} and β_{i12}; between v_{i2} and v_{i7} and hence between β_{i21} and β_{i22}, and so on).

A spatial factor approach to spatially varying coefficients also provides a reduced parameterization. For M outcomes and p predictors v_i is of dimension Mp. A simplification is to assume a single factor (or possibly factors) underlying the spatial effects β_{ikm} for predictor k. Thus let F_{ik} be a single such factor, and define for $k = 1, \ldots, p$

$$\beta_{ik1} = \lambda_{k1} \sum_{j=1}^{n} w_{ij} F_{jk}$$

$$\beta_{ik2} = \lambda_{k2} \sum_{j=1}^{n} w_{ij} F_{jk}$$

$$\vdots$$

$$\beta_{ikM} = \lambda_{kM} \sum_{j=1}^{n} w_{ij} F_{jk}$$

(8.19)

Then $F_i = (F_{i1}, \ldots, F_{ip})$ is of dimension p, and might be modelled as multivariate normal. For identifiability the loadings $\lambda_{k1}(k = 1, \ldots, p)$ may be preset (e.g. to one) leaving $p(M - 1)$ free loadings.

Finally robust spatial regression models based on non-parametric approaches may be considered. Thus the underlying v_i in (8.18c) may be drawn from a discrete mixture with a known number J of components or, as in a DPP(G_0, κ) model, an unknown and stochastically varying total of components under a baseline prior G_0 and concentration parameter κ. Under the latter option define categorical indicators for area i,

$$D_i \sim \text{Categorical}(\pi)$$

where π is of length C (C less than or equal to n), and updated using an appropriate prior such as the stick breaking prior. Associated with each cluster c is a value $V_c = (v_{c1}, v_{c2}, \ldots, v_{c,Mp})(c = 1, \ldots, C)$ drawn from the baseline prior G_0 (that might be multivariate normal or Student t of dimension Mp). Then if at a particular iteration $t = 1, \ldots, T$ the selected cluster for area j is $c_j = D_j^{(t)}$, one obtains β_{ikm} as a spatially filtered average of $v_{c_j h}$ ($h = 1, \ldots, Mp$). Thus (8.18d) is replaced by

$$\beta_{i11} = \sum_{j=1}^{n} w_{ij} v_{c_j 1} \qquad \beta_{i12} = \sum_{j=1}^{n} w_{ij} v_{c_j 4}$$

$$\beta_{i21} = \sum_{j=1}^{n} w_{ij} v_{c_j 2} \qquad \beta_{i22} = \sum_{j=1}^{n} w_{ij} v_{c_j 5} \qquad (8.20)$$

$$\beta_{i31} = \sum_{j=1}^{n} w_{ij} v_{c_j 3} \qquad \beta_{i32} = \sum_{j=1}^{n} w_{ij} v_{c_j 6}$$

The means $\{B_{11}, B_{21}, B_{31}, B_{12}, B_{22}, B_{32}\}$ are obtained by averaging the $\beta_{ikm}^{(t)}$ over areas and iterations.

Example 8.4 Varying effect of suicide risk factors   A varying regression effect model for male and female suicides in 1989–1993 in 354 English local authorities is considered. There are three predictors – a social fragmentation score, a deprivation score and a rurality score[1] – each of which has a varying predictor effect. Alternative models (A and B) as in (8.17) and (8.18) are considered as well as a spatial factor model (model C) as in (8.19). The model under (8.18) is then $y_{im} \sim \text{Poi}(E_{im}\mu_{im})$ for $m = 1, 2$ with

$$\log(\mu_{i1}) = \kappa_1 + \beta_{i11}x_{i1} + \beta_{i21}x_{i2} + \beta_{i31}x_{i3}$$
$$\log(\mu_{i2}) = \kappa_2 + \beta_{i12}x_{i1} + \beta_{i22}x_{i2} + \beta_{i32}x_{i3}$$

Under (8.17) the intercepts are omitted for identifiability. Fit is assessed using the DIC and pseudo marginal likelihood (PsML) of Gelfand (1996), while cross-validatory predictive checks for individual areas are based on Marshall and Spiegelhalter (2003). Additionally, the Moran I statistic is calculated from model residuals $y_{im} - \hat{y}_{im}$ to assess whether varying intercepts u_{im} and s_{im} are required in addition to spatially varying regression effects. Note that allowing for spatially varying regression may avoid the need for spatial errors (Fotheringham *et al.*, 2003), in the sense that model deviations are no longer spatially correlated.

With the prior in (8.17), average predictor effects B_{km} are found to be all significantly positive for males, but only social fragmentation has a clear positive effect on female suicide (Table 8.2). It is also apparent that the Moran I is not different from zero, so there is no residual spatial correlation. Mean regression effects under (8.18) are similar but the fit appears to improve slightly. The correlation between the 354 values of β_{i11} under model A and under model B is 0.75. For the five other coefficients (β_{i21} through to β_{i32}) the correlations are 0.77, 0.78, 0.80, 0.77 and 0.88. So the spatially varying regression coefficients seem to be reasonably stable over the two models. The Moran coefficient is also satisfactory under model B.

The Marshall and Spiegelhalter (2003) diagnostics show more divergencies (*P* values over 0.975 or under 0.025) for model B than model A, i.e. 35 vs. 18 in 758 observations ($N = 354$ male suicide totals, $N = 354$

---

[1] Social fragmentation combines people not married, population turnover, one-person households and 'bedsitter' (private rented) accommodation. Deprivation combines unemployment, social groups IV and V, no-car households and social renting. Rurality is based on agricultural workers and population density (negatively weighted).

Table 8.2 Mean regression effects (spatial regression heterogeneity)

	Mean regression effect	2.5%	Median	97.5%
Model A	Male, social fragmentation	0.047	0.085	0.123
	Male, deprivation	0.004	0.041	0.079
	Male, rurality	0.023	0.061	0.100
	Female, social fragmentation	0.084	0.136	0.188
	Female, deprivation	−0.042	0.005	0.053
	Female, rurality	−0.070	−0.023	0.035
	Moran (males)	−0.038	0.025	0.090
	Moran (females)	−0.023	0.033	0.091
	DIC	4573		
	d_e	295		
	PsML	−2312		
Model B	Male, social fragmentation	0.039	0.085	0.131
	Male, deprivation	0.015	0.059	0.101
	Male, rurality	0.023	0.067	0.111
	Female, social fragmentation	0.083	0.148	0.211
	Female, deprivation	−0.038	0.014	0.071
	Female, rurality	−0.070	−0.007	0.059
	Moran (males)	−0.041	0.019	0.080
	Moran (females)	−0.014	0.045	0.103
	DIC	4556		
	d_e	257		
	PsML	−2303		
Model C	Male, social fragmentation	0.059	0.100	0.140
	Male, deprivation	0.007	0.048	0.091
	Male, rurality	0.022	0.065	0.110
	Female, social fragmentation	0.087	0.140	0.199
	Female, deprivation	0.002	0.027	0.062
	Female, rurality	−0.014	0.014	0.051
	Moran (males)	−0.034	0.022	0.081
	Moran (females)	0.064	0.102	0.137
	DIC	4547		
	d_e	180		
	PsML	−2293		

female suicide totals). However a Q–Q plot of the P values for model B against the $2N$ order statistics $1/(2N + 1)$ of a uniform $(0,1)$ distribution appears satisfactory (Figure 8.1).

Model C as would be expected has fewer effective parameters and also does not lead to any major loss in goodness of fit. Both the DIC and PsML improve slightly. There are again 35 divergent areas as per

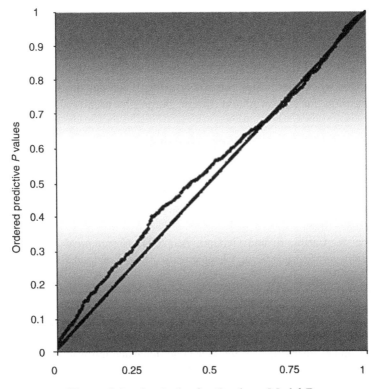

Figure 8.1 $Q-Q$ plot for P values, Model B

Marshall and Spiegelhalter (2003) and the $Q-Q$ plot is in line with a uniform distribution. However, the Moran statistic for females is no longer satisfactory.

EXERCISES

1. In Example 8.1 assess how far the model in (8.7) affects fit to the extreme areas (e.g. 18, 49) identified under the mixed ICAR model.
2. In Example 8.2 (model C) try a scale mixture version of the normal for the unstructured error

$$u_i \sim N(0, 1/(\tau_u \lambda_i))$$

with

$$\lambda_i \sim \text{Ga}(0.5\nu, 0.5\nu)$$

(equivalent to Student t with ν degrees of freedom). Assess changes in fit to the suspect cases and in overall terms.

3. In Example 8.3 try the common spatial factor model for all ages, not just ages under 65. First try a single error φ_i for all ages and then two errors φ_{i1} and φ_{i2} for ages under and over 65. Under the first option assess how far the correlation between $100 \exp(\varphi_i)$ and the IMD score changes when the model extends to all ages. Under the second option assess how far the estimated disease burden by area is affected.

REFERENCES

Anselin, L. (2001) Spatial econometrics. In *A Companion to Theoretical Econometrics*, B. Baltagi (ed.). Oxford: Basil Blackwell, 310–330.

Bailey, T. (2001) Spatial statistical analysis in health. *Cadernos de Saude Publica*, **17**, 1083–1098.

Banerjee, S., Carlin, B. and Gelfand, A. (2004) *Hierarchical Modeling and Analysis for Spatial Data*. Chapman and Hall: London/CRC Press Boca Raton, FL.

Bernardinelli, L., Clayton, D. and Montomoli, C. (1995) Bayesian estimates of disease maps: how important are priors? *Statistics in Medicine*, **14**, 2411–2431.

Besag, J. (1989) Towards Bayesian image analysis. *Journal of Applied Statistics*, **16**, 395–407.

Besag, J. and Kooperberg, C. (1995) On conditional and intrinsic autoregression. *Biometrika*, **82**, 733–746.

Besag, J., York. J. and Mollié, A. (1991) Bayesian image restoration with two applications in spatial statistics. *Annals of the Institute of Statistics and Mathematics*, **43**, 1–59.

Browne, W., Goldstein, H. and Rasbash, J. (2001) Multiple membership multiple classification (MMMC) models. *Statistical Modelling*, **1**, 103–124.

Carlin, B. and Pérez, M. (2000) Robust Bayesian analysis in medical and epidemiological settings. In *Robust Bayesian Analysis*, Lecture Notes in Statistics 152, Insua, D. R. and Ruggeri, F. (eds). Springer: New York, 351–372.

Clayton, D. (1996) Generalized linear mixed models. In *Markov Chain Monte Carlo in Practice*, Gilks, W., Richardson, S. and Spiegelhalter, D. (eds). Chapman and Hall: London, 275–301.

Clayton, D. and Kaldor, J. (1987) Empirical Bayes estimates of age-standardized relative risks for use in disease mapping. *Biometrics*, **43**, 671–681.

Congdon, P. (1997) Bayesian models for the spatial structure of rare health outcomes: a study of suicide using the BUGS program. *Journal of Health and Place*, **3**(4), 229–247.

Congdon, P. (2002) A life table approach to small area health need profiling. *Statistical Modelling*, **2**, 1–26.

Congdon, P. (2003a) *Applied Bayesian Modelling*. John Wiley & Sons: Chichester.

Congdon, P. (2003b) Modelling spatially varying impacts of socioeconomic predictors on mortality outcomes. *Journal of Geographical Systems*, **5**, 161–184.

Congdon, P. (2004) A multivariate model for spatio-temporal health outcomes with an application to suicide mortality. *Geographical Analysis*, **36**, 234–258.

Diggle, P., Moyeed, R. and Tawn, J. (1998) Model based geostatistics. *Applied Statistics*, **47**, 299–350.

Diggle, P., Ribeiro, P. and Christensen, O. (2003) An introduction to model-based geostatistics. In *Spatial Statistics and Computational Methods*, Lecture Notes in Statistics 173, Møller, J. (ed.). Springer: New York.

Fernandez, C. and Green, P. (2002) Modelling spatially correlated data via mixtures: a Bayesian approach. *Journal of the Royal Statistical Society, Series B*, **64**, 805–826.

Fotheringham, A., Charlton, M. and Brunsdon, C. (2003) *Geographically Weighted Regression*. John Wiley & Sons: Chichester.

Gamerman, D., Moreira, A. and Rue, H. (2003) Space-varying regression models: specifications and simulation. *Computational Statistics and Data Analysis*, **42**, 513–533.

Gelfand, A. (1996) Model determination using sampling-based methods. In *Markov Chain Monte Carlo in Practice*, W. Gilks, S. Richardson and D. Spiegelhalter (eds). Chapman and Hall: London, 145–161.

Gelfand, A. and Sahu, S. (1996) Identifiability, propriety, and parameterization with regard to simulation-based fitting of generalized linear mixed models. *Technical Report* 96–36, University of Connecticut, Dept. of Statistics.

Gelfand, A. and Vounatsou, P. (2003) Proper multivariate conditional autoregressive models for spatial data analysis. *Biostatistics*, **4**, 11–15.

Gelfand, A., Ghosh, S., Knight, J. and Sirmans, C. (1998) Spatio-temporal modeling of residential sales markets. *Journal of Business & Economic Statistics*, **16**, 312–321.

Gelman, A., Carlin, J., Stern, H. and Rubin, D. (2003) *Bayesian Data Analysis*, 2nd Edition. CRC Press: Boca Raton, FL.

Kelsall, J. and Wakefield, J. (1999) Discussion of 'Bayesian models for spatially correlated disease and exposure data', by Best, N., Waller, L., Thomas, A., Conlon, E. and Arnold R. In *Bayesian Statistics 6: Proceedings of the Sixth Valencia Meeting on Bayesian Statistics*, Bernardo, J., Berger, J., Dawid, A. and Smith, A. (eds). Oxford University Press: Oxford.

Knorr-Held, L. and Rasser, G. (2000) Bayesian detection of clusters and discontinuities in disease maps. *Biometrics*, **56**, 13–21.

Lagazio, C., Dreassi, E. and Biggeri, A. (2001) A hierarchical Bayesian model for the analysis of spatio-temporal variation in disease risk. *Statistical Modelling*, **1**, 17–29.

Langford, I., Leyland, A., Rasbash, J. and Goldstein, H. (1999) Multilevel modelling of the geographical distributions of rare diseases. *Journal of the Royal Statistical Society, Series C*, **48**, 253–268.

Lawson, A. and Clark, A. (2002) Spatial mixture relative risk models applied to disease mapping. *Statistics in Medicine*, **21**, 359–370.

Leroux, B., Lei, X. and Breslow, N. (1999) Estimation of disease rates in small areas: a new mixed model for spatial dependence. In *Statistical Models in Epidemiology, the Environment and Clinical Trials*, Halloran, M. and Berry, D. (eds). Springer: New York, 135–178.

LeSage, J. (2004) A family of geographically weighted regression models. In *Advances in Spatial Econometrics: Methedology Tools and Applications*, Anselin, L., Florax, J. and Rey, S. (eds). Springer: New York.

Leyland, A., Langford, I., Rasbash, J. and Goldstein, H. (2000) Multivariate spatial models for event data. *Statistics in Medicine*, **19**, 2469–2478.

MacNab, Y. (2003) Bayesian modeling of spatially correlated health service outcome and utilization rates. *Biometrics*, **59**, 305–315.

Marshall, E. and Spiegelhalter, D. (2003) Approximate cross-validatory predictive checks in disease mapping models. *Statistics in Medicine*, **22**, 1649–1660.

Militino, A., Ugarte, M. and Dean, C. (2001) The use of mixture models for identifying high risks in disease mapping. *Statistics in Medicine*, **20**, 2035–2049.

Mollié, A. (1996) Bayesian mapping of disease. In *Markov Chain Monte Carlo in Practice*, Gilks, W., Richardson, S. and Spiegelhalter, D. (eds). Chapman and Hall: London, 359–379.

Pascutto, C., Wakefield, J., Best, N., Richardson, S., Bernardinelli, L., Staines, A. and Elliott, P. (2000) Statistical issues in the analysis of disease mapping data. *Statistics in Medicine*, **19**, 2493–2519.

Sun, D., Tsutakawa, R. and Speckman, P. (1999) Posterior distribution of hierarchical models using CAR(1) distributions. *Biometrika*, **86**, 341–350.

Sun, D., Tsutakawa, R., Kim, H. and He, Z. (2000) Spatio-temporal interaction with disease mapping. *Statistics in Medicine*, **19**, 2015–2035.

Wakefield, J., Best, N. and Waller, L. (2000) Bayesian approaches to disease mapping. In *Spatial Epidemiology: Methods and Applications*, Elliott, P., Wakefield, J., Best, N. and Briggs, D. (eds). Oxford University Press: Oxford, 104–127.

Wang, F. and Wall, M. (2003) Generalized common spatial factor model. *Biostatistics*, **4**, 569–582.

CHAPTER 9

Time Series Models for Discrete Variables

9.1 INTRODUCTION: TIME DEPENDENCE IN OBSERVATIONS AND LATENT DATA

There are a range of possible methods for modelling time series of count, binary and multinomial variables with considerable overlap between methods for different types of discrete response. There are some well-defined classes of models such as integer autoregressive models, hidden Markov models and transition models for counts with gamma (i.e. conjugate) mixing. Alternatively, dependence on past observations or on lagged predictors may be handled by adapting continuous data methods.

Models with state-space features may be included within the class of dynamic generalized linear models or DGLMs (West *et al.*, 1985; Gamerman, 1998). Thus let y_t have a conditional density given state θ_t that belongs to the exponential family

$$f(y_t|\theta_t, \phi_t) = \exp[\{y_t\theta_t - b(\theta_t)\}/a(\phi_t) + c(y_t, \phi_t)]$$

with expectation $\mu_t = \mathrm{E}[y_t|\theta_t, \phi_t]$. Then with link g and a p-dimensional predictor vector X_t including an intercept such that

$$g(\mu_t) = \beta_t X_t$$

Then as for metric responses one might specify a linear first-order updating scheme

$$\beta_t = G_t\beta_{t-1} + \omega_t \qquad t = 2, \ldots, T$$

Bayesian Models for Categorical Data   P. Congdon
© 2005 John Wiley & Sons, Ltd

where ω_t has mean 0 and p-dimensional covariance matrix W, and the initial condition ω_1 has a diffuse prior.

As for continuous outcomes we expect correlations over time in discrete variables due both to data-driven dependence in an outcome over successive time points and to dependence in latent variables such as regression errors. These are sometimes known as observation- and parameter-driven models respectively (Cox, 1981; Jackman, 1998).

Neglecting correlation from either source (i.e. assuming temporal independence) may lead to mis-statement of standard errors on regression coefficients, as Poirier and Ruud (1988) demonstrate for correlated binary data modelled via a probit link. As with cross-sectional count or binomial data, there may well be departures from standard variance–mean assumptions (e.g. overdispersion or excess zeros) which also need to be allowed for if regression parameters are to be efficiently estimated (Jowaheer and Sutradhar, 2002).

Various methods allowing for autocorrelation and overdispersion exist, raising questions of model choice and diagnosis. For example, following Davis *et al.* (2000) a general approach might therefore allow both a latent autoregressive process and lagged dependence in the observations. Model discrimination procedures would decide whether one or the other or both were required for a particular data set. As Grunwald *et al.* (2000) suggest, there may be a range of diagnostic tests that can be applied within classical estimation frameworks, e.g. via parametric bootstrap methods as in Tsay (1992). Analogous procedures in Bayesian estimation via repeated sampling might involve using the predictive check techniques of Gelman *et al.* (1996). For example, if overdispersion exists in the observed count data such that $\tau_y = \text{Var}(y)/\bar{y}$ exceeds one, then the model can be checked in terms of whether it reproduces this observed feature. For samples of new data z from a model the predictive diagnostic τ_z can be compared with the observed level of overdispersion, via the check

$$P_c = \text{E}[\text{Pr}(\tau_z > \tau_y)|y] \qquad (9.1)$$

As well as reproducing the observed overdispersion, a satisfactory model will effectively eliminate error autocorrelations. As in Cameron and Trivedi (1998, p 228) one might consider autocorrelations in the Pearson residuals

$$e_t = (y_t - \mu_t)^{0.5}/\varphi_t$$

where $\varphi_t = \varphi(\mu_t, \delta_t)$ is the variance function. For example, under a gamma–Poisson mixture as in (5.1), with $y_t \sim \text{Po}(\mu_t \phi_t)$, where $\mu_t = \exp(\beta X_t)$ and

$\phi_t \sim G(\delta_t, \delta_t)$, the variance function is $\mu_t + \mu_t^2/\delta_t$. Then the lag k autocorrelation $(k = 1, \ldots, L)$ is estimated as

$$\hat{r}_k = \sum_{t=k+1}^{T} e_t e_{t-k} \Big/ \sum_{t=1}^{T} e_t^2$$

and is $N(0, 1/T)$ when the true autocorrelations $r_k(j = 1, \ldots, L)$ are zero. Equivalently $\sqrt{T}\hat{r}_k$ is $N(0,1)$ under temporal independence and the Box–Pierce statistic $T\sum_{k=1}^{L} \hat{r}_k^2$ is $\chi^2(L)$. To guard against incorrect standardization (i.e. incorrect specification of φ_t) alternative diagnostics are

$$R_k = \sum_{t=k+1}^{T} e_t e_{t-k} \Big/ \left(\sum_{t=k+1}^{T} e_t^2 e_{t-k}^2 \right)^{0.5}$$

which are $N(0,1)$ when $r_k = 0, k = 1, \ldots, L$.

9.2   OBSERVATION-DRIVEN DEPENDENCE

In this section we consider models with a clear dependence on lagged observational values, while possibly also allowing for data augmentation or introducing error effects that may also potentially be autocorrelated. For count data, the data augmentation approach is not commonly used. Instead direct Poisson sampling is typically used and a straightforward autoregressive data dependence might involve a log-link and lags in the log of the dependent variable. To avoid taking logs of zero one might (see Zeger and Qaqish, 1988) set $y_t^* = y_t + c$, where $0 < c < 1$, and a lag 1 dependence is then

$$\begin{aligned} y_t &\sim Po(\nu_t) \\ \log(\nu_t) &= \rho \log(y_{t-1}^*) + \beta x_t \end{aligned} \tag{9.2}$$

A positive ρ implies ν_t is increased as the previous outcome increases, whereas a negative ρ means ν_t falls as y_{t-1} increases. Such a model might not remove temporal correlation in the errors $y_t - \nu_t$ or standardized residuals $\eta_t = (y_t - \nu_t)/\varphi_t$ where $\varphi_t = Var(y_t|X_t)$. One might therefore add additional lags to (9.2) as in

$$\log(\nu_t) = \rho_1 \log(y_{t-1}^*) + \rho_2 \log(y_{t-2}^*) + \cdots + \rho_m \log(y_{t-m}^*) + \beta X_t$$

An apparently obvious observation-driven model for count data, namely

$$\log(\nu_t) = \beta X_t + \rho y_{t-1} \tag{9.3}$$

is stationary when $\beta = 0$ and $\rho < 0$, but implies exponential growth in the series when $\rho > 0$. In fact for constant X_t, ρ is a growth rate, since $\log(\nu_t) = \log(\nu_{t-1}) = \rho(y_{t-1} - y_{t-2})$.

A further option for count data, considered by Grunwald *et al.* (2000) and Brandt *et al.* (2000), is a conditional linear autoregessive CLAR(L) process up to lag L. In its lag 1 version the CLAR(1) model involves a lag parameter ρ, a mean parameter λ_t and additional parameters ϕ_t (e.g variances for Gaussian data). For example, with $y_t \sim \text{Po}(\nu_t)$

$$\nu_t = \rho_t y_{t-1} + \lambda_t \qquad (9.4)$$

where restrictions $\rho_t > 0$ and $\lambda_t > 0$ apply. A possible option is $\lambda_t = \exp(\beta_t X_t)$. More generally

$$\nu_t = \rho_t y_{t-1} + \varepsilon_t$$

where ε_t might be gamma distributed with mean λ_t. The lagged observation dependence might take the form of 'thinning' (see section 9.5), with

$$\nu_t = \rho_t \circ y_{t-1} + \lambda_t$$

The Gaussian AR(1) model fits into this framework by taking $\lambda_t = (1 - \rho)\mu$ and

$$y_t - \mu = \rho(y_{t-1} - \mu) + \varepsilon_t$$

where $\varepsilon_t \sim \text{N}(0, \sigma^2)$. Whether applied to metric or discrete responses, this model with $\rho_t = \rho$ has an exponential lag decay property such that $\text{corr}(y_t, y_{t-k}) = \rho^k$.

Cai *et al.* (2001) consider models for count (or ordinal categorical) data y_{t+1} where the regression parameter depends on the previously observed value y_t. Under this functional autoregressive approach, the conditional probability that $y_{t+1} = j$ given $(X_{t+1}, X_t, \ldots, y_t = i, y_{t-1} = k, \ldots)$ might be specified as $\Pr(y_{t+1} = j|\beta_{y_t}, \rho_{y_t}, X_t, y_t)$. For a Poisson density with q predictors this leads to

$$\begin{aligned} y_{t+1} &\sim \text{Poi}(\nu_{t+1}) \\ \log(\nu_{t+1}) &= \alpha_{y_t} + \beta_{1,y_t} x_{1,t+1} + \cdots + \beta_{q,y_t} x_{q,t+1} + \rho_{y_t} y_t \end{aligned} \qquad (9.5)$$

The impact of y_t on y_{t+1} is thus expressed indirectly through the intercepts and/or regression coefficients. A high correlation between successive values will lead to α_i increasing with i. Though Cai *et al.* do not use state-space priors, these may be relevant when there are many different values of $y_t = i$. Random walks adjusted for unequal differences between successive values may apply.

In each of these cases modifications may be necessary to account for overdispersion, e.g. taking counts y_t to be negative binomial rather than Poisson, or equivalently, or adopting a conjugate mixture, namely gamma–Poisson or beta–binomial. One might also add a normal or otherwise distributed error in the log or logit link (for count and binomial data respectively) to represent overdispersion.

9.2.1 Binary data and Markov Chain models

As an example of observation-driven dependence where latent data are introduced, consider a times series for a binary outcome y_t, $t = 1, \ldots, T$, where the latent variable W_t is introduced according as y_t is 1 or 0,

$$W_t \sim N(\mu_t, 1)\, I(a_t, b_t) \tag{9.6}$$

with $a_t = -\infty$ or 0 (and $b_t = 0$ or ∞) according as $y_t = 0$ or 1. Then an AR(1) dependence on previous responses could be specified

$$\mu_t = \rho W_{t-1} + \beta X_t$$

where $\rho \in [-1, 1]$ under stationarity (Beck et al., 2001).

A more extensive model proposed by Heckman (1981) allows lagged effects on W_t of W_{t-1}, $W_{t-2}, \ldots$, and of the observations y_{t-1}, $y_{t-2}, \ldots$. As for cross-sectional data, greater generality in terms of link functions, and robustness in terms of downweighting outliers, is achieved by scale mixing, whereby

$$W_t \sim N(\mu_t, V_t)$$

The differing scales V_t may be defined non-parametrically, allowing for clustering in the scales, or alternatively drawn from a positive parametric density (e.g. gamma).

If a direct likelihood model is used for binary data, a logistic model analogous to (9.3) has a sounder basis for studying Markov chain dependencies. In a first-order Markov chain with constant transition probabilities, the move between binary states y_t and y_{t+1} is governed by the 2×2 matrix

$$P = \begin{bmatrix} p_{00} & p_{01} \\ p_{10} & p_{11} \end{bmatrix}$$

where $p_{00} + p_{01} = 1$, $p_{10} + p_{11} = 1$ (Cox and Snell, 1989, p 98). Moves between states separated by K periods are governed by a matrix P^K if a first-order Markov chain is appropriate to the observations.

The distribution among states $\rho^{(t)} = (\rho_{0t}, \rho_{1t})$ at time t under a first-order Markov chain model is specified by

$$\rho^{(t)} = P^{t-1} \rho^{(1)}$$

As t increases, $\pi^{(t)}$ approaches an equilibrium satisfying

$$\rho P = \rho$$

with solution $\Pr(y_t = 1) = \rho_1 = p_{01}/(p_{01} + p_{10})$, $\rho_0 = 1 - \rho_1$. In the equilibrium, the joint density of y_{t-1} and y_t is

$$\Pr(y_{t-1} = j, y_t = k) = \rho_j p_{jk}$$

and therefore

$$E(y_{t-1}, y_t) = \rho_1 p_{11}$$
$$\mathrm{Cov}(y_{t-1}, y_t) = \rho_1 p_{11} - \rho_1^2$$

and the correlation between successive binary observations is

$$\mathrm{Corr}(y_{t-1}, y_t) = (\rho_1 p_{11} - \rho_1^2)/(\rho_0 \rho_1) = p_{11} - p_{01}$$

In second- and higher order Markov chains the probability of a move to state 1 or 2 between t and $t + 1$ depends not only on the current state but on preceding states. The dependence is on both the states at t and $t - 1$ in a second-order chain, on the states at t, $t - 1$ and $t - 2$ in a third- order chain, and so on. A first-order Markov chain stationary through time may be represented by the model

$$y_t \sim \mathrm{Bern}(\pi_t)$$
$$\mathrm{logit}(\pi_t) = \alpha + \beta y_{t-1} \tag{9.7a}$$

(see e.g. Lindsey, 1993, p 184). Higher order Markov chains will be appropriate when both higher order lags and interactions between lags in logit link regression models are significant. For example, a second-order Markov chain implies the model

$$\mathrm{logit}(\pi_t) = \alpha + \beta_1 y_{t-1} + \beta_2 y_{t-2} + \beta_3 y_{t-1} y_{t-2} \tag{9.7b}$$

When covariates X_t are available, the transition probabilities may be modelled by logistic regression with the regression coefficients specific to the previous state, analogous to (9.5). Thus

$$y_t \sim \mathrm{Bern}(\pi_t)$$
$$\mathrm{logit}(\pi_t) = \alpha[y_{t-1}] + \beta[y_{t-1}]X_t$$

So

$$p_{01}^{(t)} = \exp(\alpha_0 + \beta_0 X_t)/[1 + \exp(\alpha_0 + \beta_0 X_t)]$$
$$p_{00}^{(t)} = 1/[1 + \exp(\alpha_0 + \beta_0 X_t)]$$
$$p_{11}^{(t)} = \exp(\alpha_1 + \beta_1 X_t)/[1 + \exp(\alpha_1 + \beta_1 X_t)]$$
$$p_{10}^{(t)} = 1/[1 + \exp(\alpha_1 + \beta_1 X_t)]$$

When observations are unequally spaced a continuous time Markov process may be used. Let κ_{01} and κ_{10} be the transition intensities between states 0 and 1 and vice versa, with $\rho_1 = \kappa_{01}/(\kappa_{01} + \kappa_{10})$ and $\rho_0 = 1 - \rho_1$. Then following Cox and Snell (1989) and Jones (1993), and assuming a time gap t between observations,

$$p_{01}^{(t)} = \rho_1[1 - \exp(-\{\kappa_{01} + \kappa_{10}\}t)]$$
$$p_{10}^{(t)} = \rho_0[1 - \exp(-\{\kappa_{01} + \kappa_{10}\}t)]$$

(9.8a)

As t tends to infinity, these probabilities tend to steady-state probabilities ρ_1 and ρ_0 respectively, and the correlation at time gap t is $\exp[-(\kappa_{01} + \kappa_{10})t]$. Covariates may be introduced by setting

$$\log[\kappa_{01}(t)] = \alpha_0 + \beta_0 X_t$$
$$\log[\kappa_{10}(t)] = \alpha_1 + \beta_1 X_t$$

(9.8b)

More general regression structures for binomial and binary time series include non-parametric modelling as in Hyndman (1999). Thus for binary data, lags in y are included together with conventional linear effects in predictors z and smoothed regression effects in predictors x_1, x_2, etc.:

$$y_t \sim \text{Bern}(\pi_t)$$
$$g(\pi_t) = \gamma z + s_1(x_{1t}) + s_2(x_{2t}) + \cdots + \rho_1 y_{t-1} + \rho_2 y_{t-2} \cdots$$

Following Wood and Kohn (1998) this type of semi-parametric model may also be based on augmented data sampling, as in (9.6).

Example 9.1 Strikes and output These data (Kennan, 1985) illustrate Poisson time series regression with both lags in the response and overdispersion. They relate to monthly strikes y_t in the USA from January 1968 to December 1976 and their relation to output fluctuations (measured as the cyclical difference of output from a trend level). A baseline model is provided by a static Poisson regression without dependence on previous observed values or autocorrelated errors. Thus

$$y_t \sim \text{Poi}(\nu_t)$$
$$\log(\nu_t) = \beta_0 + \beta_1 x_t$$

with $x_t =$ output. In a two-chain run of 5000 iterations (convergent at under 500) we obtain the regression $\mu_t = \exp(1.65 + 3.09x_t)$. This model does not reproduce the observed overdispersion as shown by a predictive check. It is also subject to significant temporal correlations $\hat{r}_1, \hat{r}_2, \ldots$, etc., with a Box–Pierce statistic of 54 (for 5 degrees of freedom) based on the first five $\hat{r}_k$.

Instead consider lagged dependence on shifted values

$$y_t^* = y_t + c \qquad (0 < c < 1)$$

with the mean taking the form (Zeger and Qaqish, 1988)

$$\log(v_t) = \beta_0 + \beta_1 x_t + \rho_1 \log(y_{t-1}^*) + \rho_2 \log(y_{t-2}^*) + \cdots + \rho_L \log(y_{t-L}^*)$$

Simply taking this model with $L = 1$ and $c = 0.5$ eliminates the problem of reproducing the overdispersion. Moreover, the Box–Pierce statistic has posterior mean 7.7 and a comparison with a chi-square variable (with 5 degrees of freedom) shows no significant autocorrelation (the probability that the observed Box–Pierce exceeds the chi-square variable is 0.8). This model has a DIC of 574, with the largest deviance for y_{37} (9 strikes) following a zero count $y_{36} = 0$ in the previous month.

The fit is improved if a negative binomial model with lag 1 in $y_t^* = y_t + c$ is assumed. Thus

$$y_t \sim NB(\mu_t, \delta)$$
$$\log(\mu_t) = \beta_0 + \beta_1 x_t + \rho \log(y_{t-1}^*)$$

This gives a DIC of 544 and a 95% interval on β_1 of (0.5,4.4). However, the Box–Pierce statistic worsens under this model.

Example 9.2   Rainfall at Madison   Lindsey (1993) considers a binary time series relating to June days at Madison (Wisconsin) with precipitation recorded ($y_t = 1$) against rain-free days ($y_t = 0$). The series extends over 11 consecutive years (1961 to 1971) so there are 330 observations. The data are treated as one series, though June 1, 1962 will be separated from June 30, 1961. Assuming a first-order Markov chain yields a DIC of 414 and a transition matrix

$$P = \begin{bmatrix} 0.72 & 0.28 \\ 0.58 & 0.42 \end{bmatrix}$$

where the first row is for $y_{t-1} = 0$ and the second row is for $y_{t-1} = 1$ and the first and second columns are for $y_t = 0$ and $y_t = 1$. So the probability of rain, given the previous day is rainy, is 0.42.

A second-order Markov chain as in (9.7b) yields the following probabilities; conditional on the joint pair of states at $t - 2$ and $t - 1$:

States at		State at t	
$t - 2$	$t - 1$	0	1
0	0	0.746	0.254
0	1	0.499	0.501
1	0	0.653	0.347
1	1	0.690	0.310

This yields a slight improvement in the DIC to 412.7.

Example 9.3   England vs. Scotland internationals   To illustrate the case of unequally spaced binary data, consider England vs. Scotland internationals between 1872 and 1987, with $y_t = 1$ if England won. These occurred annually except for the two wars, so there is a gap of six years between matches in 1914 and 1920 and of eight years between 1939 and 1947. In all there are 104 observations. A possible predictor of the response is whether the match was played in England.

First consider the model in (9.8a) without using the predictor. The first year is modelled as a separate fixed effect. The estimates of κ_{01} and κ_{10} are 1.85 and 2.9 respectively and the steady-state probabilities are $\rho_0 = 0.61$ and $\rho_1 = 0.39$. There is a low correlation between successive y_t, namely 0.03. The DIC is 142.5. The second model uses the covariate as in (9.8b). The effects are not pronounced; for example, β_1 is marginally negative (median -0.5 but with a 95% interval from -1.7 to 1.1), so the probability of England not winning at t given they won at $t-1$ is lower if England is playing at home at time t.

9.2.2   Observation-driven models for individual multicategory data

For categorical data with $J > 2$ categories, models in lagged observations analogous to (9.7) involve multiple logit specifications (Kedem and Fokianos, 2002). As discussed in Chapter 6, a reference category (e.g. the first or Jth) is usually selected and choices or allocations are then relative to the reference category. Suppose the data are in dummy indicator form: $D_{tj} = 1$ if observation y_t is in the jth of the J categories and so $D_{t1} = D_{t2} = \cdots = D_{t,j-1} = D_{t,j+1} = \cdots = D_{tJ} = 0$. Thus

$$y_t \sim \text{Categorical}(p_t)$$

or equivalently

$$D_t \sim \text{M}(1, p_t)$$

where $p_t = (p_{t1}, p_{t2}, \ldots, p_{tJ})$ is of dimension J.

If the Jth category is the reference then 'own' lags in the same category and 'cross' lags in other categories may be included, as in VAR(L) models for multivariate continuous data. If lagged dependence extends beyond the first period interactions may occur, as in (9.7b), e.g. on the products $D_{t-1,j} D_{t-2,j}$ in a lag 2 model. A second-order lag model for the probabilities p_{tj} would potentially involve $J-1$ lag 1 main effects, $J-1$ lag 2 main effects and $(J-1)^2$ interaction effects. For instance, if interaction effects are excluded and a predictor x_t is also available, then a lag 2 model is

$$p_{tj} = \exp(\eta_{tj}) / \sum_k \exp(\eta_{tk}) \qquad (9.9a)$$

where $\eta_{tJ} = 0$, while for $j = 1, \dots, J - 1$

$$
\begin{aligned}
\eta_{tj} = \alpha_j + \beta_j x_t &+ [\gamma_{j11} D_{t-1,1} + \cdots + \gamma_{j1,J-1} D_{t-1,J-1}] \\
&+ [\gamma_{j21} D_{t-2,1} + \cdots + \gamma_{j2,J-1} D_{t-2,J-1}]
\end{aligned}
\qquad (9.9b)
$$

The lag coefficients γ_{jkm} model the impact on category j of the lag k in category m.

To estimate first- or higher order Markov chain models for data with $J > 2$ categories, a direct analysis via a multinomial likelihood is indicated. Under stationarity, a first-order Markov chain

$$(y_t | y_{t-1} = j) \sim \text{Mult}(1, p_j)$$

where $p_j = (p_{j1}, p_{j2}, \dots, p_{jJ})$ and $\sum_{k=1}^{J} p_{jk} = 1$. Under conjugacy, the transition probability vector in row j follows a Dirichlet prior. A first-order chain therefore involves $J(J - 1)$ free transition probabilities and a pth order chain involves $J^p(J - 1)$.

More parsimonious models have been suggested. The mixture transition distribution (MTD) model of Raftery (1985) assumes that Lth-order dependence is expressible as a mixture (specifically a linear combination) of separate impacts of each lag. So

$$
\begin{aligned}
\Pr\{y_t = k_0 | y_{t-1} &= k_1, y_{t-2} = k_2, \dots, y_{t-L} = k_L\} \\
&= \sum_{j=1}^{L} \eta_j \Pr(y_t = k_0 | y_{t-j} = k_j) = \sum_{j=1}^{L} \eta_j Q_{k_j k_0}
\end{aligned}
\qquad (9.10)
$$

where $\sum_{k=1}^{J} \eta_k = 1$. The $J \times J$ transition matrix of probabilities $Q_{k_j k_0} = \Pr(y_t = k_0 | y_{t-j} = k_j)$ pools over lags of different order. This model involves just $J(J - 1) + (L - 1)$ parameters, compared with $J^p(J - 1)$ in a pth-order Markov chain. The first-order MTD model is equivalent to a Markov chain of order 1.

9.2.3   Time series of aggregate multinomial data

Consider numbers of observations $n_t > 1$ at time t, with $y_t = (y_{t1}, y_{t2}, \dots, y_{tJ})$ a vector of counts. Then

$$y_t \sim \text{Mult}(n_t, p_t)$$

with $p_t = (p_{t1}, p_{t2}, \ldots, p_{tJ})$. One might replace the multinomial logit model in (9.9) with one involving lags in $y_{t,j}$. Extra multinomial variation would mean adding random effects ε_{tj} that might be multivariate normal of order $J - 1$.

Alternatively a conjugate prior, allowing for temporal dependence and for extra variation, is provided by the Dirichlet. Thus a temporal independence prior specifies

$$p_t \sim \mathrm{Dir}(\alpha_{t1}, \ldots, \alpha_{tJ})$$

and the α_{tj} are uncorrelated over time, a standard choice being $\alpha_{t1} = \alpha_{t2} = \cdots = \alpha_{tJ} = 1$ for all t. However, the independence prior neglects that probabilities close in time will be more similar than probabilities separated by a larger time gap. Gustafson and Walker (2003) propose a prior penalizing large changes between prior probabilities, namely

$$p(p_t|\kappa, \alpha) \propto \mathrm{Dir}(\alpha, \alpha, \ldots, \alpha) \exp\left[-\sum_{t=2}^{T}\sum_{j=1}^{J}(p_{tj} - p_{t-1,j})^2/\kappa \right]$$

Smaller κ imply greater smoothing (high autocorrelation) while as $\kappa \to \infty$ the independence prior holds. A DGLM approach is used by Cargnoni *et al.* (1997), illustrated by a single predictor application with

$$g(p_t) = \beta_{0t} + \beta_{1t}x_t$$

where g might be the multivariate logit as in (9.9), and the category-specific intercepts β_{0t} and regression slopes β_{1t} allow for cross-series pooling of strength via random walk priors such as

$$\beta_{0t} \sim N_{J-1}\left(\beta_{0,t-1}, \Sigma_0\right)$$

Example 9.4   DNA sequences   To illustrate VAR(*L*) type and MTD models for polytomous data consider the first 1000 records in the DNA sequence for the gene BNFR1 of the Epstein–Barr virus. These data involve $J = 4$ unordered categories of nucleotide (1 = adenine, 2 = guanine, 3 = cytosine, 4 = thymine) and are obtainable from the databases at the European Molecular Biology Laboratory. They are considered by Kedem and Fokianos (2002, p 116) with thymine as reference. They consider models as in (9.9b) without interactions between lags and find that the best model among a range of options up to lag $L = 4$ in $\{D_{t-k,j}, j = 1, \ldots, 3\}$ is one involving lags $k = 1$ and 3 only. Thus for $j = 1, 2, 3$

$$\eta_{tj} = \beta_{0j} + [\gamma_{j11}D_{t-1,1} + \gamma_{j12}D_{t-1,2} + \gamma_{j13}D_{t-1,3}]$$
$$+ [\gamma_{j31}D_{t-3,1} + \gamma_{j32}D_{t-3,2} + \gamma_{j33}D_{t-3,3}]$$

There are no predictors, so this particular model involves 21 parameters (three intercepts, three own lags at lag 1, three own lags at lag 3, six cross-lags at lag 1 and six cross-lags at lag 3). A two-chain run shows convergence at around 800 iterations. The deviance at the posterior mean is 2666.5 (compared with 2663.2 at the maximum likelihood estimate) giving a DIC of 2687.5.

By comparison the DIC for the lag 1 only model is 2703. Strictly this means that the model including lag 3 effects is preferred. In fact both models contain insignificant coefficients and a variable selection method may be applied to gain parsimony and obtain a more clearly identified model.

A second order mixture transition model as in (9.10) achieves a comparable DIC with that under a multinomial logit model. In a two-chain run (convergent after 500 iterations), the average deviance for this model is 2678, but since there are fewer (13) parameters the DIC is competitive at 2691. Note that a series of gamma densities is used as the prior for the elements on each row of Q rather than a Dirichlet prior. The estimated mixture proportions are $\eta = (0.79, 0.21)$ with transition matrix

$$Q = \begin{bmatrix} 0.18 & 028 & 0.37 & 0.17 \\ 0.26 & 0.29 & 0.21 & 0.23 \\ 0.22 & 0.33 & 0.33 & 0.12 \\ 0.13 & 0.26 & 0.38 & 0.23 \end{bmatrix}$$

This model was estimated without a constraint on η_1 and η_2, though an unconstrained run shows that one value is clearly higher and a constraint could be set accordingly. Higher order models may require a preset constraint in order to avoid label switching.

9.3   PARAMETER-DRIVEN DEPENDENCE VIA DLMs

Typical model forms where temporal dependence is modelled by random effects latent variables include state-space models, particularly dynamic linear models (DLMs), considered in this section, and autocorrelated error models (section 9.4).

Random effects evolution of states β_t (i.e. regression intercepts or coefficients) is the hallmark of dynamic generalized linear models, where y_t belongs to one of the exponential family of distributions

$$F(y_t | \theta_t, \phi_t) = \exp\left[\frac{y_t \theta_t - b(\theta_t)}{a(\phi_t)} + c(y_t, \phi_t)\right]$$

where a link function g relates the mean $\mu_t = E(y_t|\theta_t) = b'(\theta_t)$ to a p-dimensional regressor vector X_t, via $g(\mu_t) = \beta_t X_t$. The regression coefficients follow a Markov transition model

$$\beta_t = F_t\beta_{t-1} + \omega_t \qquad t = 2, \ldots, T$$

where F_t are known $p \times p$ matrices ($F_t = I$ is a common option) and the error is often assumed multivariate normal

$$\omega_t \sim N_p(0, W_t)$$

with $W = W_t$ usually taken for identifiability (or some other device such as discounting, as in West and Harrison, 1997). In practice not all the coefficients need be subject to random evolution.

More extensive models for the means μ_t might involve trend, seasonal and overdispersion effects, as in the model of Ferreira and Gamerman (2000) for monthly meningococcal meningitis cases in Rio de Janeiro. Specifically this is a dynamic log-linear Poisson model (see also Fahrmeier and Tutz, 2000)

$$y_t \sim Po(\nu_t)$$
$$\log(\nu_t) = \alpha_t + s_t + \eta_t + \beta_t x_t$$

with trend and seasonal (month) effects

$$\alpha_t = \alpha_{t-1} + \omega_{1t}$$
$$s_t = -(s_{t-1} + s_{t-2} + \cdots + s_{t-11}) + \omega_{2t}$$

which have errors

$$\omega_{1t} \sim N(0, W_1) \qquad \omega_{2t} \sim N(0, W_2)$$

The overdispersion error takes the form

$$\eta_t \sim N(0, W_3)$$

A dynamic binary time series model involving a trend and a nonstationary effect of the past response might be defined as

$$y_t \sim Bern(\pi_t)$$
$$logit(\pi_t) = \alpha_t + \beta_t y_{t-1}$$
$$\alpha_t = \alpha_{t-1} + \omega_{1t}$$
$$\beta_t = \beta_{t-1} + \omega_{2t}$$

For multinomial data, a dynamic version of (9.9) might involve taking random walks in intercepts α_{jt} and regression effects β_{jt} or making the cross-lag effects γ_{jkm} vary through time. Omitting lagged responses leads to a dynamic multiple logit model which for a single predictor is

$$y_t \sim \text{Categorical}\,(p_t)$$
$$p_{tj} = \exp(\eta_{tj}) \sum_j \exp(\eta_{tj})$$

where $\eta_{tJ} = 0$ and

$$\eta_{tj} = \alpha_{jt} + \beta_{jt} x_t \qquad j = 1, \ldots, J-1$$

For a DGLM for an ordered categorical response one might set

$$
\begin{aligned}
p_{t1} &= \gamma_{t1} \\
p_{tj} &= \gamma_{tj} - \gamma_{t,j-1} \qquad j = 2, \ldots, J-1 \\
\gamma_{tj} &= \text{Pr}(y_t \leq j) = F(\kappa_{tj} - \beta_t X_t) \qquad j = 1, \ldots, J-1 \\
p_{tJ} &= 1 - \gamma_{i,J-1}
\end{aligned}
$$

where F might be a normal or logistic distribution function and possibly time-varying cut points on the underlying continuous scale are constrained to satisfy

$$\kappa_{t1} < \cdots < \kappa_{t,J-1}$$

Ordinal data time-series varying coefficient regression models for unequally spaced observations $r = 1, \ldots, R$ at times $t_1, t_2, \ldots, t_R$ are considered by Kauermann (2000). Thus

$$\gamma_{tj} = \text{Pr}(y_r \leq j | x_r, t_r) = F[\kappa_j(t_r) + \beta(t_r) X_r]$$

where cut points and regression effects may vary over time via state-space models allowing for varying gaps $\Delta t_r = t_r - t_{r-1}$ between predictors (Chapter 3).

9.3.1   Measurement error models

If X_t contains only an intercept then the above DGLMs essentially represent a true state underlying a response measured with error. To exemplify this latent state approach, Carlin and Polson (1992) examine binary data for REM sleep in infants with $y_t = 1$ if in REM sleep at minute t, y_t otherwise. They suggest that underlying the observed y_t, and similarly the augmented data W_t as in (9.6), is a 'true' sleep state variable α_t. Assuming a first-order random walk in the true state one might have

$$W_t \sim \text{N}(\alpha_t, 1) \tag{9.11a}$$
$$\alpha_t \sim \text{N}(\alpha_{t-1}, \sigma_\beta^2) \tag{9.11b}$$

with the initial condition α_0 a separate fixed effect. An autoregressive dependence might also be assumed, with

$$\alpha_t \sim N(\rho\alpha_{t-1}, \sigma_\alpha^2) \qquad (9.11c)$$

A similar type of measurement error model retains binomial or Poisson sampling but assumes autoregressive or random walk dependence in a state variable defined by the mean parameters (Kitagawa and Gersch, 1996, chapter 13) or by linking transformations of them such as logit or log transforms. For example, assume a Bernoulli density for the binary variable y_t, with $y_t \sim \text{Bern}(\pi_t)$. Then, as in Cox and Snell (1989), one may let

$$\text{logit}(\pi_t) = \lambda_t$$
$$\lambda_t \sim N(\mu + \rho(\lambda_{t-1} - \mu), \tau_\lambda)$$

Kitagawa and Gersch propose first- or second-order random walks in λ_t (occasionally higher orders may be used). For example, an RW(1) normal prior has the form

$$\lambda_t \sim N(\lambda_{t-1}, \tau_\lambda) \qquad (9.12)$$

Equivalently

$$\Delta^k \lambda_t \sim N(0, \tau_\lambda)$$

where $k = 1$ for an RW(1) prior. The same process applies to binomial data with n_t subjects at risk and y_t successes, with probability π_t of success. An illustration provided by Kitagawa and Gersch involves the number of occurrences of rainfall over 1 mm in Tokyo in $T = 366$ observations. Thus 365 pairs of days and 1 single day are obtained by aggregating over two years (1983–1984). For example, the first observation combines Jan 1 1983 and Jan 1 1984. So $n_t = 2$ with the exception $n_t = 1$ for 29 February in 1984 – see Exercise 9.3.

For Poisson data the corresponding state-space model would typically involve the log transform of the Poisson mean, i.e. if

$$y_t \sim \text{Po}(\nu_t)$$

and

$$\log(\nu_t) = \kappa_t$$

then

$$\Delta^k \kappa_t \sim N(0, \tau_\kappa)$$

9.3.2 Conjugate updating

Harvey and Fernandes (1989) combine gamma–Poisson or beta–binomial mixing (as in cross-sectional data) with state-space updating through time. The observations follow a relevant exponential family density, such as Poisson, binomial and negative binomial. Changes in the underlying level or state of the process through times t are governed by sequential updating based on the observations at times t. The state process follows the relevant conjugate density (gamma for Poisson, beta for binomial and negative binomial, and so on). The impact of covariates is generally assumed fixed. Thus for count data, let

$$y_t \sim \mathrm{Po}(\lambda_t m_t)$$
$$m_t = \exp(\beta x_t)$$

Given preceding observations $D_{t-1} = \{y_{t-1}, y_{t-2}, \ldots, x_{t-1}, x_{t-2}\}$, the time-varying levels λ_t are taken to be gamma, $\lambda_{t-1}|D_{t-1} \sim \mathrm{G}(a_{t-1}, b_{t-1})$ with the subsequent period's prior being

$$\lambda_t|D_{t-1} \sim \mathrm{G}(wa_{t-1}, wb_{t-1})$$

where $w \sim U(0, 1)$ is a discount factor. Combining the likelihood after observing D_t and the prior gives, via gamma conjugacy with the Poisson, a full conditional

$$\lambda_t|D_t \sim \mathrm{G}(wa_{t-1} + y_t, wb_{t-1} + m_t)$$

To allow for observation-driven lagged impacts, a Poisson autoregressive model to lag L, or PAR(L) model, is outlined by Brandt and Williams (2001) drawing on this methodology and that of the CLAR(L) model in (9.4). Thus

$$m_t = \sum_{k=1}^{L} \rho_k y_{t-k} + \exp(\beta x_t) \tag{9.13}$$

Another model (the Poisson exponentially weighted moving average model or PEWMA; Brandt et al., 2000) includes period growth rates in the model for m_t, with

$$m_t = \exp(r_t + \beta x_t)$$

Ord et al. (1993) and Harvey and Fernandes (1989) consider the same framework for binomial–beta, NB–beta and multinomial–Dirichlet mixture models. Thus for binomial data $y_t \sim \mathrm{Bin}(n_t, \pi_t)$ and $\pi_t \sim \mathrm{Be}$

(wc_{t-1}, wd_{t-1}) where the process is initialized by vague positive fixed effects c_1 and d_1, e.g. $c_1 \sim \text{Ga}(1, 0.001)$, $d_1 \sim \text{Ga}(1, 0.001)$. Then the full conditional for π_t is

$$\pi_t | D_t \sim \text{Be}(wc_{t-1} + y_t, wd_{t-1} + n_t - y_t)$$

For multinomial data $y_t \sim M(n_t, p_t)$ the prior on p_t is Dirichlet with parameters $\{wc_{j,t-1}, j = 1, \ldots, J\}$ and the full conditional is

$$p_t | D_t \sim \text{Dir}(wc_{1,t-1} + y_{1t}, wc_{2,t-1} + y_{2t}, \ldots, wc_{J,t-1} + y_{Jt})$$

For the negative binomial

$$p(y_t | \pi_t, r) = \binom{r + y_t - 1}{y_t} \pi_t^r (1 - \pi_t)^{y_t}$$

the expectation is

$$E(y_t | \pi_t, r) = r(1 - \pi_t)/\pi_t$$

and the conjugate prior for π_t is beta, as for the binomial. However, to ensure a constant expectation the prior has the form

$$\pi_t \sim \text{Be}(\omega a_{t-1} + (1 - \omega), \omega b_{t-1})$$

Explanatory variables may be introduced by setting $r_t = \exp(\beta X_t)$.

Program 9.5   REM sleep   Carlin and Polson (1992) examine a series of 120 observations at minute intervals with $y_t = 1$ if an infant is in REM sleep at minute t, and y_t otherwise. In their model the binary data are augmented by the underlying continuous (normal) variables W_t

$$W_t | \tau^2, y_t, \alpha_t \sim N(\alpha_t, \tau^2) \, I(0, \infty) \qquad \text{if } y_t = 1$$
$$\sim N(\alpha_t, \tau^2) \, I(-\infty, 0) \qquad \text{if } y_t = 0$$
$$\alpha_t \sim N(\rho \alpha_{t-1}, \sigma^2)$$

with α_0 a separate fixed effect. One goal is to predict the next value in the series (observation y_{121} with state α_{121}). Following Carlin and Polson (1992, p 584) the following priors are assumed:

$$\alpha_0 \sim N(0, 1) \quad 1/\sigma^2 \sim \text{Ga}(3, 3) \quad 1/\tau^2 \sim \text{Ga}(0.5, 0.5) \quad \rho \sim N(0.5, 10)$$

A two-chain run of 20 000 iterations in WINBUGS13 (convergent from under 1000) estimates σ^2 as 1.2 and ρ as 0.93 (and a 95% interval 0.82 to 1). The estimated mean for α_{121} is 2.39 and the probability that y_{121} is one is estimated as 0.85. Note that the five preceding observations (y_{116} to y_{120}) are all one. Figure 9.1 plots the estimates of α_t.

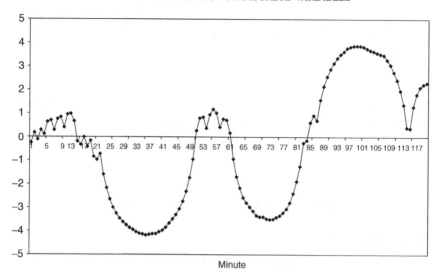

Figure 9.1   Latent states for infant REM sleep

While the preceding model may be identified with informative priors on the two variances, identifiability is more straightforward by setting $\tau^2 = 1$. This approach reduces the estimated variance of the states to 0.89 (posterior mean) but the probability that y_{121} is one remains at 0.85.

Program 9.6   Purse snatchings   Brandt and Williams (2001) analyse data on purse snatchings in the Hyde Park neighbourhood of Chicago before and after a crime prevention programme. The data are 71 counts over 28-day periods from January 1969, with the intervention commencing at the 42nd period ($x_t = 0$ for $t = 1, \ldots, 41$ and $x_t = 1$ for $t = 42, \ldots, 71$).

First a CLAR(1)-type model as in (9.13) is adopted with

$$m_t = \rho_1 y_{t-1} + \exp(\beta_0 + \beta_1 x_t)$$

with prior $\beta_1 \sim N(0, 1)$ on the intervention effect. Diffuse priors are adopted on the initial values a_1 and b_1. A two-chain run of 10 000 iterations (convergent from 2500) shows a negative mean for β_1 of -0.36, with the 95% interval entirely negative ($-0.69, -0.02$) in line with an effective intervention. The lag parameter ρ has mean 0.72. Using equation (2.17b) there are around 42 effective parameters and a DIC of 421. A predictive check on whether new data sampled from the model have a variance–mean relationship in tune with the observed data suggests that this aspect of the data is satisfactorily reproduced, with P_c in (9.1) standing at 0.26.

A standard negative binomial, with

$$P(y_t) = \frac{\Gamma(y_t + r)}{\Gamma(y_t + 1)\Gamma(r)} \left(\frac{r}{r + \mu_t}\right)^r \left(\frac{\mu_t}{r + \mu_t}\right)^{y_t}$$

and regression effects included via the CLAR(1) model

$$\mu_t = \rho_1 y_{t-1} + \exp(\beta_0 + \beta_1 x_t)$$

give a DIC of 446 and also a satisfactory predictive check (namely 0.39) in terms of reproducing the observed overdispersion. The coefficient on β_1 is still negative but the 95% interval is no longer entirely negative, namely $(-0.76, 0.01)$. Under this model $\hat{r}_1$ is estimated as -0.13, and so temporal autocorrelation at lag 1 does not appear significant ($\sqrt{T}\hat{r}_1$ is around -1.1).

Program 9.7 Non-accidental mortality in Birmingham Smith *et al.* (2000) give several alternative analyses of a series of daily deaths y_t (from causes other than accidents) among the over 65s in Birmingham, Alabama, between August 3, 1985 and December 31, 1988 (a period of 1247 days). They use Poisson regression to relate deaths to minimum daily temperature (tmin), mean specific humidity (mnsh) and PM10 readings. Their best model for the meteorological variables involves a lag 3 term in tmin, lags at 0, 1, 2 and 3 in mnsh, and lags at 0 and 1 in the square of mnsh. (A lag at 0 means the predictors of y_t include contemporaneous values of mnsh and mnsh^2.) They model the impact of PM10 as a three-day average, PMMEAN, of the readings on the previous three days (PM10_1, PM10_2 and PM10_3). They try a simple linear term in PMMEAN and find a coefficient of 0.00098 (s.d. 0.00040). They also try a seasonally varying impact of PMMEAN and find a weaker effect in summer (June, July, August) than the other seasons.

Here the lag 1 term in the square of mnsh is omitted as it proved non-significant and an intercept varying by year is adopted as there is a slight upward trend in deaths among the over 65s. To model seasonal effects (higher deaths in winter months) a monthly seasonal random effect is used (via an RW(1) model). There are four missing values in the PMMEAN series and these necessitate modelling of the explanatory variable by a first-order random walk (other interpolation methods might be used). IG(1,1) priors on the evolutionary variances in both these random walks are used in the expectation of reasonably smoothly changing series.

The most important remaining aspect of the model is representing the impact of the running PM10 mean, PMMEAN, on the death rate. Model A follows Smith *et al.* in using a seasonally varying PM10 effect. For numerical reasons, PM10 is scaled by 0.01 so the coefficients on it should be approximately 100 times those cited by Smith *et al.* The seasonal effect estimated here peaks in spring and fall and is weaker in the other seasons. The coefficients with 95% intervals for spring, summer, fall and winter are respectively 0.147 (0.027, 0.265), 0.023 (−0.136, 0.167), 0.141 (0.016, 0.262) and 0.022 (−0.115, 0.165). The DIC for this model is 16 751, which includes a component from the interpolation model for PM10.

A second model uses a cubic spline with knots at PM10 readings of 80 and 100. These were based on analysing death rates according to grouped sets of days with PM10 readings in bands 2.5 wide (these being successively 10–12.49, 12.5–14.99, etc.) with values over 120 constituting the 45th band. Then a seven-period moving average of these 45 bands gives an impression of the smooth PM10 effect, showing the death rate rising at higher values. Smith *et al.* use threshold and general additive models to demonstrate this non-linear effect. This model has a slightly worse DIC than model A (at 16 764) but shows the accelerating effect of PM10 on mortality (Figure 9.2). Figure 9.2 also shows the distorting impact of the sparse values of PM10 over 120.

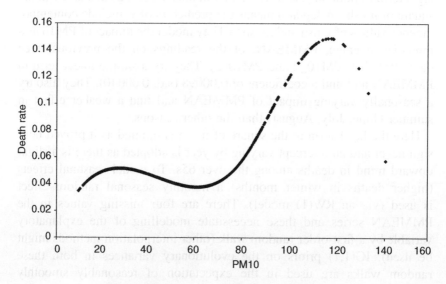

Figure 9.2 Impact of PM10 on death rate

9.4 PARAMETER-DRIVEN DEPENDENCE VIA AUTOCORRELATED ERROR MODELS

Davis *et al.* (2000) emphasize the importance of detecting autocorrelation in the regression errors, indicating where present the need for a latent series with a potentially complex ARMA structure. For simplicity let the latent series ε_t be AR(1). Then in a data augmented binary regression analysis one might have

$$W_t \sim N(\mu_t, 1)$$
$$\mu_t = \beta X_t + \varepsilon_t$$
$$\varepsilon_t \sim N(\rho \varepsilon_{t-1}, \sigma_\varepsilon^2)$$

with X_t including an intercept. Czado (2001) describes such a model with y_t defined by migraine attacks, and W_t being an unobservable latent threshold for such an attack. For a count time series one might have (Oh and Lim, 2001; Chen and Ibrahim, 2000)

$$y_t \sim Po(\nu_t)$$
$$\log(\nu_t) = \beta X_t + \varepsilon_t$$

(9.14a)

with

$$\varepsilon_t = \rho \varepsilon_{t-1} + u_t$$

(9.14b)

where $|\rho| \leq 1$ and u_t is an uncorrelated error with variance σ^2. An alternative conjugate model (Zeger, 1988; Davis *et al.*, 2000) has

$$\mu_t = \exp(\beta X_t)\eta_t$$

where η_t is gamma with mean 1. In either case the latent process introduces both autocorrelation and overdispersion in y_t, though the autocorrelation in y_t is less than or equal to that in the latent series (Chen and Ibrahim, 2000).

Oh and Lim (2001) combine this approach with data augmentation (i.e. introducing latent normal W_t underlying the observations y_t), while Chen and Ibrahim (2000) suggest this model along with methods for formally incorporating historical data into the prior for $\{\beta, \rho, \sigma_\varepsilon^2\}$. Oh and Lim mention that the cdf of a Poisson variable y_t with mean ν_t is closely approximated by

$$\Phi(A[y_t, \nu_t]) = \Phi(-3u_t^{-0.5}[(\nu_t u_t)^{1/3} - 1 + u_t/9])$$

where Φ is the standard normal cdf and $u_t = [y_t + 1]^{-1}$. Define $\eta_t = \log(\nu_t) = \beta X_t + \varepsilon_t$ as above; then for $y_t > 0$ the latent variables underlying the Poisson observations are obtained by sampling

$$W_t \sim N(\mu_t, 1) \, I(A[y_t - 1, \nu_t] + \eta_t, \, A[y_t, \nu_t] + \eta_t)$$

For $y_t = 0$ they are obtained as

$$W_t \sim N(\mu_t, 1) I(-\infty, \, A[0, \nu_t] + \eta_t)$$

To include observational lags as well as lags in the latent process one might adopt the CLAR model structure with lags in y_{t-k} and ε_{t-k}. For example, with first-order lags in both, one obtains

$$y_t \sim Po(\nu_t)$$
$$\nu_t = \rho_1 y_{t-1} + \exp(\beta X_t + \varepsilon_t) \qquad (9.15)$$
$$\varepsilon_t = \rho_2 \varepsilon_{t-1} + u_t$$

where $\rho_1 \sim U(-1, 1)$ and $\rho_2 \sim U(-1, 1)$. Zeger and Qaqish (1988) suggest an empirical error model where the error is defined by comparing the lagged and transformed responses $\log(y_{t-k}^*)$, as in (9.2), with the regression predictions βX_{t-k}. So an AR(1) model analogous to (9.14) is

$$y_t \sim Po(\nu_t)$$
$$\log(\nu_t) = \beta x_t + \rho_1(\log y_{t-1}^* - \beta X_{t-1})$$

This might be combined with lags in the $\log(y_{t-k}^*)$ themselves to give a joint observation–parameter-driven model.

Program 9.8   Purse snatchings (continued)   The purse snatchings data are reanalysed with models allowing both observational and error lags. The first model uses the transformation of Zeger and Qaqish (1988) so that

$$y_t \sim Po(\nu_t)$$
$$\log(\nu_t) = \rho_1 \log y_{t-1}^* + \beta X_t + \varepsilon_t$$
$$\varepsilon_t = \rho_2 \varepsilon_{t-1} + u_t$$

$U(-1, 1)$ priors are assumed on both correlation parameters. This model shows ρ_2 to average 0.88 and ρ_1 to be negative (mean -0.23). The Whistlestop effect disappears completely. The DIC is lower at 406 than for the model used in Example 9.6, with $d_e = 31$.

The second model uses a CLAR(1) model with AR(1) errors in the exponential component, as in (9.15). The DIC for this model is 418 and as for the first model ρ_2 is strongly significant, with mean 0.86 from the second half of a two-chain run of 20 000 iterations. The Whistlestop intervention coefficient is again not significant with a 95% interval $(-0.48, 0.68)$.

9.5 INTEGER AUTOREGRESSIVE MODELS

Integer-valued autoregressive (INAR) models for count observations y_t involve a form of survival of some of the previous counts in the series together with immigration of new units. An INAR(1) model involving y_{t-1} only assumes

$$y_t = R_t + \omega_t \qquad t > 1$$

with

$$R_t \sim \text{Bin}(\rho, y_{t-1})$$

and ω_t is a Poisson 'innovation' series with mean $(1 - \rho)\mu$. If y_{t-1} is Poisson with mean μ, then R_t is Poisson $\rho\mu$, and y_t is Poisson with mean μ (McKenzie, 1986). The initial condition (the prior for the first observation y_1) for an INAR(1) model is taken as

$$y_1 \sim \text{Po}(\mu)$$

The notation

$$R_t = \rho \circ y_{t-1}$$

is used to denote binomial thinning, and taking $0 < \rho < 1$ provides a stationarity constraint. An INAR(2) process involves thinning on y_{t-1} and y_{t-2} and replacement probabilities, ρ_1 and ρ_2. Then

$$y_t = R_{1t} + R_{2t} + \omega_t$$

where $R_{1t} \sim \text{Bin}(\rho_1, y_{t-1})$, $R_{2t} \sim \text{Bin}(\rho_2, y_{t-2})$. The innovation ω_t is then Poisson with mean $\left(1 - \sum_{k=1}^{2} \rho_k\right)\mu$. Stationarity for an INAR(p) process is defined by

$$\sum_{k=1}^{p} \rho_k < 1$$

(Cardinal *et al.*, 1999). The INAR(1) model without covariates implies $y_t \sim \text{Poi}(\nu_t)$, where $\nu_t = E(y_t | y_{t-1}) = \rho y_{t-1} + (1 - \rho)\mu$ (Winkelmann, 2000, p 203).

To allow for overdispersion both ω_t and y_1 may be negative binomial (McKenzie, 1986). Alternatively, random thinning may be assumed (Grunwald *et al.*, 2000, p 481); for example, with replacement probabilities varies over time ρ_t according to a beta density with mean ρ^* (Jowaheer and Sutradhar, 2002). Thus

$$
\begin{aligned}
y_t &= W_t + \omega_t \\
W_t &\sim \text{Bin}(\rho_t, y_{t-1}) \\
\rho_t &\sim \text{Beta}(b\rho^*, b(1 - \rho^*))
\end{aligned}
\tag{9.16}
$$

where the innovations and the first observation y_1 are Poisson or negative binomial.

Suppose predictors are available and a regression term such as $\theta_t = \exp(\beta X_t)$ is introduced. Assuming $y_1 \sim \text{Poi}(\theta_1)$ then to be fully consistent with the INAR framework such that $E(y_t) = \theta_t$, it is necessary that ω_t have mean $\lambda_t = \theta_t - \rho \theta_{t-1}$. For λ_t to be positive in turn requires θ_t to be constant or monotonically decreasing (Azzalini, 1994), so limiting the flexible modelling of time-varying predictors. Alternatively, following Brannas (1995), the innovation term ω_t may be modelled as Poisson with mean θ_t and for the first observation a pre-series latent observation y_0 might be introduced, so that for $t = 1, \ldots, T$

$$
\begin{aligned}
y_t &= \rho \circ y_{t-1} + \omega_t \\
\omega_t &\sim \text{Po}(\theta_t)
\end{aligned}
$$

Alternatively or additionally covariates Z_t may be introduced into modelling the time-varying replacement probabilities ρ_t. To ensure $0 < \rho_t < 1$ a logit or probit link might be used:

$$
\text{logit}(\rho_t) = \gamma Z_t
$$

INAR models are framed to resemble Gaussian autoregression models but can be framed in terms of conditional means (Grunwald et al., 2000). For example, survival and immigration might also be modelled directly in Poisson or NB means, such that ν_t combines thinning (e.g. $\rho \circ y_{t-1}$) with an innovation series ω_t with density confined to positive values (e.g. Poisson, negative binomial or a gamma). So

$$
\begin{aligned}
y_t &\sim \text{Po}(\nu_t) \qquad t > 1 \\
\nu_t &= \rho \circ y_{t-1} + \omega_t \\
\omega_t &\sim \text{Po}(\mu(1 - \rho)) \\
y_1 &\sim \text{Po}(\mu)
\end{aligned}
\tag{9.17}
$$

Overdispersion might be introduced by taking ω_t to be a discrete mixture with K classes, as in

$$
\omega_t \sim \text{Po}(\mu_{G_t}) \qquad G_t \sim \text{Categorical}(\pi_1, \ldots, \pi_K)
$$

by random thinning (taking ρ_t to vary randomly), or taking the innovations to be gamma distributed, as in

$$\omega_t \sim \text{Ga}(c\theta_t, c) \tag{9.18}$$

with $\theta_t = \exp(\beta X_t)$. Discrete innovation series other than Poisson are considered by McCabe and Martin (2003).

Program 9.9 Purse snatchings (continued) The purse snatchings data are reanalysed using an INAR(1) structure in Poisson means. The first model is as in (9.17) without predictors X_t and taking ρ constant. The estimated ρ is 0.42 with $\theta = 13.4$. The predictive check on overdispersion shows this not to be accounted for, with probability as in (9.1) of 0.01 and the DIC deteriorates to 432, as measured by $D(\bar{\mu}|y) + 2d_e$.

To allow for overdispersion and predictor effects, one option is random thinning as in (9.16). So

$$y_t \sim \text{Po}(\nu_t)$$
$$\nu_t = \rho_t \circ y_{t-1} + \omega_t$$
$$\rho_t \sim \text{Beta}(b\rho^*, b(1 - \rho^*))$$
$$\omega_t \sim \text{Po}(\theta_t)$$
$$\theta_t = \exp(\beta_0 + \beta_1 X_t)$$

A Be(1,1) prior on ρ^* is assumed and a Ga(0.5,0.5) prior on b. A two-chain run of 10 000 iterations shows convergence from around 2000 iterations. The 95% interval on β_1, namely $(-0.70, -0.04)$, is confined to negative values with a posterior average of -0.36; ρ^* is estimated as 0.35. The average deviance is 379 with 44 effective parameters, giving a DIC of 423. The predictive check on predictive vs. actual overdispersion is satisfactory, around 0.47.

9.6 HIDDEN MARKOV MODELS

Whereas in state-space models the unobserved series driving the observations is continuous, in hidden Markov models (HMMs) the unobserved states are discrete; see, for example, Paroli and Spezia (2002) and Leroux and Puterman (1992). In a first-order hidden Markov model (by far the most commonly used) the unobserved process S_t is in one of m categories, with movement between states at successive periods determined by a stationary one-step transition matrix $P = \{P_{ij}\}$ of order m, namely

$$P_{ij} = \Pr[S_t = j | S_{t-1} = i] \tag{9.19}$$

that involves $m(m-1)$ unknown probabilities. In a first-order model, there is one initial state S_1 which is multinomial with probabilities $q = \{q_1, \ldots, q_m\}$. The distribution among states at time t, $\rho^{(t)} = (\rho_{1t}, \ldots, \rho_{mt})$, is specified by

$$\rho^{(t)} = P^{t-1}\rho^{(1)}$$

As t increases, $\rho^{(t)}$ approaches an equilibrium satisfying

$$\rho P = \rho$$

with solution (for $m = 2$)

$$\rho_1 = p_{12}/(p_{12} + p_{21}) \qquad \rho_2 = 1 - \rho_1$$

For example, for a first-order Poisson HMM with no predictors, there are m means $\mu_1, \ldots, \mu_m$ with

$$Y_t \sim \text{Po}(\mu_{S_t})$$
$$S_t \sim \text{Categorical}(P_{S_{t-1},1:m}) \qquad t > 2$$
$$S_1 \sim \text{Categorical}(q_{1:m})$$

As for other mixture models there are possible questions of label switching, especially as the number of latent categories m increases, and identifiability constraints on the model parameters are needed. For example, in a Poisson HMM with no predictors, such a constraint is an ordering of the Poisson means (Scott, 2002, pp 342–343).

Among developments of such models are the incorporation of additive errors, possibly with state-dependent variances (de Gunst *et al.*, 2001) or other state-dependent parameters (e.g. the autocorrelations in AR error schemes). Thus for a Poisson HMM one might have

$$y_t \sim \text{Po}(\nu_t)$$
$$\nu_t = \mu_{S_t} + \varepsilon_t$$
$$\varepsilon_t = \rho_{S_t}\varepsilon_{t-1} + u_t$$

with $u_t \sim \text{N}(0, \phi_{S_t})$.

Given the underlying states S_t the observations y_t are usually taken to be mutually independent. If S_t is allocated to category k, then y_t is assigned to the kth of the m possible components of a suitable density. For count data, Poisson components might have different means (Leroux and Puterman, 1992), while for a negative binomial they might have different means and overdispersion parameters. Suppose component k of a Poisson

HMM has mean μ_k; then the likelihood is defined by the state-dependent probabilities

$$\delta_{tk} = \Pr(y_t = y | S_t = k) = \exp(-\mu_k)\mu_k^y/y!$$

and the marginal distribution of y_t is

$$\Pr(y_t = y) = \sum_k \mu_k \delta_{tk}$$

For count data, shifting between regimes with different intensities is often used as a model for overdispersion.

Example 9.10 Purse snatchings (continued)   A two-state Poisson HMM is applied to these data. Although the average deviance is higher than for some previous models at 424, the parsimony of the model (with five parameters) means that the penalized fit is relatively acceptable. The means of the two Poisson groups are 21.4 and 9.9, and the probabilities of belonging to the first higher intensity group approach 1 for periods 23–43. One might try higher numbers of states m and compare the AIC and BIC of the resulting models with the m=2 case. Scott (2002) notes that the BIC for latent Markov models may unduly penalize larger models (i.e. those with higher m).

EXERCISES

1. In Example 9.1 try a CLAR(1) model with $y_t \sim \text{Poi}(\nu_t)$

$$\nu_t = \rho y_{t-1} + \exp(\beta x_t)$$

with $\gamma > 0$ and the mixed version

$$\nu_t = \rho y_{t-1} + \varepsilon_t$$

$\varepsilon_t \sim \text{Ga}(\kappa \exp(\beta x_t), \kappa)$. How does fit compare with the models using the transform $y_t^* = y_t + c$? Try also the discrete conditioning model with mean

$$\log(\mu_{t+1}) = \alpha_{y_{t-1}} + \beta x_t$$

where α_i has 19 possible values (since y_t varies between 0 and 18). A random state-space prior (e.g. RW(1)) for α_i is one option.

2. In Example 9.3, try a model with first and second lags ($k = 1, 2$) in the Poisson and NB models with transformed observations

$$y_t^* = y_t + c \qquad (c = 0.5)$$

How does this affect fit and autocorrelation in the error terms?

3. The following data relate to the Tokyo rainfall data mentioned earlier and in the code below are modelled via a stationary binomial model. Consider instead non-stationary RW(1) and RW(2) models (with normal or other errors) in the logit of the rainfall probability and assess the gain in fit over the stationary binomial.

   model {for (t in 1:59) {y[t] ~ dbin(p,2)}

      for (t in 61:366) {y[t] ~ dbin(p,2)}

   # model for 29th February 1984

     y[60] ~ dbern(p)}

   # prior on constant binomial probability

     p ~ dbern(1,1)}

   list(y=c(0,0,1,1,0,1,1,0,0,0,0,0,0,0,1,0,0,1,1,0,1,1,0,0,0,0,0,0,0,0,1,

      0,1,0,0,0,0,0,0,0,0,0,1,1,0,0,0,2,1,0,0,0,0,1,1,0,1,0,0,0,

      0,1,1,0,0,0,0,0,0,2,0,0,1,1,0,2,1,0,1,1,1,0,1,2,0,0,1,1,1,1,1,

      2,0,0,1,1,0,1,2,0,1,1,1,0,0,1,2,1,0,2,1,0,1,0,0,0,0,0,1,0,0,

      1,1,0,0,0,1,1,0,0,0,0,0,1,1,1,2,2,0,0,0,0,0,1,1,1,0,0,0,1,0,0,

      0,0,0,1,0,1,0,0,0,2,1,1,2,1,0,1,2,2,0,2,2,1,1,1,2,2,2,1,1,0,

      0,0,1,0,1,1,1,2,1,0,1,1,0,0,2,1,1,1,1,2,2,0,1,0,0,0,0,1,1,0,0,

      0,0,0,0,0,0,0,0,0,0,0,0,1,1,1,1,1,1,0,1,2,1,0,0,0,0,0,0,0,1,0,

      1,0,0,0,1,0,1,1,1,2,0,0,0,1,2,2,0,1,1,2,2,1,0,1,1,1,1,1,1,0,

      0,0,0,0,1,0,0,0,1,0,2,1,1,0,1,1,1,0,2,1,1,1,1,0,0,0,1,0,0,0,0,

      0,0,0,0,0,1,1,0,1,1,0,0,0,0,1,1,0,0,1,1,0,0,0,1,0,0,0,0,0,0,

      0,0,0,0,0,0,0,0,0,1,0,0,0,0,1,1,1,0,0,0,0,1,0,0,0,0,0,0,1,1))

4. In Example 9.6 apply a negative binomial–beta mixture with the prior and parameterization discussed in section 9.3.2.

5. In the analysis of the data in Example 9.7, Smith *et al.* mention issues with representing overdispersion and also possible error autocorrelation. Using Example 9.8 as a template, investigate these questions.

6. In Example 9.10 try $m = 3$ and $m = 4$ in the Poisson HMM for the purse snatching data and assess fit against the $m = 2$ model via the AIC and BIC.

REFERENCES

Azzalini, A. (1994) Logistic regression and other discrete data models for serially correlated observations. *Journal of the Italian Statistical Society*, **3**, 169–179.

Beck, N., Epstein, D., Jackman, S. and O'Halloran, S. (2001) Alternative models of dynamics in binary time-series-cross-section models: the example of state failure. *Political Methodology Working Papers* – 2001 (http://polmeth.wustl.edu).

Brandt, P. and Williams, J. (2001) A linear Poisson autoregressive model: the Poisson AR(p) model. *Political Analysis*, **9**, 164–184.

Brandt, P., Williams, J., Fordham, B. and Pollins, B. (2000) Dynamic models for persistent event count time series. *American Journal of Political Science*, **38**, 823–843.

Brannas, K. (1995) Explanatory variables in the INAR(1) count data model. *Umea Economic Studies* 381.

Cai, Z., Yao, Q. and Zhang, W. (2001) Smoothing for discrete valued time series. *Journal of the Royal Statistical Society, Series B*, **63**, 357–375.

Cameron, C. and Trivedi, P. (1998) *Regression Analysis of Count Data*, Econometric Society Monograph No. 30. Cambridge University Press: Cambridge.

Cardinal, M., Roy, R. and Lambert, J. (1999) On the application of integer-valued time series models for the analysis of disease incidence. *Statistics in Medicine*, **18**, 2025–2039.

Cargnoni, C., Müller, P. and West, M. (1997) Bayesian forecasting of multinomial time series through conditionally Gaussian dynamic models. *Journal of the American Statistical Association*, **92**, 640–647.

Carlin, B. and Polson, N. (1992) Monte Carlo Bayesian methods for discrete regression models and categorical time series. In *Bayesian Statistics 4*, Bernardo, J. *et al.* (eds). Clarendon Press: Oxford, 577–586.

Chen, M.-H. and Ibrahim, J. (2000) Bayesian predictive inference for time series count data. *Biometrics*, **56**, 678–685.

Cox, D. (1981) Statistical analysis of time series: some recent developments. *Scandinavian Journal of Statistics*, **8**, 93–115.

Cox, D. and Snell, E. (1989) *Analysis of Binary Data*, 2nd Edition, Monographs on Statistics and Applied Probability, 32. Chapman and Hall: London.

Czado, C. (2001) Individual migraine risk management using binary state space mixed models. *Working Paper*, Technische Universität München, Lehrstuhl für Statistik.

Davis, R., Dunsmuir, W. and Wang, Y. (2000) On autocorrelation in a Poisson regression model. *Biometrika*, **87**, 491–505.

De Gunst, M., Künsch, H. and Schouten, J. (2001) Statistical analysis of ion channel data using hidden Markov models with correlated state-dependent noise and filtering. *Journal of the American Statistical Association*, **96**, 805–815.

Fahrmeier, L. and Tutz, G. (2000) *Multivariate Statistical Modelling Based on Generalized Linear Models*. Springer: Berlin.

Ferreira, M. and Gamerman, D. (2000) Dynamic generalized linear models. In *Generalized Linear Models: A Bayesian Perspective*, Dey, D., Ghosh, S. and Mallick, B. (eds). Marcel Dekker: New York, 57–72.

Gamerman, D. (1998) Markov chain Monte Carlo for dynamic generalised linear models. *Biometrika*, **85**, 215–227.

Gelman, A., Carlin, J., Stern, H. and Rubin, D. (1996) *Bayesian Data Analysis*. Chapman and Hall, London.

Grunwald, G., Hyndman, R., Tedesco, L. and Tweedie, R. (2000) Non-Gaussian conditional linear AR(1) models. *Australian & New Zealand Journal of Statistics*, **42**, 479–495.

Gustafson, P. and Walker, L. (2003) An extension of the Dirichlet prior for the analysis of longitudinal multinomial data. *Journal of Applied Statistics*, **30**, 293–310.

Harvey, A. and Fernandes, C. (1989) Time series models for count or qualitative observations. *Journal of Business and Economic Statistics*, **7**, 407–417.

Heckman, J. (1981) Statistical models for discrete panel data. In *Structural Analysis of Discrete Data with Econometric Applications*, Manski, C. and McFadden, D. (eds). The MIT Press: Cambridge, MA, 114–178.

Hyndman, R. (1999) Nonparametric additive regression models for binary time series. *Proceedings, 1999 Australasian Meeting of the Econometric Society, 7–9 July 1999, University of Technology, Sydney*.

Jackman, S. (1998) Time series models for discrete data: solutions to a problem with quantitative studies of international conflict. Department of Political Science, Stanford University.

Jones, R. (1993) *Longitudinal Data with Serial Correlation: A State-space Approach*. Chapman and Hall: London.

Jowaheer, V. and Sutradhar, B. (2002) Analyzing longitudinal count data with overdispersion. *Biometrika*, **89**, 389–399.

Kauermann, G. (2000) Modeling longitudinal data with ordinal response by varying coefficients. *Biometrics*, **56**, 692–698.

Kedem, B. and Fokianos, K. (2002) *Regression Models for Time Series Analysis*. John Wiley & Sons: New York.

Kennan, J. (1985) The duration of contract strikes in US manufacturing. *Journal of Econometrics*, **28**, 5–28.

Kitagawa, G. and Gersch, W. (1996) *Smoothness Prior Analysis of Time Series*. Springer: Berlin.

Leroux, B. and Puterman, M. (1992) Maximum penalized likelihood estimation for independent and Markov-dependent poisson mixtures. *Biometrics*, **48**, 545–558.

Lindsey, J. (1993) *Models for Repeated Measurements*. Clarendon Press: Oxford.

McCabe, B. and Martin, G. (2003) Coherent Bayesian predictions of low count time series. *Working Paper*, Department of Economics and Accounting, School of Management, University of Liverpool.

McKenzie, E. (1986) Autoregressive moving average processes with negative binomial and geometric marginal distributions. *Advances in Applied Probability*, **18**, 679–705.

Oh, M. and Lim, Y. (2001) Bayesian analysis of time series Poisson data. *Journal of Applied Statistics*, **28**, 259–271.

Ord, K., Fernandes, C. and Harvey, A. (1993) Time series models for multivariate series of count data. In *Developments in Time Series Analysis: In Honour of Maurice B. Priestley*, Subba Rao, T. (ed.). Chapman and Hall: London, 295–309.

Paroli, R. and Spezia, L. (2002) Parameter estimation of Gaussian hidden Markov models when missing observations occur. *Metron*, **60**, 165–181.

Poirier, D. and Ruud, P. (1988) Probit with dependent observations, *Review of Economic Studies*, **55**, 593–614.

Raftery, A. (1985) A model for higher-order Markov chains, *Journal of the Royal Statistical Society, Series B*, **47**, 528–539.

Scott, S. (2002) Bayesian methods for hidden Markov models: recursive computing in the 21st century. *Journal of the American Statistical Association*, **97**, 337–351.

Smith, R., Davis, J., Sacks, J. and Speckman, P. (2000) Regression models for air pollution and daily mortality: analysis of data from Birmingham, Alabama. *Environmetrics*, **10**, 719–745.

Tsay, R. (1992) Model checking via parametric bootstraps in time series analysis. *Applied Statistics*, **41**, 1–15.

West, M. and Harrison, J. (1997) *Bayesian Forecasting and Dynamic Models*, 2nd Edition. Springer: New York.

West, M., Harrison, J. and Migon, H. (1985) Dynamic generalised linear models and Bayesian forecasting. *Journal of the American Statistical Association*, **80**, 73–83.

Winkelmann, R. (2000) *Econometric Analysis of Count Data*, 3rd Edition. Springer: Berlin.

Wood, S. and Kohn, R. (1998) A Bayesian approach to robust nonparametric binary regression. *Journal of the American Statistical Association*, **93**, 203–213.

Zeger, S. (1988) A regression model for time series of counts. *Biometrika*, **75**, 621–629.

Zeger, S. and Qaqish, B. (1988) Markov regression models for time series: a quasi-likelihood approach. *Biometrics*, **44**, 1019–1031.

CHAPTER 10

Hierarchical and Panel Data Models

10.1 INTRODUCTION: CLUSTERED DATA AND GENERAL LINEAR MIXED MODELS

Multilevel models are appropriate for hierarchically arranged data with nested sources of variability such as patients within hospitals, progeny within sires, or pupils within classes, when incorrect inferences may be drawn if the different sources of variability are not taken into account. Analysing data in which individuals j are arranged within clusters i ($i = 1, \ldots, n; \ j = 1, \ldots, R_i$) should address the similarity (autocorrelation) between individuals in the same cluster, and also possibly varying causal impacts of predictors x_{ijk} on the outcome y_{ij} across clusters. Other forms of clustered data include panel data (repetitions of observations at times t for subjects i), with the clusters being the subjects. Repeated data may also be defined by cross-categorized factors (e.g. treatment combinations, diagnostic tests, product attributes) with the subjects being patients, quality raters, etc. For longitudinal data it is often appropriate for the autocorrelation model to be structured in time, using methods such as those in Chapter 9. If subjects are arranged spatially then both space and time structuring should be included (section 10.10).

Some of the mechanisms underlying autocorrelation within clusters of human subjects in hierarchically structured school outcome data are spelt out by Snijders and Bosker (1999, p 9) such as residence in a similar neighbourhood, attending the same school, or shared group norms. For instance, pupil exam results vary partly because of different pupil abilities, but also because of differences between classes or schools.

Bayesian Models for Categorical Data   P. Congdon
© 2005 John Wiley & Sons, Ltd

Pupils in the same class tend to perform more similarly than pupils in different classes, and pupils in one school are usually more similar to each other than to pupils in other schools. The institutional context may also impact on individual-level attainment processes; for instance, the slope relating exam performance to ability may be influenced by class size or teacher style (Aitkin and Longford, 1986). Similar contextual effects apply for health and behavioural outcomes, though there is still much debate about the relative importance of individual-level influences and those due to institutional or neighbourhood settings (Goldstein and Spiegelhalter, 1996).

In longitudinal or panel studies a collection of subjects is observed through time, so that repeated observations are clustered within subjects. For instance, a discrete outcome y_{it} may be observed at $t = 1, \ldots, T_i$ occasions for each of n subjects. While there are many overlaps with specifications involving cross-sectional clustered data, there are distinct issues that may occur, such as the modelling of growth paths through time, or the addition of lagged values ($y_{i,t-1}$, $y_{i,t-2}$, etc.) as predictors of the current response. Just as for clustered data it is likely that subjects vary in their average levels on the outcome, but they may also vary in growth rates over time, or in the impact of other time-specific predictors.

10.2　HIERARCHICAL MODELS FOR METRIC OUTCOMES

As for cross-sectional data without clustering, the linear model for metric data forms a baseline for non-metric data or is sometimes of direct relevance if, say, the Albert–Chib data augmentation method (Albert and Chib, 1993) via Gibbs sampling is applied. Consider a univariate metric outcome and a two-level hierarchy and let y_i be an $R_i \times 1$ vector $y_i = (y_{1i}, y_{2i}, \ldots, y_{R_i i})'$ with R_i being the number of repetitions within clusters or subjects $i = 1, \ldots, n$. Let X_i be an $R_i \times p$ matrix of time-varying predictors with ith row $X_{ir} = (x_{ir1}, \ldots, x_{irp})$, and $\gamma_i = \{\gamma_{i1}, \gamma_{i2}, \ldots, \gamma_{ip}\}'$ be a $p \times 1$ random regression parameter vector. Also let 1 be $R_i \times 1$. Then a two-level hierarchical model with variation in intercepts α_i and in all predictor effects has the form

$$y_i = 1\alpha_i + X_i\gamma_i + \varepsilon_i \tag{10.1}$$

The observation level error $\varepsilon_i = (\varepsilon_{i1}, \varepsilon_{i2}, \ldots, \varepsilon_{iR_i})'$ is most usually taken to follow a parametric density (e.g. normal) with constant variance $\varepsilon_{ir} \sim N(0, \sigma_\varepsilon^2)$. The cluster-specific random effects $\eta_i = (\alpha_i, \gamma_{i1}, \ldots, \gamma_{ip})$

may have independent prior densities or be associated in a multivariate prior density, e.g.

$$\eta_i \sim N_{p+1}(H, \Sigma) \tag{10.2}$$

where $H = (A, \Gamma_1, \ldots, \Gamma_q)$ represents average regression effects. These population averages typically have diffuse prior densities, e.g. the prior on the average intercept might be $A \sim N(0, V_\alpha)$ where V_α is a known large variance. Alternatively variations in η_i might be modelled in terms of regression on r cluster-level attributes W_i, e.g.

$$\eta_i \sim N_{p+1}(W_i \varphi, \Sigma)$$

where φ is of dimension $r \times (p+1)$.

Often a mixed model (Laird and Ware, 1982) with some regression parameters fixed over clusters is used, so that

$$y_i = 1\alpha + X_i\beta + Z_i b_i + \varepsilon_i \tag{10.3}$$

where X_i is an $R_i \times p$ matrix of predictors associated with fixed regression parameters β, and Z_i is an $R_i \times q$ matrix of predictors the impact of which on the outcome is expressed by a $q \times 1$ vector of random cluster-specific effects $b_i = (b_{i1}, b_{i2}, \ldots, b_{iq})'$. If the z_{ik} are a subset of $\{1, x_{i1}, \ldots, x_{ip}\}$ then the b_i may be parameterized to represent deviations from the population average regression parameters. So the total effect of the kth predictor in cluster i is $\beta_{ik} = \beta_k + b_{ik}$. b_i typically takes a parametric multivariate prior parallel to (10.2), namely

$$b_i \sim N_q(0, \Sigma_b) \tag{10.4}$$

Other options are possible: for instance, Z_i might include variables not present in X_i. Non-parametric alternatives to (10.4) are also possible, such as discrete mixtures. The precision matrix $T_b = \Sigma_b^{-1}$ of the predictor random effects under a multivariate normal prior (10.4) is typically assigned a Wishart prior $T_b \sim W(\nu_0, Q)$, though options for greater flexibility have been proposed (e.g. Boscardin and Weiss, 2004). The full conditional density of T_b is then a Wishart with shape $\nu = \nu_0 + n$ and scale

$$\left(Q^{-1} + \sum_{i=1}^{n} b_i b_i' \right)^{-1}$$

Wakefield *et al.* (1994) and Chib and Carlin (1999) note that Gibbs sampling can be used for all the parameters in this model, since all full conditionals reduce to standard densities. Despite this there may be a gain in identifiability and convergence in MCMC estimation if overlaps between the fixed and random effects specifications are avoided (Chib *et al.*, 1999). Thus suppose Z_i consists only of X variables, say $Z_i = (X_{i1}, \ldots, X_{iq})$, with randomly varying coefficients that have means $(\beta_1, \ldots, \beta_q)$. To avoid overlaps, the model is reframed so that X_i now consists only of $X_i^* = (X_{i,q+1}, \ldots, X_{ip})$, and b_i in (10.4) now has mean $(\beta_1, \ldots, \beta_q)$.

Heterogeneity in intercepts representing intersubject variations in, say, average susceptibility or ability that are not captured by measured covariates is often an issue. In many applications this is the only form of heterogeneity considered (so $q = 1$) and the linear mixed model reduces to

$$y_i = 1\alpha + X_i\beta + 1b_i + \varepsilon_i \tag{10.5}$$

where b_i might a priori be taken as $b_i \sim N(0, \sigma_b^2)$. Under the alternative parameterization above the intercept is excluded from the fixed effects and the mean of the first random effect assumes the role of the intercept (Chib and Carlin, 1999). So (10.5) becomes

$$y_i = 1b_i + X_i\beta + \varepsilon_i$$

where $b_i \sim N(\alpha, \sigma_b^2)$.

The above models may be expressed in a three-stage form which is the basis for conjugate generalized linear mixed models for discrete outcomes as discussed by Daniels and Gatsonis (1999). Thus the model of equation (10.3) may be written

$$y_{ir} \sim N(\mu_{ir}, \sigma^2)$$

where the conditional mean is $\mu_{ir} = E(y_{ir}|b_i) = \alpha + X_{ir}\beta + Z_{ir}b_i$. The second stage of this model specifies the prior density $p(b_i|\theta_b)$ for the cluster effects b_i while the third stage specifies priors on hyperparameters, $p(\theta_b)$.

10.3 HIERARCHICAL GENERALIZED LINEAR MODELS

For non-normal data (binary, categorical or count) similarly assume that the responses y_{ir} follow the exponential family density

$$p(y_{ir}|\theta_{ir})] = \exp\{[y_{ir}\theta_{ir} - \psi(\theta_{ir})]/a_{ir}(\phi) + c(y_{ir}, \phi)\}$$

The appropriate density (e.g. Poisson) from this family provides the likelihood for the observations at the first stage. Under a conjugate mixture approach (Daniels and Gatsonis, 1999), the second stage specifies the conjugate mixing density for θ_{ir} (e.g. gamma for a Poisson likelihood, beta for a binomial likelihood) in terms of a mean μ_{ir} (e.g. Poisson mean, binomial probability) and scale parameter δ_i. A mixture on the θ_{ir} at the second stage is especially relevant in the case of over-dispersion. The second stage also specifies how μ_{ir} is linked to the regression function $\eta_{ir} = \eta_{ir}(X_{ir})$, e.g.

$$g(\mu_{ir}) = X_{ir}\beta_i \qquad (10.6)$$

where the $q \times 1$ regression vectors β_i vary over clusters. The third stage specifies densities for varying effects between clusters i in terms of cluster attributes W_i for example,

$$\beta_i \sim N(W_i\varphi, \Sigma_\beta)$$

while the fourth stage specifies priors on hyperparameters.

Thus consider a conjugate Poisson–gamma mixture for repeated count data. Then one form possible is a multiplicative subject-level frailty

$$y_{ir} \sim \text{Poi}(\mu_{ir})$$
$$\mu_{ir} \sim G(\delta_i\mu_{ir}, \delta_i)$$

where δ_i might be itself assigned a Gamma prior. Then $\log(\mu_{ir}) = \beta_i X_{ir}$ and $\beta_i \sim N(W_i\phi, \Sigma_\beta)$. Related model specifications (Lee and Nelder, 2000) might specify

$$\mu_{ir} = \exp(X_{ir}\beta)\nu_i$$

where

$$\nu_i \sim G(\kappa, \kappa)$$

and the precision κ of the ν_i might be assigned its own prior. If overdispersion is still unaccounted for then a gamma distributed observa-tion-level effect ν_{ir} might be introduced, again with mean 1:

$$\mu_{ir} = \exp(X_{ir}\beta)\nu_{ir}\nu_i$$

A related model often used to analyse integer item scores (e.g. in education) is known as the Rasch Poisson count model (van Duijn and Jansen, 1995) and assumes for subject i and item r

$$\mu_{ir} = \nu_i\pi_{ir}$$

where the effects $\nu_i \sim \text{Ga}(c, c/m)$ have mean m and $\pi_{ir} \sim \text{Dir}(b_1, \ldots, b_R)$. In educational applications, the ν_i are ability parameters and the π_{ir} are difficulty parameters with prior difficulties b_r constant over subjects. The marginal likelihood here is the product of a negative binomial for the subject total $y_{i+} = \sum_r y_{ir}$ (and with parameters c and $s = c/m$) and a multinomial for y_{ir} conditional on y_{i+} with parameters

π_{ir}/π_{i+}. So the multinomial is modelling how the total score for an individual is distributed between individual items. This is also known as the Dirichlet model in consumer research (Goodhardt *et al.*, 1984). Generalizations might include making the ability mean m specific to pupil groups (e.g. $m_1, \ldots, m_G$), or relating an individual-level mean m_i to predictors via log-linear regression.

The alternative to conjugate approaches is the generalized linear mixed model involving random effects in a log- or logit linear regression term η_{ir}. For example, with $y_{ir} \sim \text{Po}(\mu_{ir})$ or $y_{ir} \sim \text{Bin}(n_{ir}, \mu_{ir})$, heterogeneity in intercepts or slopes across clusters could be modelled via

$$g(\mu_{ir}) = X_{ir}\beta + Z_{ir}b_i \qquad (10.7)$$

where X_{ir} is $1 \times p$, $Z_{ir} = (z_{ir1}, z_{ir2}, \ldots, z_{irq})$ is $1 \times q$, and $b_i = (b_{i1}, \ldots, b_{iq})'$. This form is often simpler for purposes of modelling variation between clusters in both slopes and intercepts, e.g. via a multivariate normal prior

$$b_i \sim \text{N}_q(W_i\varphi, \Sigma_b)$$

Robust alternatives might involve scale mixing or discrete mixtures. For example, a scale mixture would specify

$$b_i \sim \text{N}_q(W_i\varphi, \Sigma_b/\lambda_i)$$

where λ_i are gamma (leading to multivariate t or Cauchy distributed b_i) or exponential (leading to a double exponential b_i). One discrete mixture option (Weiss *et al.*, 1999) allows for an inflated variance (dispersion) component with

$$b_i \sim \pi\text{N}_q(W_i\varphi, \Sigma_b) + (1 - \pi)\text{N}_q(W_i\varphi, k\Sigma_b)$$

where $k = 100$, say. As for the metric model above, identifiability may be improved if X_{ir} and Z_{ir} are mutually exclusive. So if there were intercept variation between clusters (and no other form of heterogeneity, so $q = 1$) then $Z_{ir1} = 1$ and $b_i = b_{i1}$ has a non-zero mean which represents the average intercept. The X_{ir} will not then include an intercept.

Modelling intracluster correlation may remove much extra variation in the GLMM of (10.7), but, if required, an extra level of variation at the observation level may be used

$$g(\mu_{ir}) = X_{ir}\beta + Z_{ir}b_i + \varepsilon_{ir} \qquad (10.8)$$

Such extensions will generally much improve fit as measured by $D(\bar{\theta}|y)$ or $D(\bar{\mu}|y)$ but may risk overparameterizing the model. Here the prior on the ε_{ir} may need to be sufficiently informative to prevent them effectively modelling the data (Johnson and Albert, 1999).

The logit linear random effects approach generalizes to clustered ordinal and multinomial responses with J levels. Let $y_{ir} \in 1, 2, \ldots, J$ be repetitions of categorical variables for individuals or groups. Thus for an ordinal response with predictors possibly varying over repetitions, one seeks to model the cumulative probabilities

$$\gamma_{irj} = \Pr(y_{ir} \le j | X_{ir}) \tag{10.9a}$$

with probabilities $\pi_{irj} = \Pr(y_{ir} = j)$ obtained as

$$
\begin{aligned}
\pi_{ir1} &= \gamma_{ir1} \\
\pi_{irj} &= \gamma_{irj} - \gamma_{ir,j-1} & j &= 2, \ldots, J-1 \\
\gamma_{irj} &= F(\kappa_j - \mu_{ir}) & j &= 1, \ldots, J-1 \\
\pi_{irJ} &= 1 - \gamma_{ir,J-1}
\end{aligned}
\tag{10.9b}
$$

and $J - 1$ free cut point parameters κ_j. The model for the regression mean excludes an intercept but may include zero-mean random effects b_i, namely

$$\mu_{ir} = \beta_1 x_{ir1} + \beta_2 x_{ir2} + \cdots + \beta_p x_{irp} + b_i$$

or subject-varying coefficients on a q vector of predictors Z_{ir},

$$\mu_{ir} = \beta_1 x_{ir1} + \beta_2 x_{ir2} + \cdots + \beta_p x_{irp} + Z_{ir} b_i$$

For unordered categorical data with repeated observations (e.g. consumer choice data with several observations on each household or subject) alternative estimation methods are considered by Chen and Kuo (2001). For example, a random subject intercept MNL model for repeated multinomial choice data with individual specific attributes C_{ij} (e.g. Price$_{ij}$ and Quality$_{ij}$) or attributes varying over repetitions C_{ijr} (e.g. price by purchase occasion) has the form

$$\pi_{irj} = \Pr(y_{ir} = j)$$

$$= \exp(b_{ij} + \gamma_1 \text{Price}_{ijr} + \gamma_2 \text{Quality}_{ijr}) / \sum_{k=1}^{J} \exp(b_{ik} + \gamma_1 \text{Price}_{ikr}$$

$$+ \gamma_2 \text{Quality}_{ikr}) \quad j = 1, \ldots, J; r = 1, \ldots, R$$

where for identifiability one effect is zero (e.g. $b_{iJ} = 0$) and so $b_{ij}, j = 1, \ldots, J - 1$ might be taken as multivariate effects (e.g. MVN) of order $J - 1$. Random effects for other predictors can be modelled similarly: for example, γ_1 might vary over individuals, choices or both. The mixed MNL models considered in Chapter 6 are likely to be better identified if there are repeated choice observations for each subject.

10.3.1   Augmented data sampling for hierarchical GLMs

As for non-hierarchical data, models for clustered categorical data (especially binary or ordinal) may involve augmenting the data by latent continuous scales. In this case the metric forms of the mixed model (section 10.2) become relevant. Consider binary data with repetitions r within clusters i. Assume a latent metric utility y^*_{ir} that is given by

$$y^*_{ir} = X_{ir}\beta + Z_{ir}b_i + \varepsilon_{ir} \tag{10.10}$$

When $q = 1$, (10.10) reduces to a mixed intercept model

$$y^*_{ir} = X_{ir}\beta + b_i + \varepsilon_{ir} \tag{10.11}$$

The probability of an event is

$$\Pr(y_{ir} = 1) = \Pr(y^*_{ir} > 0) = \Pr(\varepsilon_{ir} > -X_{ir}\beta - Z_{ir}b_i) = 1 - F(-X_{ir}\beta - Z_{ir}b_i)$$

For F symmetric about zero (e.g. when F is the cdf of an N(0,1) variable) the last element of this equation equals $F(X_{ir}\beta + Z_{ir}b_i)$. So in this case the y^*_{ir} are obtainable by truncated normal sampling, constrained to be negative when the actual binary observation $y_{ir} = 0$, and positive when $y_{ir} = 1$. As above, robust alternatives may be obtained by scale mixing on both the model for y^*_{ir} and the model for b_i. For example, to approximate a logit link, the y^*_{ir} can be sampled from a truncated Student t density with 8 degrees of freedom, and variance $1/0.634^2$, since a t_8 variable is approximately 0.634 times a logistic variable. The scale mixture version of the Student t density then facilitates outlier detection and down-weighting of aberrant cases (see Example 10.3).

Latent scales also apply for clustered ordinal and multinomial data, and are fittable by the Albert and Chib (1993) approach. For ordinal responses a generalization of the location-scale model (section 7.4) is provided by Ishwaran and Gatsonis (2000). Thus for subject i and diagnostic tests r assume univariate ordered observations y_{ir}, underlying variables $y*$ and predictors X_{ir} and U_{ir} such that

$$\Pr(y_{ir} \le j | X_{ir}) = F\left[\frac{\kappa_j - X_{ir}\beta}{\exp(U_{ir}\delta)}\right]$$

Then $y_{ir} = j$ if $\kappa_{j-1} < y^*_{ir} \le \kappa_j$ where

$$y^*_{ir} = X_{ir}\beta + Z_{ir}\exp(U_{ir}\delta)$$
$$Z_i \sim N(0, R)$$

where R is a correlation matrix, since the variances are not identified. For clustered multinomial data, the multinomial probit can be generalized to repetitions r. Then $y_{ir} = j$ if $y^*_{irj} = \max(y^*_{ir1}, y^*_{ir2}, \ldots, y^*_{irJ})$ where one possible model might be

$$y^*_{irj} = V_{irj} + \varepsilon_{irj} = X_{ir}\beta_j + Z_{ir}b_i + C_{ij}\gamma + A_j\zeta + \varepsilon_{irj}$$

where C_{ij} are subject–choice attributes and A_j are choice-specific attributes. Identifiability involves differencing against the utility of a reference category, such as the Jth (Geweke *et al.*, 1994).

Example 10.1   Cohabitation experience   To illustrate multilevel analysis with a binary outcome consider data from the 1994 International Social Survey Program (Snijders and Bosker, 1999, chapter 14). This relates to whether 2079 Norwegian adults had lived together without being married. Age and religious adherence (religion $= 1$ if respondent attends religious service at least once a month) are the individual predictors. The experience of cohabitation and other aspects (e.g. the slope on age or the impact of religion) may differ by region, there being $i = 1, \ldots, 19$ regions varying in respondent totals R_i from 235 (Oslo) to 35 (Finmark).

The age effect over all subjects is modelled as a quadratic spline with knots at ages 30 and 40, namely

$$S(\text{Age}) = \beta_0 + \beta_1(\text{Age} - 20) + \beta_2(\text{Age} - 20)^2$$
$$+ \beta_3[(\text{Age} - 30)_+]^2 + \beta_4[(\text{Age} - 40)_+]^2$$

where for example

$$(\text{Age} - 30)_+ = (\text{Age} - 30) \quad \text{if Age} > 30$$
$$= 0 \qquad \text{if Age} \leq 30$$

In the first model (model A) the intercept varies by region. Thus $q = 1$ and $Z_{ir1} = 1$ as in (10.7), so that $y_{ir} \sim \text{Bern}(\pi_{ir})$,

$$\text{logit}(\pi_{ir}) = b_i + \beta_1(\text{Age} - 20) + \beta_2(\text{Age} - 20)^2$$
$$+ \beta_3[(\text{Age} - 30)_+]^2 + \beta_4[(\text{Age} - 40)_+]^2 + \beta_5\text{Religion}$$

and

$$b_i \sim \text{N}(\beta_0, \sigma_b^2)$$

The age effect in region i under this model is

$$S_i(\text{Age}) = b_i + \beta_1(\text{Age} - 20) + \beta_2(\text{Age} - 20)^2$$
$$+ \beta_3[(\text{Age} - 30)_+]^2 + \beta_4[(\text{Age} - 40)_+]^2$$

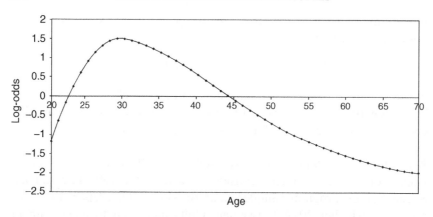

Figure 10.1 Impact of age on experience of cohabitation

Posterior means on the regression parameters $\beta = (\beta_0, \beta_1, \ldots, \beta_5)$ are obtained as $(-1.18, 0.57, -0.031, 0.025, 0.0075, -1.92)$ with the age effect when $b_i = \beta_0$ reproduced in Figure 10.1. This shows the experience of cohabitation reaching a maximum at age 30 and declining thereafter. Region effects b_i vary from -1.02 to -1.27. The DIC is 2191 with $d_e = 13$; since there are six fixed effects regression parameters plus the variance of the b_i, the 19 extra random effects have an effective dimension of 6. As an alternative predictive fit criterion, the proportion of those with cohabitation experience correctly classified using 'new data' samples is 0.60 as against 0.69 for those without experience.

A second model adds varying linear age effects over regions with a multivariate normal prior allowing for covariance between intercepts b_{i1} and age effects b_{i2}. An identity scale matrix for the Wishart prior on the inverse covariance matrix is assumed. This model has around 34.5 effective parameters, so adds considerably to complexity. In fact there is no correlation between the two region effects, so a simpler independent effects model may be more successful. There does appear to be some regional variation in the age effects with the posterior mean of the relevant parameter Σ_{b22} of the covariance matrix being 0.06, with a 95% interval $(0.03, 0.11)$. The sensitivity under this model improves slightly to around 0.605 and the DIC falls to 2188.

To illustrate the latent data technique, latent normal y^* with constant variance are sampled from a model (model C) with mean given by (10.11), i.e. varying intercepts only. This involves truncated sampling and is equivalent to a probit link, $g^{-1} = \Phi$. New latent data are sampled (without truncation) and their match with the observations assessed. The sensitivity and specificity are close to those under the direct likelihood approach (model A), namely 0.60 and 0.69. By monitoring possible outliers via Monte Carlo estimates of the CPO, the lowest CPOs are found for the youngest adults (under 18) with cohabitation experience and for three religious older adults (over 65) also with such experience.

To assess their impact on regression effects and fit a logit link is approximated by scale mixing (model D). Thus, following Albert and Chib (1993),

$$y_{ir}^* \sim N(X_{ir}\beta + b_i, 1/[\lambda_{ir}0.634^2])$$

where

$$\lambda_{ir} \sim Ga(4, 4)$$

As would be expected, the lowest weights λ_{ir} (around 0.5) are for older religious subjects with cohabitation experience; the highest weights (around 1.2) are for non-religious adults in their thirties with experience. Under model D, these subjects are less at odds with the model, with their CPOs accordingly higher than under model C. Under scale mixing the sensitivity is raised to 0.61 and the specificity to 0.70, though there are no impacts on regressor effects (e.g. in terms of their direction and significance). The effective parameter count is now 45, compared with 12 under the uniform variance model, so there is an increase in complexity, though despite this the DIC falls from 2188 to 2179.

Example 10.2 Wine bitterness A clustered ordinal data analysis considers rankings of the bitterness of wines by nine rankers on a five-point scale from 1 = not bitter to 5 = very bitter (Randall, 1989). Two wine bottles were randomly selected from products defined first by temperature at pressing, and second by skin contact at pressing. Thus each ranker rates eight bottles, two from the low-temperature–no contact category, two from the low-temperature–contact category, two from the high-temperature–no-contact category, and two from the high-temperature–contact category. Apart from the influence of these factors on ratings, it is necessary to allow for random variation between raters.

Thus the structure of (10.9) above applies to observations $r = 1, \ldots, 8$ for raters $i = 1, \ldots, 9$. A model for $\pi_{irj} = \Pr(y_{ir} = j)$ under proportional odds and without rater random effects may be defined as

$$\pi_{ir1} = \gamma_{ir1}$$
$$\pi_{irj} = \gamma_{irj} - \gamma_{ir,j-1} \qquad j = 2, \ldots, J-1$$
$$\text{logit}(\gamma_{irj}) = \kappa_j - \mu_{ir} \qquad j = 1, \ldots, J-1$$

with $J = 5$, four free cut points $\kappa_1, \ldots, \kappa_4$, and regression mean

$$\mu_{ir} = \beta_1 X_{ir1} + \beta_2 X_{ir2}$$

where $X_{ir1} = 1$ for low temperature ($= 0$ otherwise) and $X_{ir2} = 1$ for skin contact ($= 0$ otherwise). The prior (7.5) is adopted for the cut points, namely

$$\kappa_1 = \varphi_1$$

$$\kappa_j = \kappa_{j-1} + \sum_{m=2}^{j} \exp(\varphi_m)$$

with unconstrained φ_m. Model B including rater intercept variation has the form

$$\mu_{ir} = \beta_1 X_{ir1} + \beta_2 X_{ir2} + b_i$$

where the b_i are zero-mean random effects, for instance $b_i \sim N(0, \sigma_b^2)$. It is also possible to allow rater variation in the impacts of the other predictors. Labelling them as Z_{ir1} and Z_{ir2} (and retaining intercept variation) this leads to a model

$$\mu_{ir} = b_{i1} + b_{i2} Z_{ir1} + b_{i3} Z_{ir2}$$

where $b_i = (b_{i1}, b_{i2}, b_{i3})$ might be taken as multivariate normal or student t and b_{i1} has a zero mean.

A model without random intercepts gives DIC $= 185$ and $d_e = 5.9$ (a log likelihood at posterior mean of the parameters of -86.6). The mean coefficient on low temperature is -2.45 (so less bitter ratings are obtained for low-temperature pressing) while that on contact is 1.54 (so contact increases bitterness).

To allow for clustering within raters (i.e. non-independence of responses y_{ir} within raters) a normal random effect b_i is added for the model

$$\mu_{ir} = \beta_1 X_{ir1} + \beta_2 X_{ir2} + b_i$$

Because of the small sample there may be sensitivity in estimating the b_i to alternative priors on σ_b^2. Assuming

$$\tau_b = 1/\sigma_b^2 \sim Ga(1, 0.01)$$

a two-chain run of 20 000 iterations shows the most bitter ratings to be provided by rater 1 ($\bar{b}_1 = 1.05$) and the least bitter by rater 7. Allowing for such differences improves the DIC to 176.5, with $d_e = 12.1$ and log likelihood at $\bar{\theta} = (\bar{b}, \bar{\beta})$ of -76. The means on temperature and contact are enhanced to -2.75 and 1.72. An alternative prior on τ_b (only one possible option) takes

$$\eta = \tau_b/(1 + \tau_b)$$
$$\eta \sim \text{Be}(1, 1)$$

This gives a lower DIC of 170 and a wider scatter of rater effects, e.g. $\bar{b}_1 = 1.75$, and $\bar{b}_7 = -1.77$.

10.4   RANDOM EFFECTS FOR CROSSED FACTORS

The most common multilevel structure is when contextual variables are nested: for example, subjects $i = 1, \ldots, N_{jk}$ at level 1 (e.g. pupils) are nested within clusters $j = 1, \ldots, J$ at level 2 (schools) that may be further nested into level 3 aggregates $k = 1, \ldots, K$ (local education agencies); see, for example, Rasbash and Browne (2001). A random effects model can be set at each level with random effects at level j explicable by predictors at level j and k. One may also establish models for heteroscedasticity or overdispersion at level 1 with the variance of errors ε_{ijk} depending on characteristics of individuals or their contexts, e.g. $\log[\text{Var}(\varepsilon_{ijk})] = \gamma_0 + \gamma_1 X_{ijk} + \gamma_2 W_{jk}$.

However, in many situations the context involves overlapping or crossed rather than nested factors (e.g. when pupil attainment reflects both school and area of residence). Let $h = [jk]$ denote the cross-hatched factor formed by crossing levels j and k. In some applications the risk population N_h may not occur in all the combinations of contextual factors but for the moment define $H = JK$ to cover all possible combinations. Rather than let i (for pupil) range from 1 to N_h, define i to range from 1 to N where $N = \sum_h N_h$. Also let $j = j[i]$ denote the pupil's school, $k = k[i]$ denote the pupil's area of residence and $h = h[i]$ denote the crossed index $jk[i]$, with h varying from 1 to H. Then for a binary outcome with $y_i \sim \text{Bern}(\pi_i)$ possible models include a random effect h at the crossed level

$$\text{logit}(\pi_i) = \beta X_i + \gamma W_{j[i]} + \delta Z_{k[i]} + \eta_{h[i]}$$

or separate random effects u_1 and u_2 for the two crossed factors

$$\text{logit}(\pi_i) = \beta X_i + \gamma W_{j[i]} + \delta Z_{k[i]} + u_{1j[i]} + u_{2k[i]}$$

Sometimes aggregated data may be recorded for cross-hatched factors without individual information being available. For example, deaths or hospital referrals may be recorded for area of residence (j) and for the general practitioner (GP) practice (k) the patient is registered with. Let h range from 1 to $H = JK$ and let $j[h]$ and $k[h]$ denote the factors defining particular levels of the cross-hatched index $h = 1, \ldots, H$. Let y_h be counts with $y_h \sim \text{Po}(\mu_h E_h)$ with E_h being exposed to risk totals (e.g. populations living in area j and also registered with GP practice k) or expected events. Then as above there are options for modelling random effects, such as

$$\log(\mu_h) = \beta X_h + \gamma W_{j[h]} + \delta Z_{k[h]} + \eta_h \qquad (10.12)$$

where (for example)

$$\eta_h \sim \text{N}(0, \sigma_h^2)$$

This model may also be stated as

$$\log(\mu_h) = \beta X_h + \eta_h$$

where $\eta_h \sim \text{N}(\nu_h, \sigma_h^2)$ and $\nu_h = \gamma W_{j[h]} + \delta Z_{k[h]}$. Another option is separate random effects u_1 and u_2 for the two factors

$$\log(\mu_h) = \beta X_h + \gamma W_{j[h]} + \delta Z_{k[h]} + u_{1j[h]} + u_{2k[h]} \qquad (10.13)$$

where $u_{1j} \sim \text{N}(0, \sigma_1^2)$ and $u_{2k} \sim \text{N}(0, \sigma_2^2)$. An additional possibility (Congdon and Best, 2000) is to define a bivariate effect $\eta_h = (\eta_{h1}, \eta_{h2})$ with dispersion matrix Σ_η and means

$$\nu_{h1} = \gamma W_{j[h]}$$
$$\nu_{h2} = \delta Z_{k[h]}$$

with

$$\log(\mu_h) = \beta X_h + \eta_{h1} + \eta_{h2}$$
$$\eta_h \sim \text{N}_2(\nu_h, \Sigma_\eta) \qquad (10.14)$$

This structure expresses correlations in the overlapping impact of the two factors and generalizes to more than two factors. If one or more of the factors were spatially or temporally structured then one may introduce structured effects into the means. For example,

$$\nu_{h1} = \gamma W_{j[h]} + s_{j[h]}$$

where the s_j, $j = 1, \ldots, J$, are spatially structured. Unstructured effects specific to one or more factor may also be included in the means.

Example 10.3   Child respiratory hospitalizations   Child respiratory illness, including asthma incidence and hospital admissions, has been increasing in developed societies and there are well-established links with socio-economic deprivation in the area of residence, here defined by UK electoral wards (Payne *et al.*, 1993). GP practices have responsibility for registered populations that are geographically unrestricted and so will cross ward boundaries. They are responsible for referring their more ill patients to hospital, and their referral rate variation may reflect differences in the deprivation levels of their registered population but also the effectiveness of case management in primary care. So both ward (small area) and GP practice attributes may influence variation in counts of hospitalizations at the crossed factor level.

There are $j = 1, \ldots, 44$ wards and $k = 1, \ldots, 99$ practices in a study considering referrals to hospital for children aged 0–14 years with respiratory illness during the five financial years from 1 April 1991 to 31 March 1996 (Congdon and Best, 2000). Preliminary analysis of referral patterns selected combinations of wards and practices which accounted for the great majority of the at-risk population (for many combinations there are no children at risk) and $h = 1, \ldots, 345$ such combinations were identified (so $H = 345$ rather than 44×99).

Ward-level predictors included the Townsend index of neighbourhood deprivation, and a practice-level deprivation score was derived by assigning each patient affiliated to the practice with the Townsend index of their census enumeration district (which nests within wards), and then averaging over all patients. Other predictors are ward-level population density (a proxy for urban environment and traffic density), a ward-level measure of geographic access to hospitals, and an indicator of practice-level prescribing efficiency, i.e. the ratio of net costs of prophylactic asthma prescriptions to bronchodilator prescribing.

The bivariate normal effects structure in (10.14) is adopted with a Wishart $W(I, 2)$ prior assumed for Σ_η^{-1}. A spatial effect is included in the mean for wards, η_{h1}, and an unstructured effect in the mean for η_{h2}. So

$$\log(\mu_h) = \beta X_h + \eta_{h1} + \eta_{h2}$$
$$\eta_h \sim N_2(\nu_h, \Sigma_\eta)$$
$$\nu_{h1} = \gamma W_{j[h]} + s_{j[h]}$$
$$\nu_{h2} = \delta Z_{k[h]} + u_{k[h]}$$

A two-chain run of 10 000 iterations converging after 5000 shows a high negative correlation (around -0.8) between the effects. Of the five predictor effects only one is significant (ward deprivation with a positive effect, as expected). Two other regression coefficients have means in the expected direction but their credible intervals overlap zero, namely prescribing efficiency and practice deprivation. Variation between practices may in part reflect efficiency, and so of interest for assessing 'performance' are the average ranks of the practice effect u_k. For example, practice 39 has a median rank of 96 (out of 99), and practices 47 and 78 have median ranks of 92.

The average GLM deviance is 347 so the model seems to represent the dispersion structure accurately; however, d_e is relatively high at 213 (DIC $= 1952$) so more parsimonious models may be obtained. The most aberrant case according to CPO estimates is $Y_{35} = 14$ (against an expectation $E_{35} = 1.8$) and with estimated referrals lower at $\mu_{35} = 8$.

10.5   THE GENERAL LINEAR MIXED MODEL FOR PANEL DATA

Parallel to the hierarchical model for metric data (10.1) is the linear random effects model for longitudinal metric data,

$$Y_i = 1\alpha + X_i\beta + Z_ib_i + \varepsilon_i$$

where Y_i and ε_i are $T_i \times 1$, X_i is $T_i \times p$ (sometimes called a design matrix if it represents allocation of subjects to treatments or other experimental groups), and β is a $(p \times 1)$ vector of regression coefficients modelled as fixed effects. Z_i is a $T_i \times q$ matrix of predictors the impacts of which, as expressed by the $q \times 1$ vector b_i, vary between subjects.

For simplicity assume $T_i = T$ (all subjects are observed the same number of times). For categorical data $Y_i = (y_{i1}, y_{i2}, \ldots, y_{iT})'$ from the exponential family

$$p(y_{it}|\theta_{it}) = \exp\{[y_{it}\theta_{it} - \psi(\theta_{it})]/a_{it}(\phi) + c(y_{it}, \phi)\}$$

with link g, $\mu_{it} = E(y_{it}|\theta_{it}) = d\psi(\theta_{it})/d\theta_{it}$ and $Var(y_{it}|\theta_{it}) = d^2\psi(\theta_{it})/d^2\theta_{it}$. The corresponding model for the mean in a GLMM is

$$\mu_{it} = g^{-1}(\alpha + X_{it}\beta + Z_{it}b_i) \tag{10.15a}$$

The repetition over individuals and times allows for time-specific regression effects, so that with more generality

$$\mu_{it} = g^{-1}(\alpha + X_{it}\beta_t + Z_{it}b_i) \tag{10.15b}$$

For instance, a longitudinal count regression with GLMM form could take the form

$$y_{it}|b_i \sim Po(\mu_{it})$$
$$\log(\mu_{it}) = \alpha + X_{it}\beta_t + Z_{it}b_i$$

while a longitudinal logit regression with denominators N_{it} could be

$$y_{it}|b_i \sim \text{Bin}(N_{it}, \pi_{it})$$
$$\pi_{it} = \exp(\alpha + X_{it}\beta_t + Z_{it}b_i)/[1 + \exp(\alpha + X_{it}\beta_t + Z_{it}b_i)]$$

Often $q = 1$ so that there is only a varying intercept $b_i^* = \alpha + b_i$, often called a permanent effect or said to represent permanent heterogeneity. Thus

$$\mu_{it} = g^{-1}(\alpha + X_{it}\beta_t + b_i) \tag{10.16}$$

Fixed effects models have been proposed for such permanent subject parameters on the grounds of robustness, avoiding choosing a particular density for heterogeneity over subjects or the assumption that individual effects are uncorrelated with regressors (Hsiao, 1986). However, there are likely to be benefits in terms of parsimony, identifiability and 'pooling strength' by modelling heterogeneity with random effects that refer to a population-wide probability distribution. This is especially the case when the number of repetitions T_i is small and fixed effects are not well identified. As an example of parametric errors, one may assume

$$b_i \sim N_q(\Gamma, \Sigma_b)$$

with $\Gamma = (\Gamma_1, \ldots, \Gamma_q)'$ and $\Gamma_1 = 0$. Alternatively, for a random intercept model with $q = 1$, $b_i \sim N(0, \sigma_b^2)$. However, scale mixing or non-parametric error forms may be adopted in the search for greater robustness, while still pooling strength, unlike the fixed effects scheme (Ibrahim and Kleinman, 1998) – see section 10.6.

Models for longitudinal discrete data will often show overdispersion that may sometimes be resolved simply by including permanent subject effects, but may sometimes require additional random variation at unit (subject-times time) level, namely

$$\mu_{it} = g^{-1}(\alpha + X_{it}\beta_t + Z_{it}b_i + \varepsilon_{it}) \tag{10.17a}$$

where the ε_{it} are possibly normal, scale mixtures of normals, or discrete mixtures including Dirichlet process priors (Hirano, 1999). The errors ε_{it} are generally unstructured in time but time dependence may be relevant, especially for larger T. Note, however, that (10.17a) includes a form of correlation between times. Consider the case $q = 1$:

$$\mu_{it} = g^{-1}(\alpha + X_{it}\beta_t + b_i + \varepsilon_{it}) \tag{10.17b}$$

with unstructured errors ε_{it} having variance σ_ε^2. Letting the total residual be $\omega_{it} = \varepsilon_{it} + b_i$, then the correlation between ω_{it} at different periods is

$$\rho = \sigma_b^2/(\sigma_b^2 + \sigma_\varepsilon^2) \tag{10.17c}$$

Model checking for models such as (10.17) may consider Monte Carlo estimates of the CPO either at unit level (based on harmonic means of likelihoods at unit level) or at subject level (by totalling the pseudo marginal likelihoods over times for a given individual). Outliers may also be detected by directly considering estimates of the permanent subject effects $\{b_{i1}, b_{i2}, \ldots, b_{iq}\}$ and the unit-level effects ε_{it}.

If r cluster characteristics $W_i = (W_{1i}, W_{2i}, \ldots, W_{ri})'$ are thought to account for heterogeneity in intercepts or slopes over clusters, one may assume a multivariate regression for cluster effects

$$b_i = W_i \varphi + v_i$$

where the fixed effects regression parameter matrix φ is of order $r \times q$, and

$$v_i \sim N_q(0, \Sigma_b)$$

10.5.1　Time dependence

Heterogeneity between subjects, as in (10.16), or between observations as in (10.17), may operate through an autocorrelated structure in data or errors. Under augmented data sampling correlation effects may involve both the observed and augmented data. Suppose as in section 10.3 that for binary observations y_{it} latent continuous data y_{it}^* are obtained by truncated sampling (Albert and Chib, 1993). Then one might allow for lagged error effects via

$$y_{it}^* = X_{it} \beta_t + Z_{it} b_i + \eta_{it}$$
$$\eta_{it} = \rho_\eta \eta_{i,t-1} + u_{it}$$

representing (say) persistence in the impacts of unmeasured propensities in choice applications. The u_{it} are unstructured. A random walk prior on η_{it} is also possible, e.g.

$$\eta_{it} \sim N(\eta_{i,t-1}, \sigma_\eta^2)$$

True state dependence would involve a lag on y_{it} itself, and both types of dependence are included in the model

$$y_{it}^* = X_{it} \beta_t + Z_{it} b_i + \rho_Y Y_{i,t-1} + \eta_{it}$$
$$\eta_{it} = \rho_\eta \eta_{i,t-1} + u_{it}$$

Here ρ_Y measures the impact of observed choice at $t - 1$ on the current propensity.

Alternatively, following Heckman (1981), consider the intercept-only model as in (10.17b) with uncorrelated unit errors ε_{it},

$$y_{it}^* = X_{it}\beta_t + b_i + \rho_Y y_{i,t-1} + \varepsilon_{it}$$

Then a factor analytic extension to this model achieves time-varying correlation between disturbances. Let λ_t be time-varying loadings on the permanent effects, namely

$$y_{it}^* = X_{it}\beta_t + \lambda_t b_i + \rho_Y y_{i,t-1} + \varepsilon_{it} \qquad (10.18)$$

where for identifiability the b_i are taken to have known variance (e.g. $b_i \sim N(0,1)$), or one of the λ_t is constrained. If

$$\rho_t = \lambda_t^2 \sigma_b^2 / [\lambda_t^2 \sigma_b^2 + \sigma_\varepsilon^2]$$

then the correlation between total disturbances $\omega_{it} = \lambda_t b_i + \varepsilon_{it}$ at times t and s is $(\rho_t \rho_s)^{0.5}$.

For count data, sampling latent (e.g. normal) data is more problematic (though see Oh and Lim, 2001). Generalizations of the methods used in Chapter 8 may be considered to allow autoregression in Poisson or negative binomial models. For observation-driven dependence, the Zeger and Qaqish (1988) method sets $V_{it} = y_{it} + c(0 < c < 1)$, so a lag 1 dependence would be

$$y_{it} \sim Po(\mu_{it})$$

$$\log(\mu_{it}) = \rho_V \log(V_{it}) + X_{it}\beta_t + Z_{it}b_i$$

For a positive ρ_V, the means μ_{it} increase with the preceding period's count, whereas for negative ρ_V the means fall as $y_{i,t-1}$ increases. Such a model might not remove temporal correlation in residuals, defined, for example, via (Cantoni, 2004)

$$r_{it} = (y_{it} - \mu_{it})/\mu_{it}^{0.5}$$

and additional lags in V_{it} might be added or autoregression in a latent error η_{it} introduced, as above. Similarly, an extension of the conditional linear autoregessive process for count data (Grunwald *et al.*, 2000) would set

$$\mu_{it} = \gamma y_{i,t-1} + \exp(X_t\beta_t + Z_{it}b_i)$$

A factor analytic model as in (10.18) with time-varying loadings on one or more b_{ik}, $k = 1, \ldots, q$, could also be used for modelling parameter-driven autodependence. Thus with $q = 1$, and the above method for observation lags,

$$\log(\mu_{it}) = \lambda_t b_i + X_{it}\beta_t + \rho_V \log(V_{it}) + \varepsilon_{it}$$

Lee and Hwang (2000) and Diggle (1988) consider an AR(1) error component η_{ijt} in three-level longitudinal data with individuals

$j = 1, \ldots, N_i$ nested within groups $i = 1, \ldots, I$. Thus with $y_{ijt} \sim \mathrm{EF}(\mu_{ijt})$ it is now possible to let β vary over subjects as well as over times, and b_{ij} may vary by both groups and individual. For example,

$$g(\mu_{ijt}) = X_{ijt}\beta_i + Z_{ijt}b_{ij} + \eta_{ijt} + \varepsilon_{ijt}$$

where η_{ijt} are autocorrelated (e.g. AR(1) or AR(p)) and ε_{ijt} are unstructured $N(0,\sigma^2)$.

While AR(1) or random walk dependence in η_{it} or η_{ijt} is the simplest model for serial correlation, more complex models may be estimated, especially for larger data sets. As Jones (1993, p 65) points out, many panel data sets are not long enough to support complicated error structures. A general form for $\eta_i = (\eta_{i1}, \ldots, \eta_{iT})$ in a two-level model is a multinormal density of dimension T

$$\eta_i \sim N_T(0, \sigma^2 C)$$

where $\eta_i = (\eta_{i1}, \ldots, \eta_{iT})$ and the form of $C = \mathrm{Cov}(\eta_t, \eta_{t-s})$ reflects possible time dependencies. Thus an AR(1) error structure with correlation $\rho \in (-1, 1)$ gives

$$C = \begin{bmatrix} 1 & \rho & \rho^2 & \rho^3 & \cdots & \rho^{T-1} \\ \rho & 1 & \rho & \rho^2 & \cdots & \rho^{T-2} \\ \rho^2 & \rho & 1 & \rho & \cdots & \rho^{T-3} \\ \rho^3 & \rho^2 & \rho & 1 & \cdots & \rho^{T-4} \\ \vdots & & & & & \\ \rho^{T-1} & \cdots & \cdots & \cdots & \cdots & 1 \end{bmatrix}$$

Other error structures are possible (e.g. ARMA) or specialized dispersion structures such as correlation depending only on the time difference $t - s$ leading to constant parameters along every diagonal

$$C = \begin{bmatrix} 1 & \rho_1 & \rho_2 & \rho_3 & \cdots & \rho_{T-1} \\ \rho_1 & 1 & \rho_1 & \rho_2 & \cdots & \rho_{T-2} \\ \rho_2 & \rho_1 & 1 & \rho_1 & \cdots & \rho_{T-3} \\ \rho_3 & \rho_2 & \rho_1 & 1 & \cdots & \rho_{T-4} \\ \vdots & & \rho_3 & \rho_2 & \rho_1 & \\ \rho_{T-1} & \cdots & \cdots & \cdots & \cdots & 1 \end{bmatrix}$$

10.5.2   Longitudinal categorical data

Longitudinal multinomial and ordinal responses occur often for biometric or attitudinal studies where observations are classifications, ordered or unordered. Such data raise distinct modelling issues. For example, for ordinal outcomes (with J levels) thresholds on a continuous scale, possibly time specific, may be assumed to underlie observed gradings, namely $\kappa_{1t}, \kappa_{2t}, \ldots, \kappa_{J-1,t}$. Saei and McGilchrist (1998) refer to this as a time-dependent longitudinal threshold model. As for the three-level panel model for count or binary data, one may also consider modelling cluster effects b_{ij} specific to clusters i and categories j. Although it does not necessarily imply augmented data sampling, consider the observed responses Y as resulting from latent propensities or abilities Y^*. Thus for ordinal data $y_{it} = j$ if

$$\kappa_{j-1,t} < y^*_{it} < \kappa_{jt}$$

where

$$y^*_{it} = X_{it}\beta_t + Z_{it}b_i + \varepsilon_{ijt}$$

and

$$\varepsilon_{ijt} \sim N(0,1)$$

under a probit link. Departures from proportional odds would allow β_t and b_i to be group specific, with

$$y^*_{it} = X_{it}\beta_{jt} + Z_{it}b_{ij} + \varepsilon_{ijt}$$

To ensure identifiability one may either omit intercepts from X_{it} or take $\kappa_{1t} = 0$. In some panel studies changing thresholds are interpretable as changes in measurement scales or recording instruments when the focus is on estimating progress in individual ratings; for example, attempts to measure whether a mood factor is changing over time would be complicated by allowing changing scales (Steyer and Partchev, 2000).

A DGLM-based approach applicable to aggregate multinomial and ordinal data is used by Cargnoni *et al.* (1997). Here

$$Y_{it} \sim \text{Mu}(n_{it}, p_{it})$$

where $Y_{it} = (y_{it1}, \ldots, y_{itJ})$, $p_{it} = (p_{it1}, \ldots, p_{itJ})$ and $\log(p_{itj}/p_{itJ}) = \gamma_{tj} + \delta_{itj}$, where γ_{tj} are trend parameters specific to category j (e.g. national party affiliation trends), and δ_{itj} are residual trend parameters, representing area or group differentials from the overall trend. For identification $\gamma_{tJ} = \delta_{itJ} = 0$. The overall trend parameters provide a mechanism for cross-series pooling of strength. Cargnoni *et al.* assume both these sets of parameters to be MVN with an RW(1) form for δ_{itj}. They also impose the identifiability constraint

$$\delta_{Ntj} = \sum_{i=1}^{N-1} \delta_{itj}$$

In the voting trend by area example below (Example 10.5) it was found beneficial for out-of-sample forecasting (periods $T + 1$ etc.) to add a constant interaction effect λ_{ij} in the model for p_{itj}.

Autocorrelation in panel categorical data may also be modelled via hidden Markov models as in Chapter 9. For example, Böckenholt (2002) suggests a random effects model for binary panel choice data based on a latent two-state Markov chain. Preferences both in the choice and in the Markov transition behaviour are allowed to vary over individuals. Here we consider a generalized version of this approach with the y_{it} falling into one of J categories, and the latent state L_{it} falling into one of K groups. Then

$$\Pr(L_{it} = k) = q[i, L_{i,t-1}, k]$$

for $t > 1$, where

$$\log(q_{ijk}/q_{itK}) = \alpha_{jk} + \beta_{jk}F_i$$

and for identification the subject random effects F_i have known variance, and additionally $\alpha_{jK} = \beta_{jK} = 0$. Also

$$\Pr(y_{it} = j) = p[i, L_{it}, t, j]$$

where

$$\log(p_{iktj}/p_{iktJ}) = \gamma_{1kj} + \gamma_{2ij} + \gamma_{3tj}$$

and $\gamma_{1kJ} = \gamma_{2iJ} = \gamma_{3tJ} = 0$ for identification. The γ_{1kj} represent choice factors that vary according to latent group, and the subject choice random effects γ_{2ij} have dimension $J - 1$. One might as in Böckenholt (2002) model the F_i and γ_{2ij} via a J-dimensional random effects density. The final element of this model specifies the distribution between latent states at time 1: for example,

$$\Pr(L_{i1} = j) = r[i, k]$$

where

$$\log(r_{ik}/r_{iK}) = W_i\theta_k$$

where $W_i = (W_{i1}, \ldots, W_{iL})$ are fixed attributes of subject i, and $\theta_K = 0$.

Example 10.4 Respiratory symptoms in ohio children Fitzmaurice *et al.* (2003) consider binary data on wheeze symptoms among 537 Ohio children at ages 7, 8, 9 and 10 in terms of their age (A_t), maternal smoking status ($M_i = 1$ for smoking mothers) and an interaction A_tM_i. Here we contrast the effectiveness of logit and complementary log–log links (models 1 and 2 respectively) for these data using predictive criteria and model weights under parallel sampling (Congdon, 2005). The regression models in each case are defined by a single permanent subject effect b_i $(i = 1, \ldots, 537)$ and the three predictors.

To assess predictions using replicate data y_{new} from the model, sensitivity and specificity (proportions of wheeze and non-wheeze children correctly identified as such) are obtained for each age group by checking concordance between $y_{new,i,t}$ and y_{it}. A total measure of predictive accuracy in each age group is the sum of sensitivity and specificity (Youdens index ψ_t) as used by Congdon (2001a) and a 95% interval can be obtained for the difference $\psi_{1t} - \psi_{2t}$ in predictive accuracy between the models. One may also obtain posterior probabilities that model 1 provides better predictions, namely

$$\Pr(\psi_{1t} > \psi_{2t})$$

The model weights (2.19)–(2.21), DIC differences ΔDIC_{jk} and probability estimates (2.34) are also obtained.

The DICs under model 1 and 2 suggest that the complementary log–log link is preferable ($DIC_1 = 1416$ vs. $DIC_2 = 1403$). This preference is entirely due to complexity differences: model 1 has both a higher average likelihood and higher likelihood at $\bar{\theta}_1$. The effective parameters are $d_{e1} = 248.5$ and $d_{e2} = 224$ (according to the DIC method) and $d_{e1} = 492$ and $d_{e2} = 445$ according to the approximation (2.26) from Gelman *et al.* (2003). The former are used to obtain the density of $\kappa_{12}^{(t)}$ as in (2.25). Over the second half of a single chain of 10 000 iterations, covariation between $DIC(\theta_1^{(t)})$ and $DIC(\theta_2^{(t)})$ is -51. The empirical variance of $\kappa_{12}^{(t)}$ of 1975 includes this covariation but after adjustment $Var(\Delta DIC_{12})$ corresponds to

$$Var(\kappa_{12}^{(t)}) = Var(\Delta DIC_{12}) - 2Cov[DIC(\theta_1^{(t)}), DIC(\theta_2^{(t)})]$$
$$= 2(d_{e1} + d_{e2} - w_{12}) = 2(492 + 445 + 51)$$

Accordingly the difference ΔDIC_{12} of 13 compares with a standard deviation of $43 = (2.492 + 2.445)^{0.5}$, suggesting a large element of model uncertainty remains. The posterior probability that

$$\Pr[DIC(\theta_1^{(t)}) > DIC(\theta_2^{(t)})]$$

is accordingly far from decisive and is obtained as 0.62. The AIC and BIC weights include higher penalties for complexity than the DIC. The result is that the BIC decisively favours model 2 ($w_2 = 0.9997$). The Bayesian model weight estimate gives $w_2 = 0.97$, while the Akaike weight gives $w_2 = 0.8$. Likelihood weights are not penalized for complexity and very slightly favour model 1 ($w_1 = 0.55$).

Prediction-based model comparisons are also not penalized for complexity - though see Congdon (2005). The probabilities $\Pr(\psi_{1t} > \psi_{2t})$ are accordingly all around 0.5 and the differences $\psi_{1t} - \psi_{2t}$ do not seem to show a marked difference between the models.

Example 10.5 Canadian voting intentions Gustafson and Walker (2003) analyse voting intention data for $J = 4$ major Canadian parties (excluding Bloq Quebecois) over ten months in 2000 (source: www.ipsos-reid.com). March, April and June are omitted as polls were not conducted, so $T = 7$. Gustafson and Walker (2003) consider just trends in the Atlantic provinces. Areas $i = 1, \ldots, 5$ are considered here, namely Atlantic provinces, BC, Alberta, Saskatchewan/Manitoba and Ontario (Table 10.1).

Cargnoni et al. (1997) adopt an arcsin transformation to link the probabilities π_{itj} to the η_{itj}. Here the multinomial logit approach of Leonard and Hsu (1994) is adapted to this dynamic model. Additionally an RW(1) prior for both γ_{tj} and δ_{itj} is assumed together with area-specific dispersion matrices Σ_i for the δ_{itj} rather than a common Σ over all areas. Hence with $J = 4$, $T = 7$ and $N = 5$, the model is

$$\Pr(y_{it} = j) = p_{itj}$$
$$\log(p_{itj}/p_{itJ}) = \lambda_{ij} + \gamma_{tj} + \delta_{itj}$$

with $\lambda_{iJ} = \gamma_{tJ} = \delta_{itJ} = 0$. The priors on the randomly evolving parameters are

$$\gamma_{1j} \sim N(0, 1000) \qquad j = 1, \ldots, J-1$$
$$\gamma_t \sim N_{J-1}(\gamma_{t-1}, T_\gamma)$$

where $\gamma_t = (\gamma_{t1}, \gamma_{t2}, \ldots, \gamma_{t,J-1})$ and

$$\delta_{1,t} \sim N(0, 1000) \qquad\qquad j = 1, \ldots, J-1$$
$$\delta_{it} \sim N_{J-1}(\gamma_{i,t-1}, \Sigma_i) \qquad i = 1, \ldots, N-1$$

where $\delta_{it} = (\delta_{it1}, \delta_{it2}, \ldots, \delta_{it,J-1})$ and the dispersion is area specific. A Wishart prior with $J-1$ degrees of freedom and 0.0001 diagonal elements is assumed for the inverse dispersion matrices. For the interactions it is assumed that

$$\lambda_{ij} \sim N(0, 1000) \qquad i = 1, \ldots, N-1; j = 1, \ldots, J-1$$

Table 10.1   Voting intentions by province and month (excluding undecided)

	Conservative	Liberal	New Democratic	Alliance	Total
	British Columbia				
Jan	12	71	24	50	157
Feb	9	74	28	50	161
May	14	78	24	60	176
July	9	65	20	79	173
Aug	13	78	20	96	207
Sept	10	84	12	68	174
Oct	9	68	12	84	173
	Alberta				
Jan	20	40	11	50	121
Feb	21	37	11	48	117
May	20	36	11	72	139
July	11	24	8	65	109
Aug	16	42	6	86	150
Sept	12	45	6	57	120
Oct	9	38	4	75	126
	Saskatchewan/Manitoba				
Jan	11	36	25	19	91
Feb	9	42	28	26	104
May	12	43	22	32	110
July	14	26	27	39	105
Aug	8	38	23	39	108
Sept	12	48	17	24	101
Oct	4	43	26	36	109
	Ontario				
Jan	113	292	56	40	501
Feb	88	287	54	54	483
May	93	259	49	78	479
July	62	261	43	100	466
Aug	65	355	72	134	626
Sept	53	290	47	90	480
Oct	49	248	44	136	477
	Atlantic				
Jan	34	42	22	5	103
Feb	22	53	22	6	103
May	30	53	13	6	102
July	24	47	25	17	113
Aug	31	70	16	15	132
Sept	21	64	13	13	111
Oct	21	59	16	15	111

The identifiability constraints

$$\delta_{Ntj} = \sum_{i=1}^{N-1} \delta_{itj}$$

$$\gamma_{Nj} = \sum_{i=1}^{N-1} \gamma_{ij}$$

are also imposed. Additionally the variance matrices are scaled to allow for the unequal spacing of polls.

Following Gustafson and Walker (2003) one question of interest is the smoothing of 'wiggles' in the data such as the May surge in the Conservative voting intention in the Atlantic provinces. Extrapolation to future periods is also of interest. For forecasting intentions one period ahead (November, 2000) the following specification is assumed:

$$p_{i,T+1,j} = \phi_{i,T+1,j} / \sum_k \phi_{i,T+1,k}$$

$$\log(\phi_{i,T+1,j}) = \lambda_{ij} + \gamma_{T+1,j} + \delta_{iTj} \qquad j = 1, \ldots, J-1$$

$$\phi_{i,T+1,J} = 1$$

since the δ_{itj} are estimated rather imprecisely and make forecasting intervals too wide if they are extrapolated.

A two-chain run of 10 000 iterations (convergent from around 5000 iterations) shows smoothing downwards of the Conservative intention for May in the Atlantic provinces, and smoothing up of the February percentage. The forecast for the Atlantic Conservative voting proportion in November (namely $p_{5,8,1}$) is close to the October mean of 19.4% but with a wider 95% credible interval, i.e. (0.115,0.29) compared with (0.125,0.27) for October.

10.6 CONJUGATE PANEL MODELS

Section 10.5 has considered the GLMM strategies for panel models. By contrast, conjugate models for modelling both overdispersion and the permanent effects (e.g. Daniels and Gatsonis, 1999) involve Poisson–gamma mixing, beta–binomial mixing or multinomial–Dirichlet mixing. Thus gamma mixing on the underlying mean for count data as in

$$y_{it} \sim \mathrm{Po}(\lambda_{it})$$

$$\lambda_{it} \sim \mathrm{Ga}(h\mu_{it}, h)$$

$$\log(\mu_{it}) = \alpha_t + X_{it}\beta_t + Z_{it}b_i$$

with $h > 0$ preset or an extra unknown, leads to an overdispersed Poisson since

$$V(y_{it}) = \mu_{it} + \mu_{it}/h$$

Taking h_i to vary between subjects, e.g. by taking $h_i/(1 + h_i)$ to be beta distributed, leads to a subject-specific variance–mean ratio, while taking h to vary over times gives a time-specific variance–mean. Similarly for binomial data, one might specify

$$y_{it} \sim \text{Bin}(n_{it}, \pi_{it})$$
$$\pi_{it} \sim \text{Be}(r_{it}M_t, (1 - r_{it})M_t)$$
$$g(r_{it}) = \alpha_t + X_{it}\beta_t + Z_{it}b_i$$

where M_t is an extra unknown. The priors on π_{it} are independent with regard to time. An alternative is gamma or beta mixing for the intercept only (Hausman *et al.*, 1984) so that for a count response

$$y_{it} \sim \text{Po}(\delta_i\mu_{it})$$
$$\delta_i \sim \text{Ga}(h, h)$$
$$\log(\mu_{it}) = \alpha_t + X_{it}\beta_t$$

The variance is then $\mu_{it} + \mu_{it}^2/h$ and the variance–mean ratio $(1 + \mu_{it}/h)$ increases with μ_{it}. Another conjugate longitudinal model adapts the Rasch Poisson count model (Jansen, 1997) and assumes for subject i and time t

$$y_{it} \sim \text{Po}(\mu_{it})$$
$$\mu_{it} = \nu_i\gamma_{it}$$

where

$$\nu_i \sim \text{Ga}(h, h/\rho_i)$$

and

$$\gamma_{it} \sim \text{Dir}(M_{i1}, \ldots, M_{iT})$$

The ν_i are constant subject effects with $\rho_i = \exp(X_i\beta)$ possibly a function of time constant attributes. The γ_{it} are subject-specific occasion parameters that may be related to time-varying predictors. When not related to predictors the M_{it} may be assigned pooling strength priors, or taken as constant over clusters $(M_{it} = M_t)$ to represent population-wide period intensities. Or they may define a population-wide growth model, e.g.

$$\log(M_t) = \theta_0 + \theta_1 t$$

The marginal likelihood involves a negative binomial for the total count $y_{i+} = \sum_t y_{it}$ and a multinomial for y_{it} conditional on y_{i+} with parameters γ_{it}/γ_{i+}.

The impact of predictive or risk factors may modelled at individual level or via stratification. For example, suppose panel data are recorded for groups of subjects (e.g. defined by several demographic characteristics) rather than individuals. The data might be for demographic strata (e.g. age–sex–ethnic subgroups by area) and the event might be divorce, political participation, etc. For binary events, let π_{gt} be the rate at time t for subgroup g, with $y_{gt}|\pi_{gt} \sim \text{Bin}(n_{gt}, \pi_{gt})$, and let prior heterogeneity in rates $f(\pi_{gt})$ over subpopulations and periods be represented by a beta mixture. An independence prior on π would allow rates in different periods to have unrelated priors, e.g. $\pi_{gt} \sim \text{Be}(a_g, b_g)$ where a_g and b_g might be preset or extra unknowns. However, if the chance of the event is varying over time (i.e. there is non-stationarity) and rates in successive periods are correlated then successive odds ratios may be linked according to

$$[\pi_{gt}/(1 - \pi_{gt})] = \kappa_{gt}[\pi_{g1}/(1 - \pi_{g1})]$$

where 'improvement ratios' κ_{gt} are positive and

$$\pi_{g1} \sim \text{Beta}(r_g M_g, (1 - r_g)M_g)$$

where r_g is the initial rate for group g and M_g is a prior sample size (Davies *et al.*, 1982). Equivalently

$$\text{logit}(\pi_{gt}) = \log(\kappa_{gt}) + \text{logit}(\pi_{g,t-1})$$

For $\xi_{gt} = \log(\kappa_{gt})$ one might adopt state-space priors that penalize large changes, in line with an expectation of smoothly changing event rates. For Poisson data $y_{gt} \sim \text{Po}(\lambda_{gt})$, with $\lambda_{gt} \sim \text{Ga}(h\mu_{gt}, h)$, the improvement factors would be modelled via relative risks

$$\mu_{gt}/\mu_{g1} = \kappa_{gt}$$

For categorical observations (e.g. polling intentions for J political parties) for $g = 1, \ldots, N$ areas through times $t = 1, \ldots, T$ similar considerations apply. Let n_{gt} denote the total subjects (e.g. voters surveyed) in area g at time t. In this case

$$Y_{gt} \sim \text{Mu}(n_{gt}, p_{gt})$$

where $Y_{gt} = (y_{gt1}, y_{gt2}, \ldots, y_{gtJ})$, $p_{gt} = (p_{gt1}, p_{gt2}, \ldots, p_{gtJ})$. An independence prior could be

$$p_{gt} \sim \text{Dir}(\alpha_{gt1}, \ldots, \alpha_{gtJ})$$

Often it is assumed that $\alpha_{gtj} = h$ for all g, j and t with h known (e.g. $h = 1$). However, a time- and area-independent prior neglects that probabilities within the sequence for area g tend to be more similar to each other than probabilities for different areas, and that probabilities close in time for area g will be more similar than probabilities separated by a larger time gap. Gustafson and Walker (2003) propose a prior that downweights large changes between probabilities in successive periods. Specifically

$$f(p_{gt}) \propto \text{Dir}(1, 1, \ldots, 1) \exp\left[-\sum_{t=2}^{T}\sum_{j=1}^{J}(p_{gtj} - p_{g,t-1,j})^2/\lambda\right]$$

where smaller values of λ imply greater smoothing. As $\lambda \to \infty$ the independence prior with $h = 1$ is obtained. One may also adapt the improvement ratio method such that for $j = 1, \ldots, J - 1$ and $t > 1$

$$\text{logit}(p_{gtj}) = \log(\kappa_{gtj}) + \text{logit}(\pi_{g,t-1,j})$$

while $p_{g1} \sim \text{Dir}(\alpha_{g1}, \ldots, \alpha_{gJ})$.

Example 10.6 Seizure data Seizure data in $N = 59$ epileptic patients, first presented by Thall and Vail (1990), have become one of the classic panel data sets for both frequentist and Bayesian analysis. Consider the mixture model described above with $y_{it} \sim \text{Po}(\mu_{it})$ and $\mu_{it} = \nu_i\gamma_{it}$ where $\nu_i \sim \text{Ga}\ (h, h/\rho_i)$ and $\gamma_{it} \sim \text{Dir}(M_1, \ldots, M_T)$. The subject effects ν_i have means in terms of four fixed attributes

$$\rho_i = \exp(\beta X_i) = \exp(\beta_0 + \beta_1 X_{i1} + \beta_2 X_{i2} + \beta_3 X_{i3} + \beta_4 X_{i4})$$

where $X_1 =$ base seizure count, $X_2 =$ treatment (binary), $X_3 =$ treatment–base interaction and $X_4 =$ age at baseline. The γ_{it} are subject-specific occasion parameters with

$$\gamma_{it} \sim \text{Dir}(M_{i1}, \ldots, M_{iT})$$

and means $M_{it} = M_t$ that are based on a linear population growth model with $M_t = \exp(\theta_0 + \theta_1 t)$.

A two-chain run of 5000 iterations shows a significant effect of the treatment in reducing seizures (95% interval from -1.8 to -0.1) and also a significant impact of the logged baseline seizure count (95% interval from 0.6 to 1.15). Effects of X_3 on ν_i and of time in the population growth model are indicative rather than 'significant': the 95% intervals straddle zero, though most of the interval for θ_1 is negative. The DIC is estimated as 1146 ($d_e = 121$). A predictive check based on χ^2 statistics comparing actual and replicate data with μ_{it} is satisfactory (p statistic 0.44). However, outlier assessment via CPOs shows $y_{25,3}$ and $y_{39,2}$ as suspect at observation level.

10.7   GROWTH CURVE ANALYSIS

One type of panel model where random variation between subjects is a major feature is growth curve analysis for evolving biological characteristics or educational attainments. For instance, Gamerman and Smith (1996) develop a Bayesian version of the linear growth model proposed by Fearns (1975) in which a metric or discrete outcome y_{it} for subject i at time t has a mean

$$g(\mu_{it}) = b_{i1} + b_{i2}t$$

where heterogeneous intercepts b_{i1} and linear growth rates b_{i2} are modelled as exchangeable random effects. b_{i1} and b_{i2} can be taken as two sets of independent effects or to follow a bivariate (e.g. BVN) density: for example, it may be that growth rates are lower for subjects with higher initial levels ('negative feedback'), and this correlation is accommodated by a bivariate density. Greater robustness may be achieved by a mixture of bivariate normals, or by mixing the covariance with subject weights (equivalent to multivariate Student t). Factor analytic or latent growth curve models introduce factor loadings instead of functions of time (Rovine and Molenaar, 1998), e.g.

$$g(\mu_{it}) = b_{i1} + \lambda_t b_{i2}$$

where at least one of the λ_t is preset to ensure identifiability, e.g. $\lambda_1 = 1$, with λ_2 to λ_T being unknown parameters. A submodel may be used to explain intercepts b_{i1} or linear growth rates b_{i2} in terms of fixed attributes W_i of subjects.

To accommodate excess heterogeneity and/or remaining autocorrelation, one may specify

$$g(\mu_{it}) = b_{i1} + b_{i2}t + \varepsilon_{it}$$

A simple independent errors assumption $\varepsilon_i \sim N(0, \sigma_\varepsilon^2 I)$ for $\varepsilon_i = (\varepsilon_{i1}, \ldots, \varepsilon_{iT})$ may be assessed against alternatives such as $\varepsilon_i \sim N_T(0, \Sigma_\varepsilon)$, where Σ_ε is an unstructured dispersion matrix, or some specified time dependence (e.g. AR(1) dependence).

A conjugate model for growth curves might also allow for both non-parallelism and overdispersion (Jansen, 1997). Thus assume a constant subject effect ν_i and a subject–period effect γ_{it} based on a regression involving time t, with

$$\mu_{it} = \nu_i \gamma_{it}$$

where $\nu_i \sim \text{Ga}(c, c/m_i)$ and $\gamma_{it} \sim \text{Dir}(M_{i1}, \ldots, M_{iT})$. Subject means m_i might be related to fixed subject predictors W_i via a log-linear model, while growth paths are taken as non-parallel according to

$$\log(M_{it}) = \eta_0 + \eta_{1i}t$$

A frequent variation involves multilevel panel data when the subjects fall into a few subgroups $g = 1, \ldots, G$, such as treatment groups or schools (Jones, 1993; Muthén, 1997). Then the varying slopes and intercepts would have means specific to the treatment group and extra variation may also be group specific. So for groups $g = 1, \ldots, G$ and subjects $i = 1, \ldots, n_g$ within groups

$$g(\mu_{itg}) = b_{i1g} + b_{i2g}t + \varepsilon_{itg}$$

where a regression model for explaining variation in subject parameters might be

$$b_{i1g} = \delta_{01g} + \delta_{11g}W_{i1} + \delta_{21g}W_{i2} + \cdots + u_{i1g}$$
$$b_{i2g} = \delta_{02g} + \delta_{12g}W_{i1} + \delta_{22g}W_{i2} + \cdots + u_{i2g}$$

Another variation for growth curve and other panel analysis is for unequally spaced observation points (Jones and Boadi-Boateng, 1991). Thus the observations $Y_i = (y_{i1}, \ldots, y_{in_i})$ for subject i are observed at times $t_{i1}, t_{i2}, \ldots, t_{in_i}$ and the mixed intercept and slope linear growth model becomes

$$g(\mu_{ij}) = b_{i1} + b_{i2}t_{ij} + \eta_{i,t_{ij}} \qquad j = 1, \ldots, n_i$$

A lag 1 autoregressive structure in η would then account for time gaps as in

$$\eta_{i,t_{ij}} = \rho(t_{ij} - t_{i,j-1})\eta_{i,t_{i,j-1}} + u_{i,t_{ij}}$$

with $\text{Var}(\eta) = \sigma_u^2[1 - \rho^2(t_{ij} - t_{i,j-1})^2]^{-1}$

Example 10.7   Change in antisocial behaviour   The development of antisocial behaviour in adolescence has been linked *inter alia* to academic failure, home support, child gender and mother's age. Curran and Bollen (2001) use data from the US National Longitudinal Study of Youth to consider this form of growth trajectory. The antisocial behaviour scale aggregates over mother's reports on six items (e.g. child cheats or lies: 0 if not true, 1 if sometimes true, 2 if often true). The total score ranges from 0 to 12 on four measurement waves, two years apart starting in 1986, for children aged 6–8 at the first wave.

Although the six subitems are originally ordinal, the total scores are here taken as binomial from totals of 12. Predictors are W_1 for gender (1 for males), and continuous scores for home cognitive support (W_2) that are fixed over waves. The model fitted is

$$y_{it} \sim \text{Bin}(12, \pi_{it})$$
$$\text{logit}(\pi_{it}) = b_{i1} + b_{i2}\, t$$
$$b_{i1} = \delta_{01} + \delta_{11} W_{i1} + \delta_{21} W_{i2} + u_{i1}$$
$$b_{i2} = \delta_{02} + \delta_{12} W_{i1} + \delta_{22} W_{i2} + u_{i2}$$

where $u_i \sim N_2(0, \Sigma_b)$, and Σ_b is assigned a Wishart prior with identity scale matrix and 2 degrees of freedom.

The second half of a two-chain run of 20 000 iterations shows a significant positive effect of male gender on b_{i1} (with $\bar\delta_{11} = 0.45$) and a significant negative effect of support (W_2) on b_{i2} (with the posterior mean $\bar\delta_{22} = -0.028$ and a 95% interval from -0.055 to -0.005). The correlation between b_{i1} and b_{i2} has a negative mean of -0.2, but the 95% interval $(-0.48, 0.15)$ straddles zero.

To identify cases not well represented by this model CPO estimates are obtained at unit level and the log(PsML) estimates totalled over each subject's four observations. This shows subjects 137, 78 and 147 with the smallest log(PsML) values, namely -14.1, -13.6 and -12.9. Subject 137 has an erratic series of values 3, 1, 0 and 7, as does subject 147 with 0, 4, 10 and 7. These patterns are at odds with linear trends. Considering outlyingness of individual gradients b_{i2} in relation to the average growth rate $\bar b_2 = \sum b_{i2}/221$, subject 200 is identified with the lowest b_{i2} (her trajectory is 3, 0, 0, 0). Subjects 147 and 156 have the highest b_{i2} (probabilities over 0.995 of exceeding the average), though the growth for 147 is discontinuous. While a non-linear model would be infeasible with such a short series, it is relevant that the population growth is highest between periods 1 and 2 (mean 1.84 in period 2 vs. 1.49 in period 1) and slower thereafter (means 1.88 and 2.07 in the next two periods). It may therefore be relevant to try to distinguish linear growth pattern subjects from other more irregular paths.

10.8 MULTIVARIATE PANEL DATA

The above models generalize readily to more than one outcome with correlations between outcomes modelled via permanent effects or possibly observation effects. Thus for outcomes $m = 1, \ldots, M$ over time $y_{imt} \sim EF(\mu_{imt})$ a possible structure would be

$$\mu_{itm} = g^{-1}(\alpha_{tm} + X_{it}\beta_{tm} + Z_{it}b_{im} + \varepsilon_{itm})$$

where X_{it} is $1 \times p$, and Z_{it} is $1 \times q$. If unstructured in time, the unit-time–response errors $\varepsilon_{it} = (\varepsilon_{it1}, \ldots, \varepsilon_{itM})$ might be MVN, scale mixtures of MVNs, or discrete mixtures of MVNs. If autoregression is required in the ε_{itm} then a VAR(1) of order M or multivariate random walk would be relevant. For example, an RW(1) of order M is

$$\varepsilon_{it} \sim N_M(\varepsilon_{i,t-1}, T_\varepsilon)$$

The dynamic regression coefficients $\beta_{tm} = (\beta_{tm1}, \ldots, \beta_{tmp})'$ might also adopt multivariate random walk priors; an RW(1) prior with interdependence between outcomes m is

$$\beta_{tm} \sim N_p(\beta_{t-1,m}, \Sigma_m)$$

The permanent effects b_{im} are likely to be correlated over outcomes according to a covariance structure of dimension qM. For example, suppose growth curve analysis for $M = 2$ correlated responses (e.g. maths and linguistic scores) was required. Intercepts and slopes vary over subjects, so $q = 2$. Then one might specify

$$g(\mu_{it1}) = b_{i11} + b_{i12}t + \varepsilon_{it1}$$
$$g(\mu_{it2}) = b_{i21} + b_{i22}t + \varepsilon_{it2}$$

Let $d_{i1} = b_{i11}, d_{i2} = b_{i12}, d_{i3} = b_{i21}$ and $d_{i4} = b_{i22}$ and assume $d_i \sim N_4(\mu_d, \Sigma_d)$ where $d_i = (d_{i1}, d_{i2}, d_{i3}, d_{i4})$. Then negative correlations between d_{i1} and d_{i2} and between d_{i3} and d_{i4} represent negative feedback within outcomes. By contrast, positive correlations between d_{i2} and d_{i4} may be expected if y_{it1} and y_{it2} are both positive measures of ability, mood, etc. As another example, consider the mixed intercept model

$$\mu_{itm} = g^{-1}(\alpha_{tm} + X_{it}\beta_{tm} + b_{im} + \varepsilon_{itm})$$

Here the $b_i = (b_{i1}, \ldots, b_{iM})$ are outcome-specific intercepts and a multivariate prior of dimension M would be appropriate for pooling strength over the outcomes.

Multivariate longitudinal data may gain from factor analytic approaches that also pool information. For example, Steyer and Partchev (2000) consider panel multivariate ordinal data for subjects i, times t, items m and categories $j = 1, \ldots, J_m$. Then $Y_{it} = (y_{it1}, \ldots, y_{itM})$ where y_{itm} may be in one of J_m ordered categories. Their application relates to different measures of mood so a common factor ϕ_{it} pooling over the outcomes and measuring the trend in mood for subject i may be considered. If the impacts of a treatment were being monitored then it is sensible to assume constant thresholds, i.e.

$$\kappa_{mtj} = \kappa_{mj} \ (j = 1, \dots, J_m - 1)$$

on the M ordinal scales. Otherwise changes in measuring instrument are confounded with treatment impact. So

$$\Pr(y_{itm} = j) = \pi_{itmj}$$

where

$$\pi_{itm1} = \gamma_{itm1}$$
$$\pi_{itmj} = \gamma_{itmj} - \gamma_{itm,j-1} \qquad j = 2, \dots, J_m - 1$$
$$\pi_{itmJ_m} = 1 - \pi_{itm,J_m-1}$$
$$\mathrm{logit}(\gamma_{itmj}) = \kappa_{mj} - \mu_{itm} \qquad j = 1, \dots, J_m - 1$$

Regression coefficients β_{tm} on fixed attributes X_i may be specific for response and time. So the regression mean with (say) $p = 2$ fixed predictors is

$$\mu_{itm} = \lambda_m \phi_{it} + \beta_{1tm} X_{i1} + \beta_{2mt} X_{i2}$$

Also the scale of ϕ_{it} is fixed to allow λ_m to be identified. One may, for example, take the ϕ_{it} as independently N(0,1) or autocorrelated, as in

$$\phi_{it} \sim \mathrm{N}(\rho_\phi \phi_{i,t-1}, 1)$$

Factor analytic approaches may also be used in multivariate growth curve models which occur often in clinical and educational contexts. For ordinal outcomes a factor-based regression model would involve cumulative probabilities γ_{itmj} by subject i, time t, outcome m and ordered categories j. Then a logit model in γ_{itmj} might take the following form, pooling growth rate and intercept information over outcomes:

$$\mathrm{logit}(\gamma_{itmj}) = \kappa_{mj} - \lambda_{1m}\phi_{i1} - \lambda_{2m}\phi_{i2}t - X_i\beta_{tm} \qquad (10.19)$$

where $\phi_i = (\phi_{i1}, \phi_{i2})$ is a bivariate factor effect representing intercepts and growth rates. The cut points κ_{mj} are taken as fixed through time. An alternative structure might be

$$\mathrm{logit}(\gamma_{itmj}) = \kappa_{mj} - \phi_{im1} - \phi_{im2}\lambda_t - X_i\beta_{tm}$$

where ϕ_{im1} and ϕ_{im2} are (correlated) intercept and slope effects for outcomes m and some loadings λ_t are preset for identifiability (MacCallum *et al.*, 1997).

Example 10.8   Health demand   To illustrate multivariate count panel data consider the health demand data from the German Socioeconomic Panel for 1984–1994, as analysed by Riphahn *et al.* (2003). There are two responses, visits to doctor Y_1 and visits to hospital Y_2, and histories for $N = 3602$ female subjects are considered. The relevant questions were not asked in 1989–1990 and 1992–1993 so there is a maximum of seven observations per subject. While 375 subjects appear in all years, many appear only once. The total of observations for person-years ('units') is 13 083. Of interest to Riphahn *et al.* (2003) were the impacts of insurance cover after demographic influences (age, income, occupation) had been controlled for. Demand might be expected to be higher for those within the public insurance system (most of the subjects) and also among those purchasing add-on insurance (only a small minority of subjects).

There are 15 (fixed) predictors in the bivariate Poisson model of Riphahn *et al.* (2003) who assume fixed regression coefficients and permanent effects b_{im} which are uncorrelated between outcomes. Thus $y_{itm} \sim \text{Poi}(\mu_{itm})$ where

$$\log(\mu_{itm}) = \alpha_m + X_i\beta_m + b_{im} + \varepsilon_{itm}$$

where $\varepsilon_{it} \sim N_2(0, \Sigma_\varepsilon)$ but $b_{im} \sim N(0, \xi_m)$, $m = 1, 2$. Following the discussion above, possible variations on this approach would consider correlated permanent effects and time-varying regression parameters. Here we continue to assume fixed regression effects but allow in an extended model for correlated permanent effects.

The first model fitted here ignores the jointness of the outcomes. Thus all outcome-specific random effects are independent, with $\varepsilon_{itm} \sim N(0, \phi_m)$ and $b_{im} \sim N(0, \xi_m)$. The resulting DIC is 59 680 ($d_e = 9220$). The four coefficients on insurance status are all positive[1] but none have entirely positive 95% credible intervals; this may partly reflect a highly parameterized regression model, with possible scope for eliminating regressors (e.g. being married and being employed). The estimated b_{i1} and b_{i2} have a correlation of 0.19, although are a priori independent.

A modified model allows for correlation in both unit and subject effects. The DIC is reduced slightly to 59 570 ($d_e = 9154$), despite the extra parameterization, though this reduction is unlikely to be 'significant' in view of the high number of effective parameters in the two models (Congdon, 2005). The correlation between the pair of permanent

---

[1] Under model A the coefficients $\beta_{1,15}$, $\beta_{1,16}$, $\beta_{2,15}$ and $\beta_{2,16}$ have posterior means (s.d.) of 0.09 (0.05), 0.055 (0.08), 0.19 (0.14) and 0.27 (0.22). In model B the corresponding estimates are 0.14 (0.06), 0.05 (0.09), 0.27 (0.14) and 0.24 (0.22).

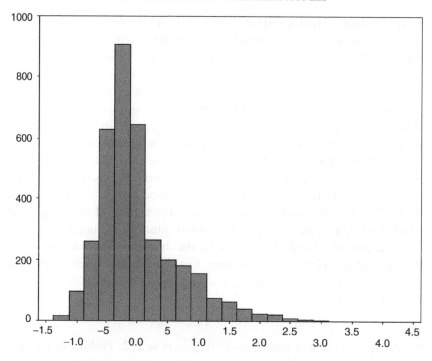

Figure 10.2 Permanent effects, hospital visits

effects is 0.40 under bivariate normality. However, there is some positive skew in the distribution of b_{i2} (hospital visits) compared with b_{i1} (doctor visits) – see Figures 10.2 and 10.3. This suggests use of a heavier tailed density than the bivariate normal for b_i, or one allowing for skewness in the hospital permanent effects (e.g. see section 3.4). Under this model the effects of public insurance have a more clearly positive impact on both types of demand.

Example 10.9 Mood change Steyer and Partchev (2000) consider a panel of length $T = 3$ for $M = 3$ ordinal mood items, each with $J_m = 5$ levels on a Lickert scale. The items are positive measures of mood, i.e. whether the subject feels 'content' (*zufrieden*), 'good' (*gut*) and 'happy' (*glücklich*). Predictors are gender (2 = F,1 = M) and age at first observation.

Two models are considered. The first (model A) is a random subject–time effects model with

$$\text{logit}(\gamma_{itmj}) = \kappa_{mj} - S_{it} - X_i\beta_m$$

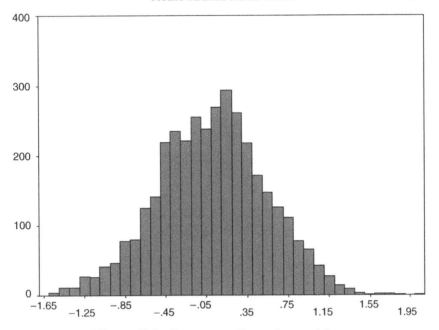

Figure 10.3 Permanent effects, doctor visits

where $X_{i1} = 1$ for females ($= 0$ for males), and $X_{i2} = $ age$/10$. The first cut point is set to zero ($\kappa_{m1} = 0$), and since there are no period intercepts, the means for each period of the subject mood scores S_{it} can be identified and represent mean population mood scores. We find from a two-chain run of 5000 iterations that the average $S_{.2}$ is highest (4.39) with a 95% interval (3.93, 4.84) compared with $S_{.1} = 4.15$ and $S_{.3} = 4.26$. As an example of a subject with major mood shift, subject 42 has item scores (5, 5, 5) at time 1 but (2, 2, 3) at time 3 with their score $S_{42,t}$ falling from 9.5 to 2.1. The estimated cut points for the three items are very similar, while the regression effects show older persons and males to have more positive mood. The DIC is 9670 with $p_e = 1275$.

The second analysis involves a factor model (model B) as in (10.19). This model includes intercepts with

$$\text{logit}(\gamma_{itmj}) = \kappa_{mj} - \alpha_t - \lambda_m \phi_{it} - X_i \beta_m$$

so that for identifiability factor scores have set means and scale, namely $\phi_{it} \sim N(0, 1)$. The loadings are constrained to positivity, namely

$$\lambda_m \sim N(1, 1)I(0,)$$

to ensure consistent labelling. Subject scores comparable with those of the first model are obtained as $S_{it} = \alpha_t + \phi_{it} \sum_{m=1}^{3} \lambda_m$.

This model reduces the DIC to 9632 with $p_e = 1283$. The estimated λ_m and their standard deviations are 2.65 (0.12), 2.59 (0.12) and 2.62 (0.12). The correlation between the two sets of S_{it} in models A and B exceeds 0.99. The most aberrant unit-level observation according to estimated CPOs is subject 123 at time 1 for item 2 (with score 5), whereas the other two items have scores of 1 recorded at this time.

10.9   ROBUSTNESS IN PANEL AND CLUSTERED DATA ANALYSIS

Models for hierarchical and clustered data sets are often applied to relatively small data sets with few replications on each subject, but involve complex random effects models or autocorrelation structures. There is therefore potential for overfitting or applying parametric structures not supported by the data. Consider the model

$$y_{it} \sim \text{EF}(\mu_{it})$$
$$g(\mu_{it}) = \alpha_t + X_{it}\beta_t + Z_{it}b_i + \varepsilon_{it} \tag{10.20}$$

A multivariate or univariate normal is the default in many studies for the vector of subject effects b_i or for unit-level errors ε_{it}. Another option in the variable intercepts model, albeit potentially heavily parameterized, is to take the b_i as fixed effects.

A model form intermediate between taking the b_i to be fixed effects and taking them to be random and fully parametric involves discrete mixtures. A flexible discrete mixture model that provides a form of averaging over different numbers of subgroups is obtained under a DPP. For example, instead of assuming N distinct b_i one may allow $J < N$ as the maximum number of possible values which in turn allows clustering into a smaller set of $J^* < J$ distinct values during estimation. The number of clusters is stochastic and its posterior density will indicate whether the assumed maximum J is sufficient. Hirano (1999) discusses non-parametric alternatives regarding ε_{it}, while Kleinman and Ibrahim (1998a, 1998b) consider DPP modelling of permanent effects.

Greater resistance to discrepant observations might also be gained by scale mixing over subjects. For example, a gamma scale mixture on normal subject effects b_i leads to a heavier tailed multivariate Student t model rather than multivariate normality. With $q = 1$, this involves

$$b_i \sim \mathrm{N}(0, \sigma_b^2/\lambda_i)$$
$$\lambda_i \sim \mathrm{Ga}(\nu/2, \nu/2)$$

where ν may be another parameter. If combined with latent data sampling as in section 10.3.1, taking ν as unknown amounts to averaging over alternative links.

Chen and Dunson (2003) point out that a parametric option such as the Wishart for Σ_b necessarily assumes heterogeneity in effects and does not include an option whereby $b_i = 0$ for all i (i.e. a homogeneous effect $\beta_{ik} = \beta_k$ of a predictor across all subjects or clusters). Suppose there are p predictors (including the intercept) and all are potentially heterogeneous, namely

$$g(\mu_{it}) = Z_{it}b_i + \varepsilon_{it}$$

where b_{i1} is a variable intercept and b_{i2}, b_{i3}, etc., are variable regression effects having unknown means. Chen and Dunson suggest taking a Choleski decomposition of Σ_b that involves expressing b_i as

$$b_i = \Lambda \Gamma c_i$$

where

$$c_{ik} \sim \mathrm{N}(0, 1) \qquad k = 1, \ldots, p$$

and

$$\Lambda = \mathrm{diag}(\lambda_1, \ldots, \lambda_p)$$

is a vector of elements $\lambda_k \geq 0$, and Γ is a lower triangular matrix with $p(p-1)/2$ unknowns. Thus $\gamma_{jj} = 1$, $\gamma_{jk} = 0$ for $k = 1, \ldots, p-1$ and for $j > k$, and the remaining terms may be assigned unrestricted normal priors (e.g. for $p = 3$ the unknowns would be $\lambda_1, \lambda_2, \lambda_3, \gamma_{21}, \gamma_{31}$ and γ_{32}). A homogeneous effect in predictors k can be allowed for by introducing a binary indicator: $A_k = 1$ if $\lambda_k = 0$ (homogeneity) and $A_k = 0$ if $\lambda_k > 0$. If $A_k = 0$ the λ_k have priors constrained to be positive, e.g. $\lambda_k \sim \mathrm{Ga}(a_k, b_k)$. The prior probability $\mathrm{P}(A_k = 0)$ may be set at 0.5 and if $A_k = 0$ at a particular iteration one sets to zero all elements $\gamma_{kj} = \gamma_{jk} = 0$, $k = 1, \ldots, p$. The proportion of iterations where $A_k = 1$ may be compared via an empirical Bayes factor with the prior proportion, to assess which predictors have heterogeneous effects across subjects or clusters.

Semi-parametric regression models for longitudinal categorical data seek to robustify the regression component by allowing for non-linear effects of one or more metric predictors (Chib and Jeliazkov, 2002). Thus instead of the standard GLMM with linear effects of X_{it}, it may be assumed that a constant non-linear smooth applies to a time-varying predictor U_{it}

$$g(\eta_{it}) = X_{it}\beta + S(U_{it}) + Z_{it}b_i + \varepsilon_{it}$$

This might apply in true panel studies where a cohort is followed through time and behaviours such as fertility or labour participation are monitored as functions of age U_{it}. Another option with time-varying U_{it} or constant U_i is a time-varying smooth such as

$$g(\eta_{it}) = X_{it}\beta + S_t(U_{it}) + Z_{it}b_i + \varepsilon_{it}$$

An example in spatial epidemiology where this might be relevant is for sets of areas where the impact of deprivation on area mortality is allowed both to be non-linear and to change between different periods. A particular application of semi-parametric modelling is when latent data are introduced as an alternative to direct binary or multinomial logit or probit analysis. Thus for binary panel data a probit link is equivalent to

$$W_{it} \sim \mathrm{N}(\mu_{it}, 1)I(A_{it}, B_{it})$$

where $A_{it} = (-\infty, 0)$ and $B_{it} = (0, \infty)$ according as $y_{it} = 0$ or 1 and

$$W_{it} = X_{it}\beta + S(U_{it}) + Z_{it}b_i$$

This type of model raises possible identification issues with a latent response and random effects for both the smooth model and the subject effects b_i.

Example 10.10 Epileptic seizures To illustrate non-parametric approaches to assessing the pattern of random effects consider again the epileptic seizure data of Example (10.6). A bivariate random effect involves the intercept and the slope over visit time (in weeks). Fixed predictors are as before: X_1 = base seizure count, X_2 = treatment (binary), X_3 = treatment–base interaction and X_4 = age at baseline. These data are subject to overdispersion as well as widely differing subject frailties, so a model such as (10.17a) with both b_i and ε_{it} modelled parametrically might be considered. One may also consider non-parametric alternatives such as a DPP model whereby the random intercepts and slopes are taken to fall into a maximum of $J \leq N$ clusters.

Here we consider a model without subject occasion errors ε_{it} but with random intercepts and slopes (b_{i1} and b_{i2}) belonging to a maximum J clusters, where J is taken as 20. The Dirichlet concentration parameter κ is taken as five, though it might also be an extra unknown; Kleinman and

Ibrahim (1998b) assume alternative preset options ($\kappa = 1.5$ and $\kappa = 100$). The base density is taken to be a bivariate normal with precision matrix P and dispersion matrix $D = P^{-1}$. Kleinman and Ibrahim assume a relatively informative Wishart prior on P (drawing on an earlier parametric GLMM analysis), with $\nu = 10$ degrees of freedom and a diagonal scale matrix S, with $S_{11} = S_{22} = 5$. Here we take $\nu = 5$ and S to be the identity matrix.

Early convergence is obtained in a two-chain run of 15 000 iterations. The average number of non-empty clusters, J^*, is 14, with posterior density (from 10 000 iterations in one chain) as in Figure 10.4. This total

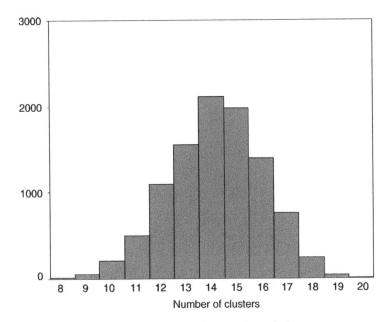

Figure 10.4   Posterior for number of clusters

is intermediate between the values 10.6 (for $\kappa = 1.5$) and 17.7 (for $\kappa = 100$) reported by Ibrahim and Kleinman (1998). Under this option for the random variation of intercepts and slopes, the average slope is negative, i.e. $\bar{b}_2 = -0.29$, but not precisely identified, with a 95% interval -0.86 to 0.25. Also there is virtually no correlation between intercepts and slopes ($D_{12} = -0.01$ with a 95% interval -0.41 to 0.34) so they might be modelled as separate DPP effects with possibly differing numbers of clusters. Both base and treatment effects are significant, however, with the latter having mean -0.89 and 95% interval $(-1.89, -0.06)$; age and the interaction effects are not significant.

A simplified model might drop one or both of the interaction and age effects. Here for illustration the age effect is modelled non-parametrically using prior (3.25), namely

$$s_t \sim N(s_{t-1}[1 + \delta_t/\delta_{t-1}] - s_{t-2}[\delta_t/\delta_{t-1}], \delta_t \tau^2)$$

There are 22 distinct age values and the smooth uses $x_t = \log(\text{age})$, with differences then being $\delta_t = x_t - x_{t-1}$. With an informative $Ga(10, 2.5)$ prior on τ^2 a quadratic rather than linear age effect is indicated, with peak influence around age 30 (Figure 10.5). The treatment effect loses

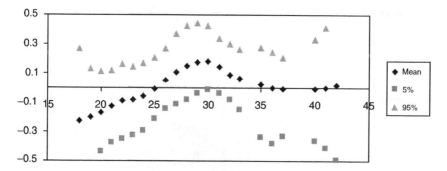

Figure 10.5   Non-linear age effect

significance under this model with a 95% interval from -1.59 to 0.22. The Gelman *et al.* (2003) complexity measure, mentioned in section 2.1, shows this model to have lower d_e (i.e. 113 with $\bar{D} = 395.4$) than the linear age model, which has $\bar{D} = 398.7$ and higher $d_e = 134$. However, inferences are likely to be sensitive to the prior on τ^2 which controls the trade-off between smoothness and fit (Chib and Jeliazkov, 2002).

10.10   APC AND SPATIO-TEMPORAL MODELS

The age-period model is a particular type of panel model that is commonly applied to death or disease counts y_{at} by age group a ($a = 1, \ldots, A$) and period t ($t = 1, \ldots, T$). An implicit cohort dimension is then obtained leading to age–period–cohort (APC) models. The usual sampling assumptions for what is actually observed, namely the y_{at}, are Poisson, in relation to person-years or expected events E_{at}, or binomial in relation to at-risk populations N_{at}. Suppose $y_{at} \sim Po(E_{at}\mu_{at})$ and that mortality or disease incidence is declining at a similar rate across all age bands. Then the mortality or disease rate trend lines for different age groups will be approximately parallel and a proportional model

$$\mu_{at} = \exp(\alpha_a) \exp(\beta_t)$$
$$\log(\mu_{at}) = \alpha_a + \beta_t$$

(10.21a)

will be appropriate. While fixed effects priors for α_a and β_t are possible, autoregressive random walk priors are preferable in terms of pooling strength between adjacent periods or age categories with similar values (Berzuini and Clayton, 1994, p 828). Possibilities include RW(1) or RW(2) models, such as

$$\alpha_a \sim N(\alpha_{a-1}, \tau_\alpha)$$

(with α_1 assumed a priori as a fixed effect) or

$$\alpha_a \sim N(2\alpha_{a-1} - \alpha_{a-2}, \tau_\alpha)$$

(with α_1 and α_2 as fixed effects). Greater robustness to shifts may be gained by scaled mixing, such as

$$\alpha_a \sim N(\alpha_{a-1}, \tau_\alpha/\omega_a)$$

where ω_a is gamma distributed. The difference $\alpha_a - \alpha_b$ is interpreted as the log of the relative risk for age group a compared with that for age group b. Since a constant may be added to all the α parameters and subtracted from all the β parameters, this model is not identified without constraints. Identifiability in a model such as (10.21a) without a constant may be gained by devices such as centring the β_t (or the α_a), or by setting one parameter to a fixed value, e.g. $\beta_1 = 0$. An alternative strategy (see Besag et al., 1995) does not impose such constraints but monitors only identifiable contrasts such as $\alpha_a - \alpha_b$ and $\beta_t - \beta_s$. If an intercept is included as in

$$\log(\mu_{at}) = \psi + \alpha_a + \beta_t$$

(10.21b)

then identifiability requires two constraints such as setting $\alpha_1 = \beta_1 = 0$ or centring both sets of parameters at each MCMC iteration.

The introduction of both age and time means that the birth cohort can be obtained as

$$c = A - a + t$$

with maximum $C = A - 1 + T$. For example, if the $T = 4$ time periods were 1983–1987, 1988–1992, 1993–1997, 1998–2002 and $A = 18$ age bands were 0–4, 5–9, 10–14, ..., 80–84 and 85+ then the first cohort ($c = 1$) would be those over 85 in 1983–1987 and the last ($c = 21$) would be 0–4 year olds in 1998–2002. Defining cohorts is simplest when age bands and periods are of the same width, though there are ways to define cohorts when widths are unequal (Knorr-Held and Rainer, 2001). A Poisson age–cohort model is then defined as

$$y_{at} \sim \text{Poi}(E_{at}\mu_{at})$$

where

$$\log(\mu_{at}) = \alpha_a + \gamma_c \qquad (10.22)$$

where the cohort effects γ_c represent factors that influence the mortality or disease incidence of a particular birth cohort throughout their lives. Identifiability in (10.22) requires centring of the γ_c or devices such as setting $\gamma_1 = 0$ or $\gamma_1 = \gamma_2$.

A model including age, period and cohort effects for data $y_{at} \sim \text{Poi}(E_{at}\mu_{at})$ then has the form

$$\log(\mu_{at}) = \psi + \alpha_a + \beta_t + \gamma_c \qquad (10.23)$$

This is an APC model and here identifiability can be problematic. It requires that all three sets of effects be centred or the use of devices such as $\alpha_1 = \beta_1 = \gamma_1 = 0$ to set the level of the three series. Additionally the relation $c = A - a + t$ introduces an extra identifiability issue and an extra constraint is needed for full identification. Often early cohort effects are poorly identified and so one might set $\gamma_1 = \gamma_2$ as well as centring all the effects at each MCMC iteration. Knorr-Held and Rainer (2000) suggest that RW(1) priors introduce a stochastic constraint that obviates the requirement for an additional formal constraint. Again a possible alternative is to summarize the model – and gauge convergence – using only identifiable parameter subsets or contrasts. These include the means μ_{at}, projections to new years (Bray, 2002), and contrasts such $\alpha_a - \alpha_b$ and $\gamma_c - \gamma_d$. In all the models (10.21)–(10.23) a zero-mean overdispersion effect may be added, so that (10.23) would become

$$\log(\mu_{at}) = \psi + \alpha_a + \beta_t + \gamma_c + \varepsilon_{atc} \qquad (10.24)$$

10.11 SPACE–TIME AND SPATIAL APC MODELS

Models defined over space and one or more time dimensions have diverse applications such as meteorology, medical imaging and disease mapping. In spatial epidemiology they may be used to monitor trends in environmental impacts (e.g. lung disease in relation to sources of airborne pollution) or trends in small-area inequalities in mortality (Congdon, 2004). Consider an application for areas i and times t, and assume

Poisson sampling of disease or death counts y_{it} in relation to expected events E_{it}. Binomial sampling might also be considered for large event totals or populations relatively small in relation to event totals; see Knorr-Held (2000) and Knorr-Held and Besag (1998).

Then one possible framework might include constant spatial and unstructured effects ('permanent area effects'), s_i and u_i, as in the cross-sectional mixed models considered in Chapter 8, combined with area-specific growth rates. Thus for $y_{it} \sim \text{Po}(E_{it}\mu_{it})$ the mixed model of Besag *et al.* (1991) might be extended as follows to include a spatially varying growth curve:

$$g(\mu_{it}) = \psi + \delta_i t + u_i + s_i \qquad (10.25)$$

where the effects δ_i may be unstructured with overall average growth rate δ. For spatially clustered trends the δ_i might be assumed to be spatially dependent (Bernardinelli *et al.*, 1995), e.g. with ICAR(ρ) form or following a robust or non-parametric option, to allow for growth discontinuities.

In addition the basic evolution through time ψ_t of the mortality or disease rate (or of a p vector of regression impacts β_t) may be specified in various ways: as independent fixed effects $\beta_t \sim N(0, \vartheta)$ with ϑ known and large, or as random effects such as an RW(1) process with

$$\beta_t \sim N_p(\beta_{t-1}, V_\beta)$$

or modelled via spline functions as in MacNab and Dean (2001). For example, with $y_{it} \sim \text{Po}(E_{it}\mu_{it})$, a time-varying intercept and regression effect model implies

$$\log(\mu_{it}) = \psi_t + s_i + u_i + \delta_i t + X_{it}\beta_t \qquad (10.26)$$

It is also possible to generalize the regression effect to be spatially as well as temporally varying, as in

$$\log(\mu_{it}) = \psi_t + s_i + u_i + \delta_i t + X_{it}\beta_{it}$$

with spatial dependence at times t and $t - 1$

$$\beta_{it} \sim N_p \left(\beta_{i,t-1} + \rho_1 \sum_{j \sim i} \beta_{jt} + \rho_2 \sum_{j \sim i} \beta_{j,t-1}, V_\beta/[2L_i + 1] \right)$$

where L_i is the number of neighbours j of area i.

More general or more heavily parameterized models may be proposed: for example, time-varying heterogeneity or spatial effects, u_{it} or s_{it} (Waller *et al.*, 1997; Carlin and Louis, 2000), autocorrelated errors ε_{it} (Congdon, 2001b), or area-specific growth functions (e.g. polynomials or spline functions), $\delta_i(t)$. In such models the addition of random effects may cause identifiability problems. Identifiability in practice may be improved by further constraints such as setting one of the u_i or s_i terms to zero.

Random effects specific to both area and time may be introduced to account for excess dispersion in relation to the Poisson or binomial. For example, Sun et al. (2000) propose a model for Poisson overdispersion including age as well as time and area dimensions

$$\log(\mu_{iat}) = \alpha_a + u_i + s_i + (\delta_i + \phi_a)t + \varepsilon_{iat}$$

which if age were omitted becomes

$$\log(\mu_{it}) = u_i + s_i + \delta_i t + \varepsilon_{it}$$

If the ε_{it} is autocorrelated in time, it may be preferable to omit the unstructured error to improve identifiability so that

$$\log(\mu_{it}) = s_i + \delta_i t + \varepsilon_{it}$$

with the error defined as

$$\varepsilon_{it} = \rho \varepsilon_{it-1} + \nu_{it}$$

for time periods $t > 1$, where $\nu_{it} \sim N(0,\tau)$, while $\varepsilon_{i1} \sim N(0,\tau/(1-\rho^2))$.

Congdon (2004) considers age–period or time–period interactions based on the Carter–Lee model (Carter and Lee, 1992). Considering age–period interactions, log relative risks for age, area and time take the form

$$g(\mu_{iat}) = \alpha_a + \beta_t + s_i + \lambda_a \kappa_t$$

with the parameters in the multiplicative function $\lambda_a \kappa_t$ constrained to gain identifiability. Lee (2000) takes the λ_a as multinomial and κ_t to sum to zero. The λ_a express variations between age groups in the adherence to the overall mortality trend represented by the κ_t parameters. If the κ_t are falling (as mortality declines) then larger λ_a indicate for which age groups mortality rates are declining most. Space–time interactions might be modelled via

$$g(\mu_{iat}) = \alpha_a + \beta_t + s_i + \eta_t b_i$$

where η_t are multinomial and represent differences between periods in the extent of spatial clustering defined by the b_i (e.g. clustering might be growing over time). Finally age–area interactions might be modelled via

$$g(\mu_{iat}) = \alpha_a + \beta_t + s_i + \zeta_a b_i$$

where the ζ_a represent age group differences in adherence to the spatial mortality regime defined by b_i. If spatial relative risks b_i are higher in deprived areas then ζ_a would be higher in those age groups (e.g. the middle-aged and children) where deprivation had the most marked mortality impact.

MacNab and Dean (2001) allow non-linear trends $S_\psi(t)$ in the level of mortality or disease or in regression impacts $S_\beta(t)$ via B-spline functions, though other forms of non-parametric regression could be used. Area–time interactions (alternatively viewed as area-specific deviations from the overall trend) may be modelled by allowing spatially varying coefficients applied to each term in a B-spline decomposition. Thus a cubic spline with a single knot would involve four polynomial functions $p_j(t), j = 1, \ldots, 4$ each of which would be premultiplied by spatial effects δ_{ji} for $i = 1, \ldots, I$ areas. The growth curve model in (10.26) with the effect of two predictors modelled via splines would become

$$g(\mu_{it}) = S_\psi(t) + S_{\beta_1}(t)X_{1t} + S_{\beta_2}(t)X_{2t} + \delta_{1i}p_1(t) + \delta_{2i}p_2(t)$$
$$+ \delta_{3i}p_3(t) + \delta_{4i}p_4(t) + u_i + s_i \tag{10.27}$$

where the smooth overall trend $S_\psi(t)$ and smooths in covariates j, namely $S_{\beta_j}(t)$, involve fixed effect parameters. For instance, a cubic spline with a single knot for $S_\psi(t)$ would take the form

$$S_\psi(t) = \psi_1 q_1(t) + \psi_2 q_2(t) + \psi_3 q_3(t) + \psi_4 q_4(t)$$

where $q_j(t)$ are polynomial functions.

The greatest generality occurs if the APC model is extended to include area i. Interactions between cohort and time, and between area and cohort, may then be relevant. Define a model with a 'main effect' spatial term s_i based on spatial adjacency and two sets of interactions as

$$\log(\mu_{iat}) = \psi + s_i + \alpha_a + \beta_t + \gamma_c + \theta_{ic} + \xi_{ct}$$

Clayton (1996, p 291) suggests a prior for such interactions based on multiplying the structure matrices underlying the joint priors in (say) cohort and area separately. Let the structure matrices of the separate area and cohort effects be denoted K_s and K_γ. Then the Kronecker product of these structure matrices $K_{s\gamma} = K_s \otimes K_\gamma$ defines the structure matrix for the joint prior and the structure of the conditional prior on θ_{ic} can then be derived. As an example of a structure matrix, an RW(1) prior in cohort effects has a structure matrix with the form

$$K_\gamma[cd] = \begin{cases} -1 & \text{if cohorts } c \text{ and } d \text{ are adjacent} \\ 0 & \text{if cohorts } c \text{ and } d \text{ are not adjacent} \\ 1 & \text{if } c = d = 1 \text{ or } c = d = C \\ 2 & \text{if } c = d = k \text{ where } k \neq 1 \text{ and } k \neq C \end{cases}$$

while an RW(2) prior has a structure matrix

$$K_\gamma = \begin{bmatrix} 1 & -2 & 1 & & & & & & & \\ -2 & 5 & -4 & & & & & & & \\ 1 & -4 & 6 & -4 & 1 & & & & & \\ & 1 & -4 & 6 & -4 & 1 & & & & \\ & & & \ddots & & & & & & \\ & & & & 1 & -4 & 6 & -4 & 1 & \\ & & & & & 1 & -4 & 6 & -4 & 1 \\ & & & & & & 1 & -4 & 5 & -2 \\ & & & & & & & 1 & -2 & 1 \end{bmatrix}$$

From Chapter 8, the prior for $s = (s_1, \ldots, s_n)$ based on adjacency is MVN with precision matrix $\tau_s K_s$ where

$$K_{s[ij]} = \begin{cases} -1 & \text{if areas } i \text{ and } j \text{ are neighbours} \\ 0 & \text{for non-adjacent areas} \\ L_i & \text{when } i = j \end{cases}$$

and L_i is the cardinality of area i (its total number of neighbours). Then an RW(1) cohort prior crossed with an ICAR(1) spatial main effect has a conditional form with variance

$$\begin{aligned} \sigma_\theta^2 / L_i & \quad \text{when } c = 1 \text{ or } c = C \\ \sigma_\theta^2 / (2L_i) & \quad \text{otherwise} \end{aligned}$$

and where the mean Θ_{ic} for θ_{ic} is defined by

$$\Theta_{i1} = \theta_{i2} + \sum_{j \sim i} \theta_{j1} / L_i - \sum_{j \sim i} \theta_{j2} / L_i$$

$$\Theta_{ic} = 0.5(\theta_{i,c-1} + \theta_{i,c+1}) + \sum_{j \sim i} \theta_{jc} / L_i - \left(\sum_{j \sim i} \theta_{j,c+1} + \sum_{j \sim i} \theta_{j,c-1} \right) / (2L_i) \qquad 1 < c < C$$

$$\Theta_{iC} = \theta_{i,C-1} + \sum_{j \sim i} \theta_{jC} / L_i - \sum_{j \sim i} \theta_{j,C-1} / L_i$$

Identifiability requires that the θ_{ic} be doubly centred at each iteration (over both areas for a given cohort c, and over cohorts for a given area i). Crossed structure matrix priors for area–time, cohort–time, age–time and age–area interactions are similarly defined.

Example 10.11 Lung cancer in Italian males Berzuini and Clayton (1994) analyse trends in lung cancer mortality in Italian males in $T = 9$ five-year periods 1944–1948, 1949–1953 through to 1984–1988. The data have $A = 13$ age bands (15–19, 20–24 through to 75–79), so there are 23 cohorts, the first being those aged 75–79 in 1944–1948 and the last being

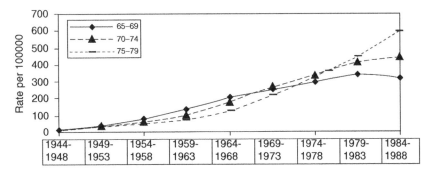

Figure 10.6 Observed trends in mortality

those aged 15–19 in 1984–1988. Figure 10.6 shows the trend in mortality for the three oldest age groups and only in the last period is there some evidence of the increase tailing off. Total deaths rose from 5789 in the first period to 97 580 in 1979–1983 and 109 170 in 1984–1988.

The model structure of (10.24) is assumed with second-order random walk priors for the age, period and cohort effects. Priors on precisions $1/\sigma^2$ for these effects follow the form adopted by Berzuini and Clayton, namely

$$\nu(\sigma_0/\sigma)^2 \sim \chi^2(\nu)$$

where σ_0 is a guess at σ (initially taken as $\sigma_0^2 = 0.01$ in the application here) and ν is the prior degrees of freedom (reflecting level of confidence in the guess); here $\nu = 2$. Identifiability constraints $\alpha_1 = \beta_1 = \gamma_1 = \gamma_2 = 0$ are applied.

A two-chain run of 2500 iterations shows convergence after 1000 iterations with median forecasts (and 50% credible intervals) for total deaths in 1989–1993 and 1994–1998 of 120 900 (111 500 to 129 800) and 131 200 (108 500 to 154 600). Taking $\sigma_0^2 = 0.001$ leads to virtually the same median for 1994–1998 but a narrower forecast interval (112 000 to 154 000). Comparable forecasts by Berzuini and Clayton (1994, Table IV) show medians 104 110 (98 500 to 110 200) and 110 700 (101 200 to 121 900) for 1989–1993 and 1994–1998 respectively.

Example 10.12 Trends in English male suicide To illustrate trends in the impact of social factors on suicide mortality, consider male suicide deaths y_{it} over 15 years between 1986 and 2000 ($t = -7, -6, \ldots, 6, 7$) in 354 English local authorities. The standard age schedule used to calculate expected suicide deaths is for England and Wales in 1993. The non-constant effects of three census-based indicators, namely social fragmentation (X_1), deprivation (X_2) and ethnicity (X_3), are modelled via cubic B splines with one interior knot at $t = 0$. A linear growth in area suicide rates to model area-level deviations from the central trend $S_\psi(t)$ is assumed, with randomly varying and spatially interdependent growth coefficients δ_i. The total area growth effect at time t is therefore $S_\psi(t) + \delta_i t$. Constant spatial intercept effects, s_i, are also included. Both s_i and δ_i follow ICAR(1) priors. So with $y_{it} \sim \text{Poi}(E_{it}\mu_{it})$

$$g(\mu_{it}) = S_\psi(t) + s_i + \delta_i t + S_{\beta_1}(t)X_{1t} + S_{\beta_2}(t)X_{2t} + S_{\beta_3}(t)X_{3t}$$

To illustrate the results, Figures 10.7 and 10.8 show the changing overall intercept $S_\psi(t)$ and the changing deprivation effect $S_{\beta_2}(t)$ respectively. The area growth coefficients δ_i vary from -0.06 to 0.037.

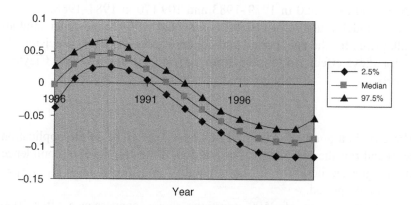

Figure 10.7 Changing interecpt

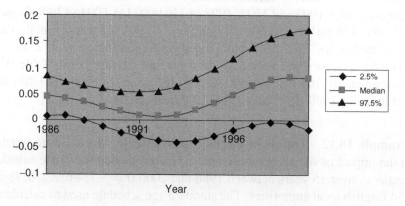

Figure 10.8 Changing deprivation effect

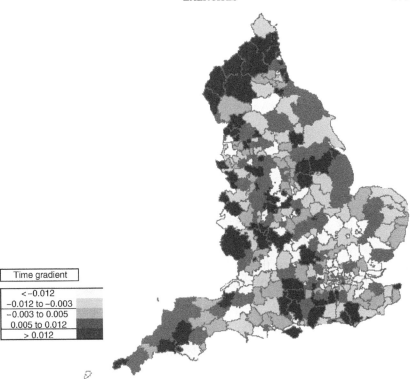

Figure 10.9   Spatially verying linear time effects

Figure 10.9 shows clustering of higher growth in parts of northern and central southern England.

The DIC is 27 128 with 454 effective parameters. With a relatively large data set such as this, some observations will not be well fit. The evidence from Monte Carlo CPO estimates is that poor fit is confined to outlying observations in the series for each area (time within area outliers) as opposed to cluster (area) level outliers. The lowest CPO is for area 65 in year 4 (Bradford in 1989) when a low death count follows a high count for the preceding year.

EXERCISES

1. In Example 10.1 try out a simple logit regression without varying region intercepts and compare its fit (e.g. via sensitivity and specificity rates estimated using replicate data as against actual data) with the varying intercepts model. Also consider a simplified model for varying

region intercepts and age effects modelled not as correlated but as independent random effects.

2. In Example 10.1 try a varying intercept model defined by a two-group discrete mixture

$$\text{logit}(\pi_{ir}) = b_{iG_i} + \beta_1(\text{Age} - 20) + \beta_2(\text{Age} - 20)^2$$
$$+ \beta_3[(\text{Age} - 30)_+]^2 + \beta_4[(\text{Age} - 40)_+]^2 + \beta_5\text{Religion}$$

where G_i is binary and the two sets of effects b_{ij} differ in mean and variance, namely

$$b_{i1} \sim N(\beta_{01}, \phi_1) \quad b_{i2} \sim N(\beta_{02}, \phi_2)$$

Either the variances or the intercepts would need to be constrained (e.g. $\beta_{01} > \beta_{02}$) to ensure identifiability.

3. In Example 10.2 try the full varying regressor effect model. With $Z_{ir1} = 1$ for low temperature and $Z_{ir2} = 1$ for skin contact, the regression model is

$$\mu_{ir} = b_{i1} + b_{i2}Z_{ir1} + b_{i3}Z_{ir2}$$

Try first taking b_{i1}, b_{i2} and b_{i3} as independent random effects and then as multivariate effects and assess gains in fit and possible identifiability issues. Are the last affected by the prior assumed for b_i?

4. In Example 10.3 try a bivariate Student t for η_h (either directly or by scale mixing) and assess global and individual observation fit against the BVN option.

5. In Example 10.4 use the latent data (Albert and Chib, 1993) method and compare (e.g. via parallel sampling) the predictive accuracy of probit and logit links. The logit link may be approximated by scale mixing or by using the logistic density to sample the latent continuous data (see Chapter 4).

6. In Example 10.4 use a complementary log–log link to compare the model fitted there (i.e. with age–maternal smoking interaction A_tM_i) with a model with Z_tM_i where $Z_t = 1$ for the last three ages and $Z_t = 0$ for children aged 7.

7. In Example 10.5 assess both fit to the observed data and out-of-sample predictive accuracy (November forecast) of assuming a common dispersion matrix Σ for the residual trend parameters δ_{itj}. Assessing the out-of-sample accuracy will require the actual November data from the Ipsos–Reid website.

8. In Example 10.6 try one of the other conjugate Poisson panel models outlined in Section 10.6 and compare its fit and implications for treatment impact with the adapted version of the Rasch Poisson model.

9. In Example 10.7 try a mixture of normals to model the growth rates, with the effects b_{i1} modelled as univariate effects independently of b_{i2}, and b_{i2} modelled as a mixture of univariate normals (growth rate subgroups)

$$b_{i2} \sim \pi N(B_{i1}, V_1) + (1 - \pi)N(B_{i2}, V_2)$$

Also the regression model in B_{i1} and B_{i2} should include different regression effects (i.e. the intercepts and the coefficients on W_1 and W_2 vary according to growth rate subgroup). For identifiability the intercepts for B_{i1} and B_{i2} may be ordered a priori.

10. In Example 10.8 try adding a skew random effect into the model for hospital visits, namely

$$\log(\mu_{it2}) = \alpha_2 + X_i\beta_2 + b_{i2} + \lambda b_{i3} + \varepsilon_{it2}$$

where $b_{i3} \sim N(0, 1)I(0,)$ and λ is a positive loading, and ascertain change in model performance.

11. In Example 10.9 try model A with period and item-specific regression effects, namely

$$\text{logit}(\gamma_{itmj}) = \kappa_{mj} - S_{it} - X_i\beta_{mt}$$

and ascertain any gain in fit. Also in model B assess gains in fit from making ϕ_{it} autoregressive, as in

$$\phi_{it} \sim N(\rho_\phi \phi_{i,t-1}, 1)$$

12. In both the models in Example 10.10 try DPP with baseline univariate normals (rather than bivariate) on b_{i1} (varying intercepts) and b_{i2} (varying time slopes). Allow the DPP to have separate clustering schemes.

13. In Example 10.12 fit a simplified model

$$g(\mu_{it}) = S_\psi(t) + s_i + \delta_{G_i}t + S_{\beta_1}(t)X_{1t} + S_{\beta_2}(t)X_{2t} + S_{\beta_3}(t)X_{3t}$$

where G_i can fall into one of three categories and $\delta_1 < \delta_2 < \delta_3$ for identifiability. Compare its fit with the spatially varying linear trend model in terms of the DIC, average deviance, and effective parameters as defined by Spiegelhalter *et al.* (2002) and Gelman *et al.* (2003).

REFERENCES

Aitkin, M. and Longford, N. (1986) Statistical modeling issues in school effectiveness studies. *Journal of the Royal Statistical Society, Series A*, **149**, 1–43.

Albert, J. and Chib, S. (1993) Bayesian analysis of binary and polychotomous response data. *Journal of the American Statistical Association*, **88**, 669–679.

Bernardinelli, L., Clayton, D. and Montomoli, C. (1995) Bayesian estimates of disease maps: how important are priors? *Statistics in Medicine*, **14**, 2411–2431.

Berzuini, C. and Clayton, D. (1994) Bayesian survival analysis on multiple time scales. *Statistics in Medicine*, **13**, 823–838.

Besag, J., York, J. and Mollie, A. (1991) Bayesian image restoration with applications in spatial statistics. *Annals of the Institute of Mathematical Statistics*, **43**, 1–59.

Besag, J., Green, P., Higdon, D. and Mengersen, K. (1995) Bayesian computation and stochastic systems. *Statistical Science*, **10**, 3–66.

Böckenholt, U. (2002) Markov models with random effects for binary panel data. *Methods of Psychological Research Online*, Special Issue.

Boscardin, W. and Weiss, R. (2004) Fitting unstructured covariance matrices to longitudinal data. *Working Paper*, UCLA Department of Biostatistics.

Bray, I. (2002) Application of Markov chain Monte Carlo methods to projecting cancer incidence and mortality by APC models. *Journal of the Royal Statistics Society, Series C*, **51**, 151–164.

Cantoni, E. (2004) A robust approach to longitudinal data analysis. *Canadian Journal of Statistics*, **32**, 169–180.

Cargnoni, C., Müller, P. and West, M. (1997) Bayesian forecasting of multinomial time series through conditionally Gaussian dynamic models. *Journal of the American Statistical Association*, **92**, 640–647.

Carlin, B. and Louis, T. (2000) *Bayes and Empirical Bayes Methods for Data Analysis*, 2nd Edition. Chapman and Hall: London CRC Press: Boca Raton, FL.

Carter, L. and Lee, R. (1992) Modeling and forecasting US sex differentials in mortality. *International Journal of Forecasting*, **8**, 393–411.

Chen, Z. and Dunson, D. (2003) A Bayesian approach for assessing heterogeneity in generalized linear models. *Working Paper*, National Institute of Environmental Health Sciences.

Chen, Z. and Kuo, L. (2001) A note on the estimation of the multinomial logit model with random effects. *The American Statistician*, **55**, 89–95.

Chib, S. and Carlin, B. (1999) On MCMC sampling in hierarchical longitudinal models. *Statistics and Computing*, **9**, 17–26.

Chib, S. and Jeliazkov, I. (2002) Semiparametric hierarchical Bayes analysis of discrete panel data with state dependence and serial correlation. *Working Papers*, Olin School of Business, Washington University in St Louis.

Chib, S., Greenberg, E. and Winkelmann, R. (1999) Posterior simulation and Bayes factors in panel count data models. *Journal of Econometrics*, **86**, 33–54.

Clayton, D. (1996) Generalized linear mixed models. In *Markov Chain Monte Carlo in Practice*, Gilks, W., Richardson, S. and Spiegelhalter, D. (eds). Chapman and Hall: London.

Congdon, P. (2001a) Predicting adverse infant health outcomes using routine screening variables: modelling the impact of interdependent risk factors. *Journal of Applied Statistics*, **28**, 183–197.

Congdon, P. (2001b) Bayesian models for suicide monitoring. *European Journal of Population*, **15**, 1–34.

Congdon, P. (2004) Modelling trends and inequality in small area mortality. *Journal of Applied Statistics*, **31**, 603–622.

Congdon, P. (2005) Bayesian predictive model comparison via parallel sampling. *Computational Statistics and Data Analysis*, **48**, 735–753.

Congdon, P. and Best, N. (2000) Small area variation in hospital admission rates: adjusting for referral and provider variation. *Journal of the Royal Statistical Society, Series C*, **49**, 207–226.

Curran, P. and Bollen, K. (2001) The best of both worlds: combining autoregressive and latent curve models. In *New Methods for the Analysis of Change*, Collins, L. and Sayer, A. (eds). American Psychological Association: Washington, DC, 105–136.

Daniels, M. and Gatsonis, C. (1999) Hierarchical generalized linear models in the analysis of variations in health care utilization. *Journal of the American Statistical Association*, **94**, 29–42.

Davies, R., Crouchley, R. and Pickles, A. (1982) Modelling the evolution of heterogeneity in residential mobility. *Demography*, **19**, 291–299.

Diggle, P. (1988) An approach to the analysis of repeated measurements. *Biometrics*, **44**, 959–971.

Fearns, T. (1975) A Bayesian approach to growth curves. *Biometrics*, **62**, 89–100.

Fitzmaurice, G., Laird, N. and Ware, J. (2003) *Applied Longitudinal Analysis*. John Wiley & Sons: New York.

Gamerman, D. and Smith, A. (1996) Bayesian analysis of longitudinal data studies. In *Bayesian Statistics 5*, Bernardo, J., Berger, J., Dawid, A. and Smith, A. (eds). Oxford University Press: Oxford, 587–597.

Gelman, A., Carlin, J., Stern, H. and Rubin, D. (2003) *Bayesian Data Analysis*, 2nd Edition. CRC Press: Boca Raton, FL.

Geweke, J., Keane, M. and Runkle, D. (1994) Alternative computational approaches to inference in the multinomial probit model. *Review of Economics and Statistics*, **76**, 609–632.

Goldstein, H. and Spiegelhalter, D. (1996) League tables and their limitations. Statistical issues in comparisons of institutional performance. *Journal of the Royal Statistical Society, Series A*, **159**, 385–443.

Goodhardt, G., Ehrenberg, A. and Chatfield, C. (1984) The Dirichlet: a comprehensive model of buying behaviour. *Journal of the Royal Statistical Society, Series A*, **147**, 621–655.

Grunwald, G., Hyndman, R., Tedesco, L. and Tweedie, R. (2000) Non-Gaussian conditional linear AR(1) models. *Australian & New Zealand Journal of Statistics*, **42**, 479–495.

Gustafson, P. and Walker, L. (2003) An extension of the Dirichlet prior for the analysis of longitudinal multinomial data. *Journal of Applied Statistics*, **30**, 293–310.

Hausman, J., Hall, H. and Griliches, Z. (1984) Econometric models for count data with an application to the patents-R&D relationship. *Econometrica*, **52**, 909–938.

Heckman, J. (1981) Statistical models for discrete panel data. In *Structural analysis of discrete data with econometric applications*, Manski, C. and McFadden, D. (eds). The MIT Press: Cambridge, MA, 114–178.

Hirano, K. (1999) A semiparametric model for labor earnings dynamics. In *Practical Nonparametric and Semiparametric Bayesian Statistics*, Dey, D., Mueller, P. and Sinha, D. (eds). Springer: New York.

Hsiao, C. (1986) *Analysis of Panel Data*. Cambridge University Press: Cambridge.

Ibrahim, J. and Kleinman, K. (1998) Semiparametric Bayesian methods for random effects models. In *Practical Nonparametric and Semiparametric Bayesian Statistics*, Dey, D., Mueller, P. and Sinha, D. (eds). Springer: New York.

Ishwaran, H. and Gatsonis, C. (2000) A general class of hierarchical ordinal regression models with applications to correlated ROC analysis. *Canadian Journal of Statistics*, **28**, 731–750.

Jansen, M. (1997) Applications of Rasch's Poisson counts model to longitudinal count data. In *Applications of Latent Trait and Latent Class Models in the Social Sciences*, Rost, J. and Langeheine, R. (eds). Waxmann Münster: New York, 380–389.

Johnson, V. and Albert, J. (1999) *Ordinal Data Modeling*. Springer: New York.

Jones, R. (1993) *Longitudinal Data with Serial Correlation: A State-space Approach*. Chapman and Hall: London.

Jones, R. and Boadi-Boateng, F. (1991) Unequally spaced longitudinal data with AR(1) serial correlation. *Biometrics*, **47**, 161–175.

Kleinman, K. and Ibrahim, J. (1998a) A semiparametric Bayesian approach to the random effects model. *Biometrics*, **54**, 921–938.

Kleinman, K. and Ibrahim, J. (1998b) A semi-parametric Bayesian approach to generalized linear mixed models. *Statistics in Medicine*, **17**, 2579–2596.

Knorr-Held, L. (2000) Bayesian modelling of inseparable space-time variation in disease risk. *Statistics in Medicine*, **19**, 2555–2567.

Knorr-Held, L. and Besag, J. (1998) Modelling risk from a disease in time and space. *Statistics in Medicine*, **17**, 2045–2060.

Knorr-Held, L. and Rainer, E. (2001) Projections of lung cancer mortality in West Germany: a case study in Bayesian prediction. *Biostatistics*, **2**, 109–129.

Laird, N. and Ware, J. (1982) Random-effects models for longitudinal data. *Biometrics*, **38**, 963–974.

Lee, J. and Hwang, R. (2000) On estimation and prediction for temporally correlated longitudinal data. *Journal of Statistical Planning and Inference*, **87**, 87–104.

Lee, R. (2000) The Lee-Carter method for forecasting mortality, with various extensions and applications. *North American Actuarial Journal*, **4**, 80–93.

Lee, Y. and Nelder, J. (2000) Two ways of modelling overdispersion in non-normal data. *Journal of the Royal Statistical Society, Series C*, **49**, 591–598.

Leonard, T. and Hsu, J. (1994) The Bayesian analysis of categorical data – a selective review. In *Aspects of Uncertainty: A Tribute to DV Lindley*, Freeman, P. and Smith, A. (eds). John Wiley & Sons: Chichester.

MacCallum, R., Kim, C., Malarkey, W. and Kiecolt-Glaser, J. (1997) Studying multivariate change using multilevel models and latent curve models. *Multivariate Behavioral Research*, **32**, 215–253.

MacNab, Y. and Dean, C. (2001) Autoregressive spatial smoothing and temporal spline smoothing for mapping rates. *Biometrics*, **57**, 949–956.

Muthén, B. (1997) Latent variable modeling of longitudinal and multilevel data. In *Sociological Methodology*, Raftery, A (ed.). Blackwell: Boston, MA, 453–480.

Oh, M. and Lim, Y. (2001) Bayesian analysis of time series Poisson data. *Journal of Applied Statistics*, **28**, 259–271.

Payne, J., Coy, J., Milner, P. and Patterson, S. (1993) Are deprivation indicators a proxy for morbidity? A comparison of the prevalence of arthritis, depression, dyspepsia, obesity and respiratory symptoms with unemployment rates and the Jarman score. *Journal of Public Health Medicine*, **15**, 161–170.

Randall, J. (1989) The analysis of sensory data by generalized linear models. *Biometrical Journal*, **31**, 783–791.

Rasbash, J. and Browne, W. (2001) Modelling non-hierarchical structures. In *Multilevel Modelling of Health Statistics*, Leyland, A. and Goldstein, H. (eds). John Wiley & Sons: Chichester, 93–105.

Riphahn, R., Wambach, A. and Million, A. (2003) Incentive effects in the demand for health care: a bivariate panel count data estimation. *Journal of Applied Econometrics*, **18**, 387–406.

Rovine, M. and Molenaar, P. (1998) The covariance between level and shape in the latent growth curve model with estimated basis vector coefficients. *Methods of Psychological Research Online*, http://www.ppm.ipn.uni-kiel.de/mpr/issue5.

Saei, A. and McGilchrist, C. (1998) Longitudinal threshold models with random components. *Journal of the Royal Statistical Society, Series D*, **47**, 365–375.

Snijders, T. and Bosker, R. (1999) *Multilevel Analysis: An Introduction to Basic and Advanced Multilevel Modeling*. Sage: London.

Spiegelhalter, D., Best, N., Carlin, B. and van der Linde, A. (2002) Bayesian measures of model complexity and fit. *Journal of the Royal Statistical Society, Series B*, **64**, 583–639.

Steyer, R. and Partchev, I. (2000) Latent state-trait modelling with logistic item response models. In *Structural Equation Modeling: Present and future*, Cudeck, R., Du Toit, S. and Sörbom, D. (eds). Scientific Software International: Chicago, 481–520.

Sun, D., Tsutakawa, R., Kim, H. and He, Z. (2000) Bayesian analysis of mortality rates with disease maps. *Statistics and Medicine*, **19**, 2015–2035.

Thall, P. and Vail, S. (1990) Some covariance models for longitudinal count data with overdispersion. *Biometrics*, **46**, 657–671.

Van Duijn, M. and Jansen, M. (1995) Modeling repeated count data: some extensions of the Rasch Poisson counts model. *Journal of Educational & Behavioral Statistics*, **20**, 241–258.

Wakefield, J., Smith, A., Racine-Poon, A. and Gelfand, A. (1994) Bayesian analysis of linear and non-linear population models by using the Gibbs Sampler. *Applied Statistics*, **43**, 201–221.

Waller, L., Carlin, B., Xia, H. and Gelfand, A. (1997) Hierarchical spatio-temporal mapping of disease rates. *Journal of the American Statistical Association*, **92**, 607–617.

Weiss, R., Cho, M. and Yanuzzi, M. (1999) On Bayesian calculations for mixture likelihoods and priors. *Statistics in Medicine*, **18**, 1555–1570.

Zeger, S. and Qaqish, B. (1988) Markov regression models for time series: a quasi-likelihood approach. *Biometrics*, **44**, 1019–1031.

CHAPTER 11

Missing-Data Models

11.1 INTRODUCTION: TYPES OF MISSING DATA

Many data sets resulting from surveys or longitudinal investigations (e.g. clinical follow-up studies) are subject to missing data. Cross-sectional surveys may be subject to refusal by some subjects to answer certain questions or items (this is known as item non-response) or be subject to complete missingness on a subject (unit non-response). Panel studies in clinical and other applications are subject to intermittently missing responses or permanent loss from observation (known as attrition) (Ibrahim *et al.*, 2001a). Let X generically denote dependent or predictor variables. Permanent loss to observation results in what is known as a monotone form of missing data since if X_{it} is observed then $X_{i,t-1}, X_{i,t-2}, \ldots$ are necessarily observed, while if X_{it} is missing then $X_{i,t+1}, X_{i,t+2}, \ldots$ are necessarily also missing. In cross-sectional (e.g. survey) applications monotone non-response would occur in the context of two items if education (say) was answered whenever income was, but for some subjects education would be recorded though income was missing. Panel studies and time series, such as those in climatology or environmetrics, may be subject to intermittent non-monotone missing data. For example, limitations of measurement instruments mean that values below detection limits are left censored (Hopke *et al.*, 2001).

Apart from whether missingess is monotone, a further question is whether the chance of a missing value is related to the value that would have been observed had response actually happened (Little and Rubin, 1987). Corresponding to whether data X_{ij} on variable j for subject i are observed or not, one may define binary indicators which count as part of the observations. Thus $R_{ij} = 1$ (for X_{ij} missing) and $R_{ij} = 0$ (for X_{ij}

Bayesian Models for Categorical Data P. Congdon
© 2005 John Wiley & Sons, Ltd

observed); if X_{ijt} is time varying then $R_{ijt} = 1$ (for X_{ijt} missing) and $R_{ijt} = 0$ (for X_{ijt} observed). The missingness indicators may be multi-nomial rather than binary when there is more than one type of missing data. For example, in the case of multilevel data (e.g. pupils within schools) the non-response may be at level 1 (pupils) or at level 2 (schools).

If the probability that X_{ij} is missing (i.e. the probability that R_{ij} is one when there is only one type of missingness) is governed by missingess at random then $\pi_{ij} = \Pr(R_{ij} = 1)$ may depend on values of observed variables (e.g. observed Y_i, or X_{ik} where $k \neq j$) but not on the values of possibly missing variables including X_{ij} itself. By contrast, when the chance of non-response is related to the values of missing variables, then response is called non-ignorable. An example might be when response to an income question depends on the level of income, or in a panel study if the chances of dropout were related to treatment allocation or response. For instance, dropout from clinical intervention studies may be greater for patients adversely affected by a treatment, or a measure of mood may be more likely to be missing when the subject is in a poor mood (Ibrahim et al., 2001a). The parameters of a missingness model may be only weakly identified by the data and so sensitivity analysis may be applied on the parameters of the central model (relating Y to X) according to alternative parameterizations of the model for $R \mid Y, X$ (Roy and Lin, 2002).

The opposite possibility is known as missingess completely at random (MCAR): none of the data collected or missing are relevant to explaining the chance of missingness; the regression for $\Pr(R_{ij}) = 1$ is independent of all predictors X and outcomes Y. Only if the MCAR assumption holds is it valid to use the strategy known as complete case analysis: that is, excluding from the analysis any observation with missing values.

Where data are missing, whether by missingness at random (MAR) or non-ignorable mechanisms, the analysis may be set up such that values for the missing data are sampled at each MCMC iteration. In non-ignorable models such a sampling scheme is a necessary feature of the model. However, approximations to such repeated sampling imputation of missing data may be obtained by a small finite set of imputations under the multiple imputation (MI) method (Rubin, 1987; Schafer, 1997). Letting the missing data be denoted X_{mis}, and the observed data X_{obs}, multiple imputation requires a sample of m values of $X_{i,\text{mis}}$ (for subjects $i = 1, \ldots, n$) from the conditional density $p(X_{\text{mis}} \mid X_{\text{obs}})$. These samples replace or 'fill in' the missing data to make m versions of the complete data and then standard complete data analysis is applied m times. Inferences on parameters or derived quantities are obtainable by averaging over the m complete data analyses.

11.2 DENSITY MECHANISMS FOR MISSING DATA

When missingness is at random, Rubin (1976) shows that likelihood-based inference does not require a model for the missing-data mechanism (see also Laird, 1988). By contrast, for non-ignorable non-response it is necessary to set up a model for the response mechanism as well as for the data themselves. As an example, suppose for a cross-sectional sample that there is a binary outcome Y, a single binary covariate X, and that Y is fully observed but X possibly missing. Then set $R = 1$ if X is missing and $R = 0$ if X is observed. One possible approach, known as a selection model, considers the joint density as

$$p(Y, X, R | \eta, \beta, \theta) = p(R | Y, X, \eta) p(Y | X, \beta) p(X | \theta) \qquad (11.1a)$$

Since X is binary, $f(X | \theta)$ reduces for each individual to a Bernoulli density

$$X_i \sim \text{Bern}(\theta) \qquad (11.1b)$$

and $p(Y | X, \beta)$ can be modelled using a logit, probit or complementary log–log link, g:

$$Y_i \sim \text{Bern}(\pi_i^Y)$$
$$g(\pi_i^Y) = \beta_1 + \beta_2 X_i \qquad (11.1c)$$

If the chance that X is missing is possibly related to observed values of X, but not to missing values of X, then the model need go no further and $p(R | Y, X, \eta)$ need not be specified. However, if there is non-ignorable missingness, i.e. if $\Pr(R_i = 1)$ is related to $X_{i,\text{mis}}$ as well as $X_{i,\text{obs}}$, then a further model stage is required. For instance, one might specify

$$R_i \sim \text{Bern}(\pi_i^R)$$
$$\text{logit}(\pi_i^R) = \eta_1 + \eta_2 Y_i + \eta_3 X_i \qquad (11.1d)$$

where predictor values X_i include imputations under (11.1b). A sensitivity analysis in this simple example might consider whether setting η_3 or η_2 to zero affects the β coefficients.

In panel data subject to attrition but not intermittent missing data, non-ignorable missingness means the chance of a dropout at time t may be related to the possibly missing value of y_{it} at that time. Ignorable missingness by contrast might include impacts of lagged (and observed) dependent variable values such as $y_{i,t-1}$ on $\Pr(R_{it} = 1)$ but not the contemporaneous value (Diggle and Kenward, 1994). So non-ignorable missingness would imply $\eta_2 \neq 0$ in a dropout model such as

$$R_{it} \sim \text{Bern}(\pi_{it}^R) \qquad (11.2a)$$

$$\text{logit}(\pi_{it}^R) = \eta_1 + \eta_2 y_{it} + \eta_3 X_{it} + \eta_4 y_{i,t-1} \qquad (11.2b)$$

where y_{it} is an additional unknown imputed under the model when $R_{it} = 1$.

Another possible factoring of the joint density in (11.1a) leads to what is known as the pattern mixture model approach to incomplete data (Little, 1993). Let Y be subject to missingness and X be fully observed; then the model for R_Y precedes that for Y itself, as in

$$p(Y, X, R | \eta, \beta, \theta) = p(Y | R_Y, X, \beta) p(R_Y | X, \eta) p(X | \theta) \qquad (11.3a)$$

The regression analysis for Y involves both the missingness indicators R and the substantive influences X which are the focus of interest, and the marginal expectation $E(Y_i | X_i)$ is obtainable by averaging over the probabilities π_i^R (Horton and Fitzmaurice, 2002). Practical strategies for pattern mixture modelling involve simplifying identifiability constraints (Molenberghs et al., 2002) such as delimiting a relatively small number of non-response patterns. For example, for $T = 3$ observation points in a panel data problem, the possible sequences are OOO (all three values of Y observed), OOM, OMO, MOO, OMM, MOM, MMO and MMM. While it is possible to include the complete non-response pattern MMM given information on some predictors, they are often ignored in practice (Hedeker and Gibbons, 1997). Hence one is left with seven response patterns and the probability $p(Y | R, X, \beta)$ can be modelled by introducing an extra categorical predictor R_i for missingness status (with seven levels). Random effects parameters (e.g. variance or dispersion matrices for permanent subject effects in GLMMs) can also be distinguished by subject response category.

If X is not subject to missingness it may be taken as known rather than random, and the joint density (11.3a) becomes

$$p(Y, X, R | \eta, \beta) = p(Y | R, X, \beta) p(R_Y | W, \eta) \qquad (11.3b)$$

One might therefore have a multinomial logit model for R (with seven categories in the example just quoted) using predictors W that may include a subset of X, and a main regression for Y in which R is introduced as a categorical predictor. Alternatively, interpretability may be gained by simply reducing non-response patterns to a binary index $R_i = 1$ for completers (those with the OOO pattern), and $R_i = 0$ for all non-completers regardless of pattern (those with one of the OOM, OMO, MOO, OMM, MOM, MMO patterns). This option makes it easy to include in the Y regression model interactions $X \times R$ between substantive factors (e.g. treatment status and time) and response status. Another simplification of the response patterns might be a trichotomy distinguishing full response from monotone missingness (OOM and OMM) and from intermittent nonmonotone missingness (MOM, MOO, OMO and MMO).

Example 11.1   NIMH schizophrenia collaborative study   As an illustration of a pattern mixture approach consider clinical trial data on 437 patients observed at one or more of seven possible occasions (weeks from start of study) as in Table 11.1. The outcome y_{it} is ordinal with $J = 7$

Table 11.1   Observed sample totals at each of seven observation points

	Weeks from start of study							Total sample
	0	1	2	3	4	5	6	
Drug	327	321	9	287	9	7	265	329
Placebo	107	105	5	87	2	2	70	108

levels of increasing severity. Not all subjects were measured at each time point. It can be seen that by six weeks there is substantial dropout. A simple pattern mixture model simply defines $R_i = 1$ or 0 according as a measurement is taken at week 6.

In Hedeker and Gibbons (1997), a GLMM as in Chapter 10 is applied, with time transformed to the square root of weeks. The predictors X_{it} are $X_1 =$ treatment status (drug vs. placebo) and $X_2 =$ time–treatment interaction, while the random effects predictors are $Z_1 =$ intercept and $Z_2 =$ time. Then a model ignoring missingness $P(y_{it}|\beta, b_i, X_{it}, Z_{it})$ can be contrasted with one that takes it into account via the mechanism $P(y_{it}|\beta, b_i, X_{it}, Z_{it}, R_i)P(R_i|W, \eta)$ as in (11.3b). Hedeker and Gibbon do not explicitly model $P(R_i|W_i, \eta)$, though interactions in the model for y_{it} achieve this indirectly. The relevant interactions are between dropout status and treatment, between drop-out and time, and between dropout and the time–treatment interaction. One might instead have a separate logit regression of R on treatment status.

With a subject–time ordinal response the relevant multinomial probability vector is denoted $\pi_{it} = (\pi_{it1}, \ldots, \pi_{itJ})$ and a model ignoring dropout is

$$y_{it} \sim \text{Categorical}(\pi_{it})$$
$$\pi_{itj} = \gamma_{itj} - \gamma_{it,j-1} \quad j = 2, \ldots, J - 1$$
$$\text{logit}(\gamma_{itj}) = \kappa_j - (\beta_1 X_{i1} + \beta_2 X_{i2} + b_{i1} + b_{i2}Z_{i1})$$

where κ_j are cut points on an underlying continuous scale. The random effects $b_i = (b_{i1}, b_{i2})$ are assumed to be bivariate normal with inverse covariance $\sum_b^{-1}$ following a Wishart prior with identity scale matrix. Because of the varying intercept b_{i1}, it is assumed that $\kappa_1 = 0$.

This is a highly parameterized model and convergence may be slow. The first model (model A ignoring missingness) is run for 20 000 iterations and results from the second half shows no clear 'main effect' β_1 of treatment, but a treatment–time effect β_2 that is significantly negative. Those on the drug regime decline more rapidly in terms of symptom severity. Note, though, that all patients seem to show an improvement over time (the median of the time effects b_{i2} is -0.52). Intercepts and random time paths are negatively correlated.

The pattern mixture model (model B) is then fitted with a dropout status main effect and interactions between dropout and treatment, dropout and time, and dropout and time × treatment. To aid identifiability the covariance matrix $\sum_b$ from model A is used to set an informative Wishart prior on the MVN precision matrix for b_i.

The DIC fits (Table 11.2) are very similar between the models but the drug–time interaction is amplified in model B, with median -2.81 instead of -1.63. Dropout is more common among those in the treatment group with higher scores and among those within the treatment group with worsening symptoms (the dropout–drug–time interaction is positive). However, dropouts in general, regardless of treatment group, tend to have lessening symptoms (the dropout × time effect is negative).

Table 11.2 Parameter estimates in with- and without-dropout models

Parameter	Without dropout (model A)			Allowing for dropout (model B)		
	2.5%	Median	97.5%	2.5%	Median	97.5%
Intercept	7.46	8.15	8.78	7.87	8.57	9.10
Time	−0.78	−0.52	−0.27	−0.28	0.03	0.36
κ_2	1.91	2.26	2.61	1.96	2.29	2.64
κ_3	3.46	3.87	4.28	3.52	3.91	4.30
κ_4	5.20	5.68	6.16	5.30	5.73	6.15
κ_5	7.36	7.91	8.50	7.47	7.96	8.43
κ_6	10.79	11.54	12.46	10.95	11.61	12.19
Drug	−0.09	0.48	1.00	−0.85	−0.09	0.79
Drug*time	−2.04	−1.63	−1.33	−3.49	−2.81	−2.14
Dropout				−1.44	−0.57	0.14
Dropout*time				−1.23	−0.75	−0.36
Dropout*drug				−0.10	0.89	1.94
Dropout*drug*time				0.75	1.41	2.22
Variance Intercepts	2.61	3.63	5.48	2.38	3.53	4.46
Covariance time and intercepts	−1.10	−0.46	−0.04	−1.03	−0.55	−0.10
Variance time	0.76	1.17	1.63	0.82	1.17	1.61
DIC	4278			4271		
d_e	521			510		

11.3 AUXILIARY VARIABLES

Often survey data contain auxiliary information that may provide information on (i.e. act as a surrogate for) incompletely observed predictors X or responses Y (Ibrahim *et al.*, 2001b; Horton and Laird, 2001). Auxiliary information might be stratifier variables (e.g. area of residence, demographic variables) available even in the event of complete unit non-response. Another possibility when a covariate X is expensive to measure is to obtain a proxy A for all subjects while the gold standard X is collected only for a small subsample. Consider the case of missing data on one or more predictors X. If the probability of response $\Pr(R_i = 1)$ on X is related to the possibly missing values X_i the complete selection model likelihood may be factored as

$$p(R_X, Y, A, X | \eta, \beta, \theta) = p(R_X | Y, A, X, \eta) p(Y | X, A, \beta) p(X, A | \theta) \qquad (11.4a)$$

Simplifications occur if missingness on X is at random rather than non-ignorable and if Y is conditionally independent of A given X. If the latter condition holds then $p(Y|X, A, \beta) = p(Y|X, \beta)$ and the auxiliary variable is not a significant predictor of Y when X is given. If additionally missingness in X depends only on observed data, then the model may be reduced to

$$p(R_X, Y, A, X|\beta, \theta) = p(Y, A, X|\beta, \theta) = p(Y|X, \beta)p(X, A|\theta) \qquad (11.4b)$$

Note that an alternative factorization if Y and A are not conditionally independent is (Horton and Laird, 2001)

$$p(Y, A, X|\delta, \beta, \theta) = p(A|Y, X, \delta)p(Y|X, \beta)p(X|\theta) \qquad (11.4c)$$

where $p(X|\theta)$ is usually just a model for the missing covariate(s), with

$$p(Y, A, X|\delta, \beta, \theta) = p(A|Y, X, \delta)p(Y|X, \beta)p(X_{\mathrm{mis}}|\theta, X_{\mathrm{obs}})$$

If additionally missingness on X is non-ignorable,

$$
\begin{aligned}
p(R_X, Y, A, X|\eta, \delta, \beta, \theta) &= p(R_X|Y, A, X, \eta)p(A|Y, X, \delta) \\
&\times p(Y|X, \beta)p(X_{\mathrm{mis}}|\theta, X_{\mathrm{obs}})
\end{aligned}
\qquad (11.4d)
$$

Example 11.2   Mental health service use   To illustrate the value of auxiliary data in improving the model for missing X and hence the prediction of Y (which is completely observed) consider data from Horton and Laird (2001) on a binary measure Y of mental health service use by 2486 children. Predictors include socio-economic variables: gender ($X_1 = 1$ for boy), age group ($X_2 = 1$ for older age group), ethnicity of child, namely whether black ($X_3 = 1$ or 0) or Hispanic ($X_4 = 1$ or 0), and also whether from a single-parent family ($X_5 = 1$ or 0). Also available is a parental measure of child pathology (the Child Behavior Checklist, CBCL) and a teacher measure of child pathology (the Teacher Report Form, TRF). The CBCL is considered an auxiliary variable for the TRF which is subject to 43% missingness and is considered one of the X variables ($X_6 = $ TRF). All other X variables and Y are fully observed.

Fitzmaurice *et al.* (1996) conclude that $\Pr(R = 1)$ for the variable TRF ($R = 1$ if TRF is missing, 0 otherwise) is not related to TRF, i.e. missingess is at random. However, for illustrative purposes we allow, as in (11.4d), for the possibility of non-ignorable non-response by including a logit model for R. An informative N(0, 1) prior is assumed for the impact of TRF on logit[$\Pr(R = 1)$], in line with a 95% expectation of an odds ratio between 0.14 and 7. A vague prior leads to lack of identifiability.

The remainder of the model uses the factorization as in (11.4c) that does not necessarily assume conditional independence of Y and A given X, so that conditional independence can be assessed by whether Y has a direct effect on A. This is the likelihood assumed here with the additional component $f(R_X|Y, A, X, \eta)$.

Thus the model has the following components: a model for missingness on TRF

$$R_i \sim \text{Bern}(\pi_{1i})$$

$$g_1(\pi_{1i}) = \eta_X X_{\text{obs},i} + \eta_6 \text{TFR}_i + \eta_7 \text{CBCL}_i + \eta_8 Y_i$$

where $\eta_X = (\eta_0, \dots, \eta_5)$ and $X_{\text{obs},i} = (1, X_1, X_2, X_3, X_4, X_5)'$; a model for the auxiliary variable CBCL (the parent pathology report)

$$\text{CBCL}_i \sim \text{Bern}(\pi_{2i})$$

$$g_2(\pi_{2i}) = \delta_X X_{\text{obs},i} + \delta_6 \text{TFR}_i + \delta_7 Y_i$$

a model for Y

$$Y_i \sim \text{Bern}(\pi_{3i})$$

$$g_3(\pi_{3i}) = \beta X_{\text{obs},i} + \beta_6 \text{TFR}$$

where $\beta = (\beta_0, \dots, \beta_5)$; and a model for values taken by the covariate (teacher's pathology report) subject to missingness

$$\text{TRF}_i \sim \text{Bern}(\pi_{4i})$$

$$g_4(\pi_{4i}) = \theta X_{\text{obs},i}$$

A two-chain run of 5000 iterations (convergent from 1000) shows that TRF is not a significant influence on $\Pr(R_i = 1)$ so the MAR assumption on TFR is confirmed. The coefficient on TRF in the model for Y has mean 1.32 (s.d. 0.17) which compares closely with the estimate of 1.30 obtained by Horton and Laird (2001, p 39) under their MLA model. A higher coefficient on TFR is obtained if it is erroneously assumed that Y and A are conditionally independent. The lack of conditional independence is apparent in the impact of Y in the logit model for the auxiliary variable CBCL, with a coefficient of 1.17 (s.d. 0.13).

11.4 PREDICTORS WITH MISSING VALUES

Consider data without auxiliary variables but with several covariates (as well as possibly Y also) subject to missing values. The joint density has the form

$$p(Y, X, R_Y, R_X | \eta, \beta, \theta) = p(R_X, R_Y | Y, X, \eta) p(Y | X, \beta) p(X | \theta)$$

with one possible conditional sequence giving

$$p(Y, X, R_Y, R_X | \eta, \beta, \theta) = p(R_Y | R_X, Y, X, \eta_Y)$$
$$p(R_X | Y, X, \eta_X) p(Y | X, \beta) p(X | \theta)$$

Assume for simplicity that only X values are subject to missingness so that

$$p(Y, X, R_X | \eta, \beta, \theta) = p(R_X | Y, X, \eta) p(Y | X, \beta) p(X | \theta)$$

The incompletely observed covariates $X_{i,\text{mis}} = (X_{i1}, \ldots, X_{im})$ may be categorical $\{X_{i1}, \ldots, X_{ik}\}$ and continuous $(X_{i,k+1}, \ldots, X_{im}\}$ and there is a question of specifying both the joint distribution of $X_{i,\text{mis}} = \{X_{i1}, \ldots, X_{im}\}$ and the joint density of the covariate missingness indicators $R_i = \{R_{i1}, \ldots, R_{im}\}$. The fully observed covariates are $X_{i,\text{obs}} = \{X_{i,m+1}, \ldots, X_{ip}\}$.

Ibrahim *et al.* (1999) propose the joint density of the covariates subject to missingness specified as a series of one-dimensional conditional distributions, so that

$$p(X_{i1}, \ldots, X_{im} | \theta\} = p(X_{im} | X_{i,m-1}, \ldots, X_{i1}, \theta_m\} \ldots p(X_{i2} | X_{i1}, \theta_2) p(X_{i1} | \theta_1)$$
$$(11.5a)$$

Alternative conditioning sequences may be tried as part of a sensitivity analysis. Possible approaches for modelling the R_i include a joint log-linear model for $p(R_i | Y_i, X_i, \eta)$ with $X_i = (X_{i,\text{mis}}, X_{i,\text{obs}})$ as predictors, or equivalently a multinomial model with all possible classifications of non-response as categories (Schafer, 1997, chapter 9). For example, if X_{mis} contains two variables then there are four possible combinations of R_1 and R_2. However, the joint density can also (Ibrahim *et al.*, 1999) be specified as a series of conditional distributions

$$p(R_{i1}, \ldots, R_{im} | \eta, X_i, Y_i\} = p(R_{im} | R_{i,m-1}, \ldots, R_{i1}, \eta_m, X_i, Y_i)$$
$$\ldots p(R_{i2} | R_{i1}, \eta_2, X_i, Y_i) p(R_{i1} | \eta_1, X_i, Y_i)$$
$$(11.5b)$$

What (11.5a) and (11.5b) mean in practice may be illustrated with the case of two incompletely observed continuous variables $\{X_{i1}, X_{i2}\}$, X_{i3} fully observed (continuous or binary), and two incompletely observed binary variables X_{i4}, X_{i5}. Suppose also that Y is fully observed. Suppose the joint density

$$p(X_{i2} | X_{i1}, \theta_2) p(X_{i1} | \theta_1)$$

is specified as a series of univariate normals. This is equivalent to a bivariate normal

$$p(X_{i1}, X_{i2} | \theta_1, \theta_2)$$

which is the first and second lowest stages combined in (11.5a). Conditional on partially missing $\{X_{i1}, X_{i2}\}$ and the fully observed X_{i3}, a binary regression may be used for $\pi_{4i} = \Pr(X_{i4} = 1)$ with

$$g[\pi_4(X_{i4}|X_{i1}, X_{i2}, X_{i3}, \theta_4)] = \theta_{40} + \theta_{41}X_{i1} + \theta_{42}X_{i2} + \theta_{43}X_{i3}$$

Note that it is not necessary to model the distribution of X_{i3}, since it is always observed; X_{i3} can be conditioned on in the sense that the means μ_{i1} and μ_{i2} of X_{i1} and X_{i2} are functions of X_{i3}. Finally a regression for X_{i5} would be of the form

$$g[\pi_5(X_{i5}|X_{i1}, X_{i2}, X_{i3}, X_{i4}, \theta_5)] = \theta_{50} + \theta_{51}X_{i1} + \theta_{52}X_{i2} + \theta_{53}X_{i3} + \theta_{54}X_{i4}$$

Note that other orders of conditioning are possible: one might also start with $p(X_{i4}|\theta_1)$ then model $p(X_{i5}|\theta_2, X_{i4})$ then $p(X_{i1}|X_{i5}, X_{i4}, \theta_3)$, etc. A sensitivity analysis would assess the impact of alternative sequences on the β parameters in the regression of Y on X.

For non-ignorable non-response, one allows the probability of missingness, such as $\Pr(R_{i1} = 1)$, to depend on missing values of the same variable (X_{i1}), other variables subject to missingness (X_{i2}, X_{i4}, X_{i5}), the response and fully observed covariates, as well as earlier R_{ik} in the conditional sequence. So a full model for the missingness of X_{i1} might be

$$g(\Pr[R_{i1} = 1]) = \eta_{11} + \eta_{12}X_{i1} + \eta_{13}X_{i2} + \eta_{14}X_{i3} + \eta_{15}X_{i4} + \eta_{16}X_{i5} + \eta_{17}Y_i \tag{11.5c}$$

The logit model for R_{i2} given R_{i1}, $p(R_{i2}|R_{i1}, \eta_2)$, is then

$$\begin{aligned} g(\Pr[R_{i2} = 1]) = {} & \eta_{21} + \eta_{22}X_{i1} + \eta_{23}X_{i2} + \eta_{24}X_{i3} + \eta_{25}X_{i4} \\ & + \eta_{26}X_{i5} + \eta_{27}Y_i + \eta_{28}R_{1i} \end{aligned}$$

and so on for $\Pr(R_{i4} = 1)$ conditional on R_{i1} and R_{i2}, and $\Pr(R_{i5} = 1)$ conditional on R_{i1}, R_{i2} and R_{i4}.

Note, though, that such models may be poorly identified and that parsimonious models (and/or informative priors) may be needed for identifiability in practice (Fitzmaurice *et al.*, 1996; Ibrahim *et al.*, 2001a, p 558). The usual predictor selection methods may be used to obtain parsimonious missingness models containing only significant predictors. Missingness may then be judged random or non-ignorable depending on which predictors are included in the parsimonious model. For example, if $\eta_{12} = \eta_{13} = \eta_{15} = \eta_{16} = 0$ in the model (11.5c) for R_{i1} then missingess in X_{i1} is at random (MAR) since it depends only on fully observed data. If also $\eta_{14} = \eta_{17} = 0$ then missingess in X_{i1} is completely at random (MCAR).

Example 11.3 Foreign language attainment Schafer (1997) presents data from a cross-sectional study of college achievement in foreign language ($n = 279$ subjects). The measure of success Y is an ordinal final grade, GRD, here reduced to a binary variable ($1 =$ grade A, 0 otherwise), since 126 of 232 complete responses on GRD were grade A. A new instrument for predicting success, the Foreign Language Attitude Scale or FLAS, is being compared with a well-established instrument, the Modern Language Aptitude Test (MLAT).

Of the available predictors, three continuous and one binary variable are used here. As well as the continuous $X_3 =$ FLAS, which is completely observed, the two other partially missing continuous scales are $X_1 =$ MLAT (missing for 49 subjects) and college grade point average ($X_2 =$ CGPA), missing for 34 subjects. The binary predictor X_4 is one if three or more prior language courses have been taken, and zero for two or less prior courses (this variable is missing for 11 subjects). We consider a selection model allowing for non-ignorable response in both X and Y.

As one possible conditional sequence, the first stage in the modelling of the $X_{i,\text{mis}} = (X_{i1}, X_{i2}, X_{i4})$ specifies the means μ_{i1} of X_{i1} as a function of the fully observed covariate X_{i3}. Then the means of X_{i2} are modelled in terms of X_{i1} and X_{i3}, while the logit regression for X_4 is in terms of X_1, X_2 and X_3. Thus

$$X_{i1} \sim N(\mu_{i1}, \omega_1) \qquad\qquad \mu_{i2} = \theta_{21} + \theta_{22}X_{i1} + \theta_{23}X_{i3}$$
$$\mu_{i1} = \theta_{11} + \theta_{12}X_{i3} \qquad\qquad X_{i4} \sim \text{Bern}(\pi_{i4})$$
$$X_{i2} \sim N(\mu_{i2}, \omega_2) \qquad\qquad g(\pi_{i4}) = \theta_{41} + \theta_{42}X_{i1} + \theta_{43}X_{i2} + \theta_{44}X_{i3}$$

Y and three of the four covariates, namely $\{X_1, X_2, X_4\}$, are subject to missing values. The conditional sequence assumed for the missingness indicators in both Y and X is

$$p(R_{iY}|R_{i1}, R_{i2}, R_{i4}, X_i, Y_i) p(R_{i4}|R_{i2}, R_{i1}, X_i, Y_i)$$
$$\times\, p(R_{i2}|R_{i1}, X_i, Y_i) p(R_{i1}|X_i, Y_i)$$

For example, the most extensive model for missingess on the response, namely for $\pi_{iY} = \Pr(R_{iY} = 1)$, is

$$g_Y(\pi_{iY}) = \eta_{Y1} + \eta_{Y2}Y_i + \eta_{Y3}X_{i1} + \eta_{Y4}X_{i2} + \eta_{Y5}X_{i3} + \eta_{Y6}X_{i4}$$
$$+ \eta_{Y7}R_{i1} + \eta_{Y8}R_{i2} + \eta_{Y9}R_{i4}$$

while the model for $\pi_{i4} = \Pr(R_{i4} = 1)$ is

$$g_4(\pi_{i4}) = \eta_{41} + \eta_{42}Y_i + \eta_{43}X_{i1} + \eta_{44}X_{i2} + \eta_{45}X_{i3} + \eta_{46}X_{i4} + \eta_{47}R_{i1} + \eta_{48}R_{i2}$$

The coefficients on η_{Y2} and η_{46} (i.e. where missingness is related to possibly missing values for the same binary variable) are taken as relatively informative, namely N(0,1), in order to achieve stable convergence. Even under these priors, convergence is relatively slow, after around 2000 iterations with two chains, with the coefficients in the model for R_{i1} showing delayed convergence.

Using 3000 more iterations, results for β in the central model (of ultimate substantive interest) show linguistic success positively related to both language aptitude scales, FLAS and MLAT, and also to college grade, but not to number of prior courses. The coefficients and 95% intervals for this model are given in Table 11.3.

Table 11.3 FLAS main model, posterior summary

	Mean	S.d.	2.5%	97.5%
β_1 (constant)	−9.0	1.95	−13.4	−5.6
β_2 (MLAT)	0.11	0.03	0.05	0.17
β_3 (CGPA)	1.07	0.45	0.28	2.02
β_4 (FLAS)	0.034	0.011	0.013	0.059
β_5 (Prior courses)	0.29	0.31	−0.30	0.89

Results from the logit regression for $\Pr(R_{i1} = 1)$ suggest that missingess on $X_1 = $ MLAT is informative: missingness is lower as MLAT increases, the relevant coefficient (eta1[4] in the code) having mean −0.57 and 95% interval $(−1.45, −0.19)$. Missingness on X_1 is also positively linked to Y itself. By contrast, missingness on $X_2 = $ CGPA is positively related to X_2 and negatively to Y. Missingess on X_4 and Y appears to be at random though the models are evidently overparameterized: there is only one coefficient (on X_1) in the model for $\Pr(R_{iY} = 1)$ that has a 95% interval confined to negative or to positive values.

Accordingly a second analysis includes regressor selection in the model for $\Pr(R_{iY} = 1)$. With prior probabilities of regressor inclusion of 0.5, two variables have posterior probabilities of inclusion above 0.5, namely Y and X_1. A reduced model just including these predictors for R_Y (and without predictor selection) still shows X_1 as the only significant predictor, so the missingness on Y is confirmed as MAR.

11.5 MULTIPLE IMPUTATION

The multiple imputation approach to missing data (see e.g. Faris *et al.*, 2002; Rubin, 1987) applies whether missingess is ignorable or non-ignorable, though is most typically applied in conjunction with an MAR assumption. Suppose X is subject to missingness. Under a MAR model, a small number m of samples of the missing data $X_{\text{mis},i,j}$ $j = 1, \ldots, m$, $i = 1, \ldots, n$, are drawn from the predictive density

$$P(X_{\text{mis}}|X_{\text{obs}}) = \int p(X_{\text{mis}}|X_{\text{obs}}, \gamma)p(\gamma|X_{\text{obs}})\mathrm{d}\gamma$$

After filling in the missing values m complete data sets are obtained. These data sets may be analysed by Bayesian or classical methods to yield separate estimates $\beta^{(1)}, \ldots, \beta^{(m)}$ of the parameters β involved in the $P(Y|X, \beta)$ model. The complete data analysis might be a logistic regression (Y on predictors X) whereas the imputation model might involve imputation of missing continuous predictors X_{mis} which figure among those in a logistic regression of Y on $X = (X_{\text{mis}}, X_{\text{obs}})$.

In the case of m separate Bayesian analyses, $\beta^{(1)}, \ldots, \beta^{(m)}$ are obtained as posterior means. Suppose also that the posterior variances of the kth parameter in each set $\{\beta_k^{(1)}, \ldots, \beta_k^{(m)}\}$ are $V_{1k}, \ldots, V_{mk}$ respectively. Then a combined estimate of the relevant regression coefficient is simply the average $\bar{\beta}_k$ over the m analyses. The average within-imputation variance of $\bar{\beta}_k$ is estimated as

$$\bar{V}_k = \sum_{j=1}^{m} V_{jk}/m$$

while the estimated between-imputation variance is

$$B_k = \sum_{j=1}^{m} (\beta_k^{(m)} - \bar{\beta}_k)^2/(m-1)$$

So the estimated total variance of the combined estimate $\bar{\beta}_k$ is

$$B_k(1 + 1/m) + \bar{V}_k \tag{11.6}$$

With several continuous X variables, one or more subject to incomplete observation, a commonly used imputation model is the multivariate normal under an MAR assumption (Schafer, 1997). This might be used after suitably transforming some of the original variables to reduce skewness. With several discrete variables one might use the same principle within the link function (Aitchison and Ho, 1989); for example, for J count variables Y_{ij}, take

$$Y_{ij} \sim \text{Po}(\exp(u_{ij}))$$

where

$$u_i \sim \text{N}_J(\mu_u, \Sigma_u)$$

This type of imputation model may be most appropriate when there is a relatively small fraction of missing data, and when between-variable correlations outweigh within-variable correlations such as may occur in panel data. When within-variable correlation is important the imputation method should include correlation in the observed data or error series (e.g. autocorrelation in the outcomes Y_{it} in a panel data situation).

A specific form of multiple imputation is based on the propensity score approach (Lavori *et al.*, 1995). At its simplest this involves predicting 'scores' $e_i = \Pr(R_i = 1)$ using a logistic regression on fully observed variables. Suppose $R_i = 1$ for a subject with X_2 missing and $R_i = 0$ with X_2 observed; suppose also that X_1 is fully observed and assists in predicting $\Pr(R_i = 1)$ in a logistic regression.

Then one would stratify subjects on the basis of the scores e_i. For example, split the sample into $g = 1, \ldots, G$ groups according to the quartiles of e_i. Then within each of those G groups suppose there were R_g respondents and Q_g non-respondents. In line with the 'approximate Bayesian bootstrap procedure' of Rubin (1987) one randomly selects Q_g values of X_2 (with replacement) from among those R_g subjects with X_2 observed. The sampled values are imputed to the non-respondent subjects in the same propensity score group. This would be repeated in line with MI principles to provide m copies of the 'complete' data.

In panel studies permanent loss of subjects is common (i.e. attrition or monotone missing data). For example, suppose there are two variables X_{i1t} and X_{i2t} and that waves 1 and 2 are fully observed but at wave 3 some subjects drop out. Then one might predict $e_{i3} = \Pr(R_{i3} = 1)$ using the fully observed data (X_{i1t}, X_{i2t}) for $t = 1, 2$ and impute missing scores for both predictors (X_{i13} and X_{i23}) according to the propensities e_{i3}. Define X_{3m}^C as the complete data resulting from the mth imputation. Then a regression to predict the missingness scores $e_{i4} = \Pr(R_{i4} = 1)$ at wave 4 might be based on the fully observed data at times 1 and 2 combined with the average 'complete' data $\bar{X}_3^C$ at time 3.

Example 11.4 Attitude to low-tar cigarettes This example uses data from Schafer (1999) on answers to the question 'The new low-tar cigarettes aren't going to hurt me' before and after a behavioural intervention; binary data for 882 subjects form responses y_1 (before) and y_2 (after). Data are further classified by grade 6 vs. grade 8 (see Table 11.4). There are no completely missing responses (non-response at both waves) but partial missingness only (wave 1 or wave 2). Effectiveness of the intervention may be gauged by reduction in the percentage agreeing with the question in the experimental group as compared with the control group.

Table 11.4 Success of behavioral intervention

Experimental group Agree before	Agree after Grade 6				Grade 8			
	N	Y	NA	Total	N	Y	NA	Total
N	61	18	4	83	91	12	9	112
Y	55	70	14	139	35	19	3	57
NA	12	20	0	32	28	9	0	37
Grand total	128	108	18	254	154	40	12	206

Control group Agree before	Agree after Grade 6				Grade 8			
	N	Y	NA	Total	N	Y	NA	Total
N	69	16	7	92	100	23	11	134
Y	13	18	1	32	26	34	0	60
NA	24	12	0	36	37	31	0	68
Grand total	106	46	8	160	163	88	11	262

The analysis is carried out for individuals and rather than multiple imputation (e.g. selecting a small number of sets of infill values for $y_{\text{mis},1}$ and $y_{\text{mis},2}$) we use MAR imputation using known grades and intervention groups, on the basis of repeated MCMC sampling. Let $G_i = 1$ for experimental group, $G_i = 2$ for control group and $C_i = 1$ for grade 6, and $C_i = 2$ for grade 8. The model is

$$y_{ij} \sim \text{Bern}(\pi_{ij}) \qquad j = 1, 2$$
$$\text{logit}(\pi_{i1}) = \beta_0 + \beta_1 \delta(C_i = 1) + \beta_2 \delta(G_i = 1) + \beta_3 \delta(C_i = 1, G_i = 1)$$
$$\text{logit}(\pi_{i2}) = \gamma_0 + \gamma_1 \delta(C_i = 1) + \gamma_2 \delta(G_i = 1) + \gamma_3 \delta(C_i = 1, G_i = 1)$$

where $\delta(A) = 1$ if A is true. It would also be possible to include y_{i1} in the model for π_{i2}. A non-ignorable model would include missingness models $p(R_1|X, Y_1)$ and $p(R_2|X, Y_1, Y_2)$ where $X = (C, G)$. The percentage reductions Δ_j for experiment group grades 6 and 8 under the model used are -0.169 (0.044) and -0.127 (0.045), while for control group grades 6 and 8 they are 0.046 (0.053) and 0.035 (0.043).

Schafer (1999) takes four sets of multiply imputed data, and bases inferences on Δ_j on the four 'complete' data sets, namely $Y_{\text{comp},1} = (Y_{\text{obs}}, Y_{\text{mis},1})$, $Y_{\text{comp},2} = (Y_{\text{obs}}, Y_{\text{mis},2})$, $Y_{\text{comp},3} = (Y_{\text{obs}}, Y_{\text{mis},3})$ and $Y_{\text{comp},4} = (Y_{\text{obs}},$

$Y_{\text{mis},4}$). The estimated Δ_j, obtained by averaging over four complete data analyses and using (11.6) for estimating combined precision, are -0.173 (0.039), -0.152 (0.039), 0.041 (0.046) and 0.028 (0.040).

11.6 SEVERAL RESPONSES WITH MISSING VALUES

One may envisage adapting the strategies outlined in sections 11.2–11.5 to multivariate or panel responses. For a repeated univariate or multi-variate response Y_{it} at times t one may extend the non-ignorable model in (11.2) to include several lags as well as contemporary effects, but such a model may be subject to collinearity. With multivariate panel responses Y_{itm} for outcomes $m = 1, \ldots, M$ one conceivably should allow for cross-lags and cross-contemporary effects between variables. For example, for $M = 3$, a non-ignorable missingess model might imply both contemporary effects on $\Pr(R_{itm} = 1)$ of each of $\{Y_{it1}, Y_{it2}, Y_{it3}\}$ and lagged effects of each of $\{Y_{i,t-1,1}, Y_{i,t-1,2}, Y_{i,t-1,3}\}$ giving 21 parameters in addition to the intercepts.

Alternatively, as in Roy and Lin (2002), one might propose $Q < M$ more underlying factors $\{F_{it1}, F_{it2}, \ldots, F_{itQ}\}$ both to model the correlation between the observations Y_{itm} and to include in a less heavily parameterized missingness model. Thus for metric or discrete outcomes Y_{itm} from an exponential family density with known scale, and $Q = 1$, one may set

$$R_{itm} \sim \text{Bern}(\pi_{itm})$$
$$g_R(\pi_{itm}) = \eta_{m1} + \eta_{m2}F_{it} \tag{11.7a}$$

$$Y_{itm} \sim \text{EF}(\mu_{itm})$$
$$g_Y(\mu_{itm}) = \beta_{0m} + \beta_{1m}F_{it} + b_{im} \tag{11.7b}$$

where b_{im} are permanent effects in the mth variable and where the factors are defined in terms of fixed effects applied to a $p \times 1$ covariate vector X_{it} and random subject effects applied to a $q \times 1$ covariate vector Z_{it}. For example,

$$F_{it} = X_{it}'\gamma_t + Z_{it}'\delta_i + v_{it} \tag{11.7c}$$

where γ_t is $p \times 1$; the δ_i might be multivariate normal of order q. If the Z_{it} and X_{it} are coincident then the means of δ_i are zero. For identifiability $v_{it} \sim \text{N}(0, 1)$ and X_{it} and Z_{it} should not include a constant if there are already constants in (11.7b).

The missingness model (11.7a) may additionally involve lagged F_{it} as well as lagged and contemporary covariates. Note that missingness is likely to be at unit (subject) level, either attrition or intermittent, and consist of total missingness at a given t for subject i on all responses Y_{itm}. Thus rather than variable-specific missingness one may consider a global missingness indicator $R_{it} = 1$ if at wave t subject i drops out or is missing, and $R_{it} = 0$ if the subject stays under observation or is subsequently observed. Then a non-ignorable model for $\pi_{it} = \Pr(R_{it} = 1)$ and a single covariate might be

$$g_R(\pi_{it}) = \eta_0 + \eta_1 F_{it} + \eta_2 F_{i,t-1} + \eta_3 X_{i,t-1} \qquad (11.7d)$$

Example 11.5 considers this approach for intermittent loss to observation rather than permanent dropout, including intermittent item rather than unit non-response.

Example 11.5 Diabetic patient biochemical markers To illustrate the use of latent (factor) variables in modelling missingness (Roy and Lin, 2002) consider data from Crowder and Hand (1990, Table 2.1) on the evolution of two biochemical markers in 27 patients. $T = 12$ measures are obtained on cyclic guanosine monophosphate ($Y_1 = $ cyclic GMP) and cyclic adenosine monophosphate ($Y_2 = $ cyclic AMP) in the patients. These are converted to ordinal scales, with $J_1 = 5$ levels on GMP (under 5, 5–9.99, 10–14.99, 15–19.99, over 20) and $J_2 = 6$ levels on AMP (under 6, 6–11.99, 12–17.99, 18–23.99, 24–29.99, 30+). The patients are classified into four groups, denoted by subject indicators G_i, namely 1 = normal control; 2 = control diabetic; 3 = diabetic with hypertension; 4 = diabetic with postural hypertension.

Missingness in these data is mostly intermittent rather than permanent (only subjects 10 and 27 fit the 'lost to follow-up' description). Also, while most of this missingness is at subject level (i.e. when Y_{it1} is missing so also is Y_{it2}) there are instances of item non-response (e.g. $Y_{19,1,2}$ is missing whereas $Y_{19,1,1}$ is not).

In the first model for these data the underlying means of the ordinal variables, $Y_{itm}(m = 1, 2; i = 1, \ldots, 27; t = 1, \ldots, 12)$, are explained by the factor scores F_{it} and constant factor loadings β_m. So for responses $m = 1, 2$ and with $\pi_{itm} = (\pi_{itm1}, \ldots, \pi_{itmJ_m})$

$$Y_{itm} \sim \text{Categorical}(\pi_{itm})$$

$$\pi_{itmk} = \gamma_{itmk} - \gamma_{itm,k-1} \qquad k = 2, \ldots, J_m - 1$$

$$\text{logit}(\gamma_{itmk}) = \kappa_{mk} - \beta_m F_{it}$$

$$F_{it} \sim \text{N}(\mu_{it}, 1)$$

$$\mu_{it} = \gamma_{G_i} + \delta_i$$

$$\delta_i \sim \text{N}(0, \sigma_\delta^2)$$

where the γ coefficients (for patient group) are subject to a corner constraint and an $N(1,1)$ prior assumed for the second factor loading β_2 (cf. Johnson and Albert, 1999, p 197), with β_1 preset as one. The latter constraint improves identifiability and convergence, and means that the F_{it} describe change at individual level in a latent factor underlying high scores in both GMP and AMP. Additionally, for identification of δ_i the first cut point κ_{m1} is set to zero. The missingness mechanism involves the F_{it} as follows:

$$R_{itm} \sim \text{Bern}(\pi_{itm})$$
$$\text{logit}(\pi_{itm}) = \eta_{0m} + \eta_{1m}F_{it}$$

A two-chain run of 5000 iterations (convergent from 500) gives a mean β_2 of 0.66 (95% interval from 0.50 to 0.86) so the underlying factor F_{it} increases with both Y_1 and Y_2. The δ scores range from -3.2 (subject 19, who belongs to the lowest category of Y_1 and Y_2 on most readings) to 5.5 (for subject 26 who records especially high scores on Y_2). There is some evidence that missingness is non-ignorable in that η_{11} is biased towards negative values with mean -0.38 and 95% interval from -0.73 to -0.10: patients with lower GMP readings tend to have higher rates of missing data. The DIC is estimated at 1633 with $d_e = 114$.

A second model introduces correlated subject–variable intercept effects b_{im} and time-specific loadings β_{2t} (with $\beta_{1t} = 1$ for identification). This is an example of a permanent subject effect in a multivariate setting (see section 10.8). Thus

$$\text{logit}(\gamma_{itmk}) = \kappa_{mk} - \beta_{mt}F_{it} - b_{im}$$

This model considerably improves fit (DIC $= 1363$) at the cost of a slight increase in d_e to 135. Imputations under the two models are similar (Table 11.5). The coefficient η_{11} is still negative but with with mean -0.26 and 95% interval from -0.54 to -0.03.

11.7   NON-IGNORABLE NON-RESPONSE MODELS FOR SURVEY TABULATIONS

Survey or census tabulations often include large proportions of non-response. Bayesian models have been proposed for different types of missing data situation in such tabulations. These include differential non-response among survey subgroups (Stasny, 1991), partially observed $R \times C$ cross-tabulations (Rosen *et al.*, 2001), and missingness on one or more factors in a multidimensional tabulation.

Table 11.5 Diabetic markers, imputations of missing ordinal data

Missing datum	Model 1			Model 2		
	2.5%	Median	97.5%	Mean	2.5%	97.5%
$Y_{1,9,1}$	1	2	4	1	2	4
$Y_{1,11,1}$	1	2	4	1	2	5
$Y_{2,11,1}$	2	3	5	1	3	5
$Y_{3,11,1}$	2	3	5	1	3	5
$Y_{4,11,1}$	2	3	5	1	3	5
$Y_{10,11,1}$	1	2	4	1	2	4
$Y_{10,12,1}$	1	2	4	1	2	4
$Y_{12,11,1}$	1	2	4	1	2	5
$Y_{13,11,1}$	1	2	3	1	2	4
$Y_{16,1,1}$	1	1	1	1	1	2
$Y_{23,11,1}$	1	1	3	1	2	3
$Y_{25,5,1}$	1	2	3	1	2	3
$Y_{27,11,1}$	1	1	1	1	1	2
$Y_{27,12,1}$	1	1	1	1	1	2
$Y_{1,9,2}$	1	3	6	1	3	5
$Y_{1,11,2}$	1	3	6	1	3	6
$Y_{2,11,2}$	2	3	6	2	4	6
$Y_{3,9,2}$	2	3	6	1	3	5
$Y_{3,11,2}$	2	3	6	1	3	6
$Y_{4,11,2}$	1	3	6	1	2	4
$Y_{8,11,2}$	1	2	4	1	3	5
$Y_{10,11,2}$	1	3	5	1	3	6
$Y_{10,12,2}$	1	3	5	1	3	6
$Y_{12,11,2}$	1	3	6	1	2	4
$Y_{13,11,2}$	1	3	5	1	3	6
$Y_{14,1,2}$	1	2	3	1	1	3
$Y_{16,1,2}$	1	1	3	1	1	3
$Y_{19,1,2}$	1	1	3	1	1	3
$Y_{23,11,2}$	1	2	4	1	2	4
$Y_{25,5,2}$	1	2	5	1	2	4
$Y_{27,11,2}$	1	1	3	1	1	3
$Y_{27,12,2}$	1	1	3	1	1	3

11.7.1 The differential non-response model

In one scenario, the tabulated outcome is presented for subgroups, defined possibly by combinations of fully observed categorical variables such as age group and gender (as in Park and Brown, 1994), or by known survey stratifiers such as urban or rural residence (as in Stasny, 1991). Non-response rates will vary according to subgroup. Under a hierarchical model one may pool over subgroups via exchangeable priors to improve both the estimates of probabilities (binomial, multinomial) of the outcome and the estimated non-response probabilities in separate subgroups. The form of hierarchical model for the non-response probabilities is defined by the form of missingess. Taking non-response probabilities to come from a single distribution, regardless of the possibly missing outcome, is consonant with missingness at random. By contrast a non-ignorable missingness model involves allowing differential non-response according to the category of the outcome (Little and Gelman, 1999).

Suppose that the survey outcome is binary and that the tabulation is in terms of $k = 1, \ldots, K$ groups defined by observed subject variables or by survey stratifiers. All subjects in subgroup k are taken to have the same probability $p_{ik} = p_k$ of the outcome, with prior density for p_k a beta distribution with parameters m_1 and m_2.

Suppose (following a selection model approach) that the chances of non-response in group k are affected by the occurrence or otherwise of the binary survey outcome. Let $\pi_{ik1} = \pi_{k1} = \Pr(R_{ik} = 1 | Y_{ik} = 1)$ denote the probability that subject i in group k is a non-respondent given that he/she is positive on the outcome. The probability that subject i in group k is a non-responder given that he/she is negative on the outcome is $\pi_{ik0} = \pi_{k0} = \Pr(R_{ik} = 1 | Y_{ik} = 0)$. The total probability of non-response (i.e. that $R_{ik} = 1$) involves the possible combinations of outcome $\Pr(Y_{ik})$ and of non-response conditional on outcome:

$$\begin{aligned} \Pr(R_{ik} = 1) &= \Pr(R_{ik} = 1 | Y_{ik} = 1)\Pr(Y_{ik} = 1) \\ &\quad + \Pr(R_{ik} = 1 | Y_{ik} = 0)\Pr(Y_{ik} = 0) \qquad (11.8a) \\ &= \pi_{k1}p_k + \pi_{k0}(1 - p_k) \end{aligned}$$

For the conditional non-response probabilities one may adopt beta priors

$$\pi_{k0} \sim \text{Beta}(\alpha_0, \beta_0) \quad \pi_{k1} \sim \text{Beta}(\alpha_1, \beta_1)$$

with a default choice being $\alpha_0 = \beta_0 = \alpha_1 = \beta_1 = 1$.

Suppose there are U_k non-respondents in the kth group to the binary outcome, with likelihood element as in (11.8a). Further suppose there are S_k respondents with an observed positive outcome $(Y = 1)$ and T_k with an observed negative outcome $(Y = 0)$. The likelihood elements for the latter two responder groups are respectively

$$\Pr(R_{ik} = 0 | Y_{ik} = 1)\Pr(Y_{ik} = 1) = (1 - \pi_{k1})p_k \qquad (11.8b)$$

and

$$\Pr(R_{ik} = 0 | Y_{ik} = 0)\Pr(Y_{ik} = 0) = (1 - \pi_{k0})(1 - p_k) \qquad (11.8c)$$

So the total likelihood involves terms (11.8a) to (11.8c) – see Stasny (1991) for an empirical Bayes approach to this model. Congdon (2001) outlines a fully Bayes approach to this situation and extensive modelling options are considered by Nandram and Choi (2002a; 2002b). The crime data used by Stasny are in Table 11.6. The domains are based on urban vs.

Table 11.6 National crime survey data

Domain	Responders reporting victimization (S_k)	Responders crime free (T_k)	Non-responders (U_k)	Total sample
UCL	156	555	104	815
UCH	95	364	73	532
UIL	162	557	101	820
UIH	72	262	36	370
UNL	92	297	79	468
UNH	15	40	9	64
RIL	11	36	7	54
RIH	10	105	20	135
RNL	35	274	32	341
RNH	79	413	64	556

rural neighbourhood (U vs. R), city, other incorporated or not corporated place (C/I/N), and high or low neighbourhood poverty (H/L). Combinations of rural and city are excluded, so that only 10 of the 12 factor combinations actually exist.

The estimates of the outcome probabilities p_k may be distorted by differential missingness between the groups. To model the total number of positives on Y one may introduce, as a latent variable, the number V_k among the total non-responders U_k who are positive on Y, leaving $U_k - V_k$ non-responders who have $Y = 0$. Then from (11.8a),

$$V_k \sim \text{Bin}(U_k, \rho_k) \qquad (11.9a)$$

with

$$\rho_k = \pi_{k1} p_k / [\pi_{k0}(1 - p_k) + \pi_{k1} p_k]$$

From the expression of the likelihood involving (11.8a)–(11.8c) one may obtain the full conditional densities of the outcome rate and the conditional response probabilities as

$$p_k | V_k, \pi_{k0}, \pi_{k1} \sim \text{Beta}(V_k + S_k + m_1, T_k + U_k - V_k + m_2) \qquad (11.9\text{b})$$

$$\pi_{k1} | p_k, V_k \sim \text{Beta}(V_k + \alpha_1, S_k + \beta_1) \qquad (11.9\text{c})$$

$$\pi_{k0} | p_k, V_k \sim \text{Beta}(U_k - V_k + \alpha_0, T_k + \beta_0) \qquad (11.9\text{d})$$

The above model is for non-ignorable response. By contrast, under MAR one takes the missingness probabilities independent of outcome that $\pi_{k0} = \pi_{k1} = \pi_k$. With a $\text{B}(\alpha, \beta)$ prior on these probabilities their full conditional density is

$$\pi_k | p_k, V_k \sim \text{Beta}(U_k + \alpha, S_k + T_k + \beta)$$

When Y is multinomial (with $J > 2$ categories) rather than binomial, one may modify the non-ignorable approach so that

$$p_k \sim \text{Dirch}(A_{k1}, \ldots, A_{kJ})$$

where $p_k = (p_{k1}, \ldots, p_{kJ})$ and $A_{kj} = V_{kj} + S_{kj} + m_j$ and non-response probabilities are updated according to

$$\pi_{kj} \sim \text{Beta}(V_{kj} + \alpha_j, S_{kj} + \beta_j) \qquad (11.10\text{a})$$

The stratum-specific non-respondents who are latent positives on Y are updated according to

$$V_k \sim \text{Mult}(U_k, \rho_k) \qquad (11.10\text{b})$$

where $V_k = (V_{k1}, \ldots, V_{kJ})$, $\rho_k = (\rho_{k1}, \ldots, \rho_{kJ})$ and $\rho_{kj} = \pi_{kj} p_{kj} / \sum_j \pi_{kj} p_{kj}$.

11.7.2 Alternative parameterizations of the differential non-response model

Little and Gelman (1999) and Nandram and Choi (2002a) consider reparameterizations of the differential non-response model when the outcome is binary. For groups $k = 1, \ldots, K$, Little and Gelman (1999) consider the ratios

$$R_k = \pi_{k1} / (\pi_{k1} + \pi_{k0})$$

together with the overall non-response rate

$$\pi_k = (\pi_{k0} + \pi_{k1})/2$$

so that $\pi_{k1} = 2R_k \pi_k$, $\pi_{k0} = 2(1 - R_k)\pi_k$. Thus the parameters $\{\pi_{k1}, \pi_{k0}, p_k\}$ are replaced by $\{\pi_k, R_k, p_k\}$. Setting a prior on the R_k, for instance a beta with mean 0.5, is consonant with a non-ignorable missingness model. Nandram and Choi (2002a) adopt a similar model with

$$\pi_{k1} = \gamma_k \pi_{k0} \tag{11.11a}$$

though their model is set out in terms of probabilities of response by group rather than non-response.

To assist in identification when the data may contain little information on conditional non-response, Little and Gelman argue that in most surveys the R_k should vary less than the p_k. They advocate drawing on previous surveys to assess prior variability in the R_k. Nandram and Choi (2002a) suggest the uniform shrinkage prior of Albert (1988) for π_{k1} and π_{k0} in order to stabilize their estimation. They adopt a parameterization of the beta $Be(\alpha, \beta)$ in terms of $\alpha = \tau\mu$, $\beta = \tau(1 - \mu)$ so that τ and μ are ortho-gonalized. Then the full conditional densities for p_k and π_{k0} may be parameterized as

$$p_k | V_k, \pi_{k0}, \pi_{k1} \sim \text{Beta}(V_k + S_k + \tau_1 \mu_1, T_k + U_k - V_k + \tau_1(1 - \mu_1)) \tag{11.11b}$$

$$\pi_{k0} | p_k, V_k \sim \text{Beta}(U_k - V_k + \tau_2 \mu_2, T_k + \tau_2(1 - \mu_2)) \tag{11.11c}$$

The shrinkage prior is

$$\tau_j = 1/Z_j^{0.5} - 1$$

where $Z_j \sim U(0, 1)$, so that values on the boundary of the parameter space are discouraged. For $\mu_j(j = 1, 2)$ a $Be(1,1)$ prior is assumed, while for the scaling factors γ_k in (11.11a) the shrinkage prior is adopted again. Specifically a truncated gamma, namely

$$\gamma_k \sim \text{Ga}(\nu, \nu)I(\,, 1/\pi_{i0})$$

is used, to ensure that π_{i1} is bounded above by one.

Example 11.6   National crime survey   Consider the Stasny (1991) data in Table 11.6. These data show not only varying victmization rates (among respondents) but non-response differing by stratum, with response rates apparently higher in lower crime areas. One therefore seeks stratum-specific and global estimates of victimization that adjust for non-response. There is a subsidiary goal via (11.9b) to pool strength over domains and stabilize estimates in small domains, such as RIL and UNH.

The empirical Bayes method used by Stasny gives estimates of π_{k0} in (11.9d) that are constant over domains ($\pi_{k0} = 93.7\%$ for all k), while the π_{k1} vary slightly from 67.9% to 69.4%; adjusted victimization rates vary from 16.6% to 30.5%. To be consistent with Stasny (1991) the rates π_{k1} and π_{k0} in Program 11.6 relate to response rather than non-response.

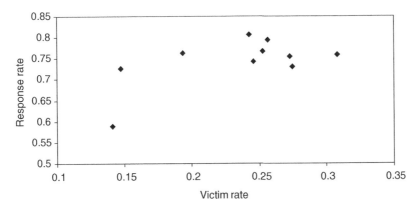

Figure 11.1 Response rate among victims

The fully Bayes conditional updating densities in (11.9) show slightly less smoothing of the victimization rates than under the EB model, with a range from 14.4% in RIH to 30.5% in UNH. The response rates π_{k0} among the crime free vary from 88% to 94% and among the victimized from 59% in RIH to 82% in UIH. Figures 11.1 and 11.2 show the relationships between p_k and π_{k1} and between p_k and π_{k0} respectively: Figure 11.2 suggests a negative relationship between p_k and π_{k0}.

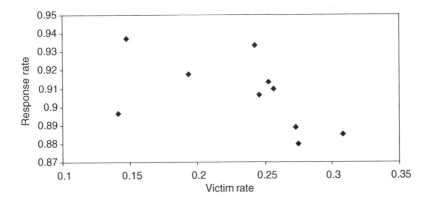

Figure 11.2 Response rate among crime free

11.7.3 Ecological inference: imputing missing interior data in R × C tables

Rosen *et al.* (2001) and King *et al.* (1999) consider a situation that often occurs in political science, i.e. inference about the cells of a population cross-tabulation from information on marginal totals. Since the cells within the cross-tabulations provide more information on individual behaviour than the marginal totals, they can be seen as relevant to the ecological inference problem (i.e. of inferring individual behaviour from aggregate data). The analysis of the binomial case when the marginal variables have only two categories (e.g. the observations are the marginal totals of a 2×2 table when there are two variables) shows overlaps with the differential response model of section 11.7.1. Consider observations for a set of electoral areas on voting vs. ethnicity as follows: the total electorate E_i eligible to vote (the grand total in the table), numbers actually voting V_i vs. those not voting $E_i - V_i$, and percentages black x_i and white, namely $(1 - x_i)$, from census data. Equivalently, numbers of black and white are $E_i x_i$ and $E_i (1 - x_i)$. Then the probability of voting θ_i in area i can be written as

$$\Pr(\text{vote}) = \Pr(\text{vote}|\text{black})\Pr(\text{black}) + \Pr(\text{vote}|\text{white})\Pr(\text{white})$$

or

$$\theta_i = \pi_i^b x_i + \pi_i^w (1 - x_i)$$

where π_i^b and π_i^w are unknown probabilities from the underlying 2×2 cross-tabulation. Then the observed data have model

$$V_i \sim \text{Bin}(\theta_i, E_i)$$

while the latent numbers of voting blacks and voting whites are sampled as

$$V_i^b \sim \text{Bin}(\rho_i, V_i)$$
$$V_i^w \sim \text{Bin}(1 - \rho_i, V_i)$$

where $\rho_i = \pi_i^b x_i / \theta_i$. The unknown percentages are updated as

$$\pi_i^b \sim \text{Beta}(V_i^b + a_b, V_i^w + b_b)$$
$$\pi_i^w \sim \text{Beta}(V_i^w + a_w, V_i^b + b_w)$$

where $\{a_b, b_b, a_w, b_w\}$ may themselves be assigned priors. Thus King et al. (1999, p 72) use E(2) priors for these parameters.

In the more general case each marginal of the table can have more than two categories. For example, for each of $i = 1, \ldots, N$ electoral regions, two variables are observed from different sources and cross-tabulation is not possible. Thus the numbers t_{ic} voting for parties $c = 1, \ldots, C$ are provided by electoral returns, while fractions x_{ir} of the voting age population who are in social classes or ethnic groups $r = 1, \ldots, R$ are from the census. The interest is in unobserved quantities such as the proportions π_{irc} of people in social class r and area i who vote for different parties c.

Assume that one or more variables Z_{ij} are available for each region that are relevant to the voting choice (e.g. local unemployment rates). Missing data for sets of areas may also be explained in part by their spatial structure in terms of adjacency or area centroids (Haneuse and Wakefield, 2004). Then the sampling model for the observed data is

$$t_{i,1:C} \sim \text{Mult}(V_i, \theta_{i,1:C})$$

where V_i is the total of voters, and taking the x_{ir} as known constants

$$\theta_{ic} = \sum_{r=1}^{R} \pi_{irc} x_{ir}$$

A Dirichlet prior on the π_{irc} is assumed with parameters that may involve a regression on relevant covariates. With one such covariate, the Dirichlet weights are

$$\alpha_{ir1} = d_r \exp(\gamma_{r1} + \beta_{r1} Z_i)$$
$$\alpha_{ir2} = d_r \exp(\gamma_{r2} + \beta_{r2} Z_i)$$

$$\vdots$$

$$\alpha_{ir,C-1} = d_r \exp(\gamma_{r,C-1} + \beta_{r,C-1} Z_i)$$
$$\alpha_{irC} = d_r$$

The parameters d_r will typically be assigned gamma or exponential priors.

The work of Haneuse and Wakefield (2004) focuses on 2×2 tables for a set of constituencies and on the marginal totals, namely Democrat and Republican votes (columns) and black vs. white voters (rows). Only the marginal totals are known (and taken from different sources). Letting x_i be the percentage black in area i, the probability θ_i of voting Republican can be written

$$\text{Pr(vote Republican)} = \text{Pr(vote Republican|black)Pr(black)}$$
$$+ \text{Pr(vote Republican|white)Pr(white)}$$

or

$$\theta_i = \pi_i^b x_i + \pi_i^w (1 - x_i)$$

where, as above, π_i^b and π_i^w are unknown race-specific probabilities of voting Republican from the underlying 2×2 cross-tabulation. Haneuse and Wakefield follow King *et al.* (1999) in taking the x_i as known constants (not stochastic); this assists in identification of the unknown π_i probabilities. They estimate these probabilities using the spatial structure of the areas in a mixed model (see Chapter 8):

$$\text{logit}(\pi_i^{\text{w}}) = \mu_{\text{w}} + u_{\text{w}i} + s_{\text{w}i}$$
$$\text{logit}(\pi_i^{\text{b}}) = \mu_{\text{b}} + u_{\text{b}i} + s_{\text{b}i}$$

where u are unstructured and s are spatial errors, e.g. $s_i \sim \text{ICAR}(1)$.

Example 11.7   Simulated voting by social class   As an example of this framework we generate data for $N = 150$ areas, $R = 3$ social classes and $C = 3$ political parties. This example follows some of the aspects set out in an example of Rosen *et al.* (2001) who were interested in assessing the proportions of different social classes voting for the National Socialists in the 1932 German elections. Specifically a covariate for each area is generated as N(0,1) and 'important covariate' coefficients β_{rc} follow those in Rosen *et al.* (2001); their substantive interest was in the effect of area unemployment Z on the pattern of voting by class.

Total voters V_i are generated from a Poisson–gamma mixture with mean 10 000. Proportions $\{0.4, 0.4, 0.2\}$ are assumed to generate the social class proportions x_{ir}, via a Dirichlet with parameters $\{4, 4, 2\}$.

To estimate the model from the simulated data we first assume $d_r \sim \text{Exp}(\lambda)$ where λ is unknown. An E(1) prior is assumed on λ, and N(0,10) priors are assumed for β_{rc} and γ_{rc}. It is necessary to model the Dirichlet prior on the π_{irc} using a series of gamma densities (a Dirichlet on $\pi_{1:C}$ with weights $\alpha_1, \alpha_2, \ldots, \alpha_C$ may be sampled by first taking $y_j \sim G(\alpha_j, 1)$ for $j = 1, \ldots, C$ and then setting $\pi_j = y_j / \sum_j y_j$). Following Rosen *et al.* (2001) the x_{ir} are taken as known rather than stochastic.

The second half of a two-chain run of 20 000 iterations reproduces the parameter values assumed in the simulation (Table 11.7).

Table 11.7   Posterior parameter estimates for simulated voting by social class

Parameter	Assumed value	Posterior mean	Standard deviation
γ_{11}	0.3	0.51	0.22
γ_{12}	0.4	0.58	0.14
γ_{21}	-0.5	-0.20	0.32
γ_{22}	0.9	1.09	0.16
γ_{31}	-0.8	-1.22	0.21
γ_{32}	-1.4	-1.12	0.20
β_{11}	1.6	1.50	0.16
β_{12}	2.2	2.12	0.17
β_{21}	-1	-0.88	0.21
β_{22}	-1.6	-1.71	0.13
β_{31}	-1.2	-1.14	0.20
β_{32}	-1.4	-0.86	0.16

11.7.4   Cross-tabulations with missing data on any factor

The general situation here is that a cross-tabulation involves K categorical variables $Y_1, Y_2, Y_3, \ldots, Y_K$ with $I_1, I_2, I_3, \ldots, I_K$ levels. Incompleteness at subject level may occur on just one of these, or on one or more in combination – with the worst case being subjects with complete non-response on all K variables. Consider for illustration a two-way table with binary variables Y_1 and Y_2.

The possible missingness patterns can be summarized as a two-way table with supplementary margins - see Jansen *et al.* (2003), Baker *et al.* (1992) and Molenberghs *et al.* (1999). Thus Jansen *et al.* consider binary data on treatment side effect and duration of follow-up (Table 11.8).

Table 11.8   Side effects by duration: numbers of subjects in different response groups

Side effect	Completely observed duration		Side effect observed duration missing	Duration observed side effect missing		Duration missing side effect missing
	Below 4 years	Above 4 years		Below 4 years	Above 4 years	
Yes	89	13	26	2	0	14
No	57	65	49			

Let $n_{jkr_1r_2}$ denote a cross-classification both by response category ($r_1 = 1$ if Y_1 observed, 0 if Y_1 missing and $r_2 = 1$ if Y_2 observed, 0 if Y_2 missing) and by the categories j and k of Y_1 and Y_2 themselves. Thus Baker *et al.* (1992) consider a study on infant birth-weight (low, normal) and maternal risk Y_2 (high, low) with observations (nine counts of mother) as follows:

> a 2×2 table for both variables observed $\{n_{jk11}\}$
> a 1×2 vector with only maternal risk observed $\{n_{.k01}\}$
> a 2×1 vector with only child birth-weight observed $\{n_{j.10}\}$
> a scalar for neither observed $\{n_{..00}\}$

In the second case the four underlying 'complete data' cells $\{n_{jk01}\}$ have to be estimated, in the third case the cells $\{n_{jk10}\}$ have to be estimated, and so on.

Following Jansen *et al.* (2003) the probability $\pi_{jkr_1r_2}$ of belonging to a particular category of the unobserved full data may be specified as a selection model

$$\pi_{jkr_1r_2} = q_{r_1r_2|jk}\, p_{jk} \qquad (11.12a)$$

where

$$p_{jk} = \theta_{jk}/\sum_j\sum_k \exp(\theta_{jk}) \qquad (11.12b)$$

is the model for $\{Y_1, Y_2\}$ and

$$q_{r_1r_2|jk} = \exp[\beta_{jk}(1-r_1) + \alpha_{jk}(1-r_2) + \gamma(1-r_1)(1-r_2)]/ \\ [1 + \exp(\beta_{jk}) + \exp(\alpha_{jk}) + \exp(\alpha_{jk} + \beta_{jk} + \gamma)] \qquad (11.12c)$$

specifies missingness given $Y_1 = j$ and $Y_2 = k$. Thus β_{jk} only operates when $r_1 = 0$, α_{jk} only when $r_2 = 0$, etc. When Y_1 and Y_2 are binary the probabilities of the observed nine counts are then given by π_{1111}, π_{1211}, π_{2111}, π_{2211}, $\pi_{.101}$, $\pi_{.201}$, $\pi_{1.10}$, $\pi_{2.10}$ and $\pi_{..00}$ where for example

$$\pi_{..00} = \pi_{1100} + \pi_{1200} + \pi_{2100} + \pi_{2200}$$

For identifiability, it is necessary to set the α_{jk} or β_{jk} equal to each other (e.g. $\alpha_{jk} = \alpha$), or equal either for all j or for all k (e.g. $\alpha_{jk} = \alpha_j$). Thus identifiable models are as follows:

(1) $\alpha_{jk} = \alpha,\ \beta_{jk} = \beta$
(2) $\alpha_{jk} = \alpha,\ \beta_{jk} = \beta_j$
(3) $\alpha_{jk} = \alpha_k,\ \beta_{jk} = \beta$
(4) $\alpha_{jk} = \alpha,\ \beta_{jk} = \beta_k$
(5) $\alpha_{jk} = \alpha_j,\ \beta_{jk} = \beta$

Model (1) represents ignorable missingness since the chance of an observation missing does not depend on the unobserved level of Y_1 or Y_2. Models (2) and (3) mean missingness on one variable is ignorable but missingness on the other variable depends on the outcome of the former. Models (4) and (5) mean missingness on one variable is ignorable but missingness on the other variable depends on its own outcome.

An alternative parameterization is given by Baker *et al.* (1992). Thus a log-linear model for the means $\mu_{jk11} = m_{jk}$ of the fully observed subtable is

$$\log(m_{jk}) = \lambda_0 + \lambda_{1j} + \lambda_{2k} + \lambda_3 + \lambda_4 + \lambda_{12jk} + \lambda_{13j} + \lambda_{14j} + \lambda_{23k} + \lambda_{24k} + \lambda_{34}$$

where λ_{1j}, λ_{2k} and λ_{12jk} are conventional log-linear effects but parameters with subscripts 3 or 4 involve missing data. Thus λ_3 is the main effect for Y_1 missing, λ_4 is the main effect for Y_2 missing, and λ_{34} is the effect for both missing. The model for data with Y_1 missing has means $\mu_{jk01} = m_{jk}\alpha_{jk}$ where

$$\log(\alpha_{jk}) = -2[\lambda_3 + \lambda_{13j} + \lambda_{23k} + \lambda_{34}]$$

The model for data with Y_2 missing has means $\mu_{jk10} = m_{jk}\beta_{jk}$ where

$$\log(\beta_{jk}) = -2[\lambda_4 + \lambda_{14j} + \lambda_{24k} + \lambda_{34}]$$

and the model for data with Y_1 and Y_2 missing has means $\mu_{jk00} = m_{jk}\alpha_{jk}\beta_{jk}\gamma$ where

$$\log(\gamma) = 4\lambda_{34}$$

Priors are placed on α_{jk}, β_{jk} and γ, subject to alternative restrictions as in (1)–(5) above for identification.

Example 11.8   Psychiatric study   Consider the data in Table 11.8 and the selection model specified by (11.12). Only one option for defining the missingness model (11.12c) is considered, namely the ignorable missing model $\alpha_{jk} = \alpha$, $\beta_{jk} = \beta$ (Baker *et al.*, 1992, p 646). The data model in (11.12b) involves three θ parameters: one for $j = 1, k = 1$, one for $j = 1, k = 2$ and one for $j = 2, k = 1$. N$(0, 1)$ priors are assumed for α, β and θ_1, θ_2, θ_3.

The resulting DIC is 52 with 5.1 parameters, and the model estimates are summarized in Table 11.9. The negative credible interval for θ_3 shows lesser side effects for patients at longer durations. Other identifiable models are considered by Jansen *et al.* (2003, Table 3). A sensitivity analysis would consider alternative missingness models and the impact on the θ parameters.

11.8   RECENT DEVELOPMENTS

The models considered above have focused on true missing data as opposed to latent variables (e.g. several manifest variables may be an indicator for a latent class as in the case of diagnosis when there is no gold standard test). They have in some cases considered simple binary situations (e.g. in sections 11.7.2 and 11.7.4), though extensions to multinomial and ordinal situations may be obtained. Furthermore situations where missingness includes both categorical and metric data have not been discussed – for

Table 11.9 Ignorable model summary

Node	2.5%	Median	97.5%
α	−4.23	−3.57	−3.06
β	−1.31	−1.04	−0.80
γ	0.94	1.72	2.54
n_{1111}	68.88	81.37	95.75
n_{2111}	46.82	58.69	71.40
n_{1211}	7.55	12.94	20.40
n_{2211}	53.36	65.15	78.55
n_{1110}	22.57	28.79	35.92
n_{2110}	15.46	20.71	26.64
n_{1210}	2.54	4.57	7.51
n_{2210}	17.31	23.09	29.46
n_{1101}	1.16	2.30	3.87
n_{2101}	0.83	1.64	2.83
n_{1201}	0.16	0.35	0.73
n_{2201}	0.92	1.82	3.15
n_{1100}	2.46	4.48	7.48
n_{2100}	1.73	3.23	5.45
n_{1200}	0.33	0.70	1.38
n_{2200}	1.96	3.58	6.01
θ_1	−0.05	0.22	0.51
θ_2	−0.44	−0.11	0.23
θ_3	−2.20	−1.62	−1.10

example, see Schafer (1997) on the general location model. Recent developments of the location model are considered by Peng *et al.* (2003). Another area where Bayesian extensions are possible is in semi-parametric modelling of the non-response mechanism (Scharfstein and Irizarry, 2003). Recent book-length treatments of missing data include Allison (2002), Gelman and Meng (2004) and Little and Rubin (2002).

EXERCISES

1. In Example 11.1 try a selection model (for non-monotone missingness) rather than a pattern mixture model (see data set C in Program 11.1 with missing data explicit). Relate $\Pr(R_{it})$ to $Y_{i,t-1}$ and Y_{it} via a logit regression ($R_{it} = 1$ if Y is NA) and assess whether a non-ignorable model is supported. For example, see equation (11.2b). Note that a model without random intercepts and slopes (on weeks$^{0.5}$)

may be simpler to fit. So one may assess whether there is a loss of fit in making this simplification.

2. In Example 11.2 it may not necessarily be that missingness on the covariate $X_6 = $ TRF is related to CBCL and TRF independently but to discrepancies between the ratings. One reason for missingness of TRF is that parents refused to give permission for school information to be obtained (Fitzmaurice *et al.*, 1996, p 104). In the model for $\Pr(R_i = 1)$ where $R_i = 1$ if TRF is missing, substitute a variable for the difference between TRF and CBCL to replace the separate effects of these two predictors and assess whether missingness is non-ignorable on this basis. Also assess whether missingness is non-ignorable if CBCL is excluded as a predictor in the logit regression for R.

3. In Example 11.3 compare the inferences on the main substantive question (the impact on whether GRD=A of FLAS, CGPA, MLAT and prior courses) as obtained from the models described with those based on a 'complete case' analysis. The latter involves excluding any subjects with missing (NA) values on any X or on Y.

4. In Example 11.3 replicate the analysis undertaken but with Y treated as ordinal rather than reduced to binary.

5. In Example 11.4 apply formal multiple imputation based on taking samples of missing Y_1 and Y_2 at iterations 1000 and 2000 in a two-chain run of Program 11.4 (this involves using the state-space command in WINBUGS). Try estimating the Δ_j and their standard errors from the corresponding four 'complete' data sets.

6. In Example 11.4 modify the existing program to allow non-ignorable missingness by making the probability of response ($R_j = 1$ if Y_j is observed) depend on the outcome as well as treatment and grade, so there are two missingness models, $p(R_1|X, Y_1)$ and $p(R_2|X, Y_1, Y_2)$ where $X = (C, G)$. How does this affect estimates of the Δ_j?

7. In Example 11.5 apply the model

$$\text{logit}(\gamma_{i1jk}) = \kappa_{jk} - \beta_{1j}F_{i1}$$
$$\text{logit}(\gamma_{itjk}) = \kappa_{jk} - \beta_{1j}F_{it} - \beta_{2j}F_{i,t-1} \quad t > 1$$

and compare the imputations with those obtained under the two models described.

8. In Example 11.5 apply the models already in the program but with the original continuous response data on Y_1 and Y_2 (included in the

programs) formed into ordinal variables with more categories (e.g. eight on both).

9. In Example 11.6 introduce a prior (e.g. $E(2)$) on the α_j and β_j parameters in (11.9c) and (11.9d) and assess any impacts on inferences.

10. In Example 11.8 try the missingness model with $\alpha_{jk} = \alpha$, $\beta_{jk} = \beta_k$ (i.e. missingness on Y_2 depends on its level).

REFERENCES

Aitchison, J. and Ho, C. (1989) The multivariate Poisson-log normal distribution. *Biometrika*, **76**, 643–651.

Albert, J. (1988) Bayesian estimation of Poisson means using a hierarchical log-linear model. In *Bayesian Statistics 3*, Bernardo, J., DeGroot, M., Lindley, D. and Smith, A. (eds). Oxford University Press: New York, 519–531.

Allison, P. (2002) *Missing Data*. Sage: Thousand Oaks, CA.

Baker, S., Rosenberger, W. and DerSimonian, R. (1992) Closed form estimates for missing counts in two-way contingency tables. *Statistics in Medicine*, **11**, 643–657.

Congdon, P. (2001) *Bayesian Statistical Modelling*. John Wiley & Sons: Chichester.

Crowder, M. and Hand, D. (1990) *Analysis of Repeated Measures*. Chapman and Hall: London.

Diggle, P. and Kenward, M. (1994) Informative drop-out in longitudinal data analysis. *Applied Statistics*, **43**, 49–93.

Faris, P., Ghali, W., Brant, R., Norris, C., Galbraith, P. and Knudtson, M. (2002) Multiple imputation versus data enhancement for dealing with missing data in observational healthcare outcome analyses. *Journal of Clinical Epidemiology*, **55**, 184–191.

Fitzmaurice, G., Laird, N. and Zahner, G. (1996) Multivariate logistic models for incomplete binary response. *Journal of the American Statistical Association*, **91**, 99–108.

Gelman, A. and Meng, X.-L. (eds) (2004) *Bayesian Analysis, Causal Inference, and Missing Data*. John Wiley & Sons: New York.

Haneuse, S. and Wakefield, J. (2004) Ecological inference incorporating spatial dependence. Chapter 13 in *Ecological Inference: New Methodological Strategies*, King, G., Rosen, O. and Tanner, M. (eds). Cambridge University Press: Cambridge.

Hedeker, D. and Gibbons, R. (1997) Application of random-effects pattern-mixture models for missing data in longitudinal studies. *Psychological Methods*, **2**, 64–78.

Hopke, P., Liu, C. and Rubin, D. (2001) Multiple imputation for multivariate data with missing and below-threshold measurements: time series concentrations of pollutants in the Arctic. *Biometrics*, **57**, 22–33.

Horton, N. and Fitzmaurice, G. (2002) Maximum likelihood estimation of bivariate logistic models for incomplete responses with indicators of ignorable and non-ignorable missingness. *Applied Statistics*, **51**, 281–295.

Horton, N. and Laird, N. (2001) Maximum likelihood analysis of logistic regression models with incomplete covariate data and auxiliary information. *Biometrics*, **57**, 34–42.

Ibrahim, J., Lipsitz, S. and Chen, M.-H. (1999) Missing covariates in generalized linear models when the missing data mechanism is non-ignorable. *Journal of the Royal Statistical Society Series B*, **61**, 173–190.

Ibrahim, J., Chen, M. and Lipsitz, S. (2001a) Missing responses in generalized linear mixed models when the missing data mechanism is non-ignorable. *Biometrika*, **88**, 551–564.

Ibrahim, J., Lipsitz, S. and Horton, N. (2001b) Using auxiliary data for parameter estimation with non-ignorably missing outcomes. *Applied Statistics*, **50**, 361–373.

Jansen, I., Molenberghs, G., Aerts, M., Thijs, H. and Van Steen, K. (2003) A local influence approach applied to binary data from a psychiatric study. *Biometrics*, **59**, 410–419.

Johnson, V. and Albert, J. (1999) *Ordinal Data Modelling*. Springer: New York.

King, G., Rosen, O. and Tanner, M. (1999) Binomial-beta hierarchical models for ecological inference. *Sociological Methods and Research*, **28**, 61–90.

Laird, N. (1988) Missing data in longitudinal studies. *Statistics in Medicine*, **7**, 305–315.

Lavori, P., Dawson, R. and Shera, D. (1995) A multiple imputation strategy for clinical trials with truncation of patient data. *Statistics in Medicine*, **14**, 1913–1925.

Little, R. (1993) Pattern-mixture models for multivariate incomplete data. *Journal of the American Statistical Association*, **88**, 125–134.

Little, R. and Rubin, D. (1987) *Statistical Analysis with Missing Data*. John Wiley & Sons: New York.

Little, R. and Rubin, D. (2002) *Statistical Analysis with Missing Data*, 2nd Edition. John Wiley & Sons: New York.

Little, T. and Gelman, A. (1999) Modelling differential nonresponse in sample surveys. *Sankhya*, **B60**, 101–126.

Molenberghs, G., Goetghebeur, E., Lipsitz, S. and Kenward, M. (1999) Non-random missingness in categorical data: strengths and limitations. *American Statistician*, **53**, 110–118.

Molenberghs, G., Michiels, B., Verbeke, G. and Curran, D. (2002) Strategies to fit pattern-mixture models. *Biostatistics*, **3**, 245–265.

Nandram, B. and Choi, J. (2002a) A Bayesian analysis of a proportion under non-ignorable nonresponse. *Statistics in Medicine*, **21**, 1189–212.

Nandram, B. and Choi, J. (2002b) Hierarchical Bayesian nonresponse models for binary data from small areas with uncertainty about ignorability. *Journal of the American Statistical Association*, **97**, 381–388.

Park, T. and Brown, M. (1994) Model for categorical data with nonignorable response. *Journal of the American Statistical Associationi*, **89**, 44–52.

Peng, Y., Little, R. and Raghuanthan, T. (2003) An extended general location model for causal inference from data subject to noncompliance and missing values. *Working Paper Series*, No. 7, University of Michigan, Department of Biostatistics.

Rosen, O., Jiang, W., King, G. and Tanner, M. (2001) Bayesian and frequentist inference for ecological inference: the R × C case. *Statistica Neerlandica*, **55**, 134–156.

Roy, J. and Lin, X. (2002) Analysis of multivariate longitudinal outcomes with non-ignorable dropouts and missing covariates: changes in methadone treatment practices. *Journal of the American Statistical Association*, **97**, 40–52.

Rubin, D. (1976) Comparing regressions when some predictor variables are missing. *Psychometrika*, **43**, 3–10.

Rubin, D. (1987) *Multiple Imputation for the Non-response in Surveys*. John Wiley & Sons: New York.

Schafer, J. (1997) *Analysis of Incomplete Multivariate Data*. Chapman and Hall: London.

Schafer, J. (1999) Multiple imputation: a primer. *Statistical Methods in Medical Research*, **8**, 3–15.

Scharfstein, D. and Irizarry, R. (2003) Generalized additive selection models for the analysis of studies with potentially non-ignorable missing outcome data. *Biometrics*, **59**, 601–613.

Stasny, E. (1991) Hierarchical models for the probabilities of a survey classification and nonresponse: an example from the National Crime Survey. *Journal of the American Statistical Association*, **86**, 296–303.

Index

Bayesian Models for Categorical Data   P. Congdon
© 2005 John Wiley & Sons, Ltd

Index compiled by Geoffrey Jones

WILEY SERIES IN PROBABILITY AND STATISTICS

ESTABLISHED BY WALTER A. SHEWHART AND SAMUEL S. WILKS

Editors
David J. Balding, Peter Bloomfield, Noel A. C. Cressie, Nicholas I. Fisher, Iain M. Johnstone, J. B. Kadane, Geert Molenberghs, Louise M. Ryan, David W. Scott, Adrian F. M. Smith, Jozef L. Teugels
Editors Emeriti
Vic Barnett, J. Stuart Hunter, David G. Kendall

The *Wiley Series in Probability and Statistics* is well established and authoritative. It covers many topics of current research interest in both pure and applied statistics and probability theory. Written by leading statisticians and institutions, the titles span both state-of-the-art developments in the field and classical methods.

Reflecting the wide range of current research in statistics, the series encompasses applied, methodological and theoretical statistics, ranging from applications and new techniques made possible by advances in computerized practice to rigorous treatment of theoretical approaches.

This series provides essential and invaluable reading for all statisticians, whether in academia, industry, government, or research.

ABRAHAM and LEDOLTER · Statistical Methods for Forecasting
AGRESTI · Analysis of Ordinal Categorical Data
AGRESTI · An Introduction to Categorical Data Analysis
AGRESTI · Categorical Data Analysis, *Second Edition*
ALTMAN, GILL, and McDONALD · Numerical Issues in Statistical Computing for the Social
    Scientist
AMARATUNGA and CABRERA · Exploration and Analysis of DNA Microarray and Protein Array
    Data
ANDĚL · Mathematics of Chance
ANDERSON · An Introduction to Multivariate Statistical Analysis, *Third Edition*
*ANDERSON · The Statistical Analysis of Time Series
ANDERSON, AUQUIER, HAUCK, OAKES, VANDAELE, and WEISBERG · Statistical Methods
    for Comparative Studies
ANDERSON and LOYNES · The Teaching of Practical Statistics
ARMITAGE and DAVID (editors) · Advances in Biometry
ARNOLD, BALAKRISHNAN, and NAGARAJA · Records
*ARTHANARI and DODGE · Mathematical Programming in Statistics
*BAILEY · The Elements of Stochastic Processes with Applications to the Natural Sciences
BALAKRISHNAN and KOUTRAS · Runs and Scans with Applications
BARNETT · Comparative Statistical Inference, *Third Edition*
BARNETT · Environmental Statistics: Methods & Applications
BARNETT and LEWIS · Outliers in Statistical Data, *Third Edition*
BARTOSZYNSKI and NIEWIADOMSKA-BUGAJ · Probability and Statistical Inference
BASILEVSKY · Statistical Factor Analysis and Related Methods: Theory and Applications
BASU and RIGDON · Statistical Methods for the Reliability of Repairable Systems
BATES and WATTS · Nonlinear Regression Analysis and Its Applications
BECHHOFER, SANTNER, and GOLDSMAN · Design and Analysis of Experiments for Statistical
    Selection, Screening, and Multiple Comparisons
BELSLEY · Conditioning Diagnostics: Collinearity and Weak Data in Regression

*Now available in a lower priced paperback edition in the Wiley Classics Library.

BELSLEY, KUH, and WELSCH · Regression Diagnostics: Identifying Influential Data and Sources of Collinearity

BENDAT and PIERSOL · Random Data: Analysis and Measurement Procedures, *Third Edition*

BERNARDO and SMITH · Bayesian Theory

BERRY, CHALONER, and GEWEKE · Bayesian Analysis in Statistics and Econometrics: Essays in Honor of Arnold Zellner

BHAT and MILLER · Elements of Applied Stochastic Processes, *Third Edition*

BHATTACHARYA and JOHNSON · Statistical Concepts and Methods

BHATTACHARYA and WAYMIRE · Stochastic Processes with Applications

BILLINGSLEY · Convergence of Probability Measures, *Second Edition*

BILLINGSLEY · Probability and Measure, *Third Edition*

BIRKES and DODGE · Alternative Methods of Regression

BLISCHKE and MURTHY (editors) · Case Studies in Reliability and Maintenance

BLISCHKE and MURTHY · Reliability: Modeling, Prediction, and Optimization

BLOOMFIELD · Fourier Analysis of Time Series: An Introduction, *Second Edition*

BOLLEN · Structural Equations with Latent Variables

BOROVKOV · Ergodicity and Stability of Stochastic Processes

BOULEAU · Numerical Methods for Stochastic Processes

BOX · Bayesian Inference in Statistical Analysis

BOX · R. A. Fisher, the Life of a Scientist

BOX and DRAPER · Empirical Model-Building and Response Surfaces

*BOX and DRAPER · Evolutionary Operation: A Statistical Method for Process Improvement

BOX, HUNTER, and HUNTER · Statistics for Experimenters: An Introduction to Design, Data Analysis, and Model Building

BOX and LUCEÑO · Statistical Control by Monitoring and Feedback Adjustment

BRANDIMARTE · Numerical Methods in Finance: A MATLAB-Based Introduction

BROWN and HOLLANDER · Statistics: A Biomedical Introduction

BRUNNER, DOMHOF, and LANGER · Nonparametric Analysis of Longitudinal Data in Factorial Experiments

BUCKLEW · Large Deviation Techniques in Decision, Simulation, and Estimation

CAIROLI and DALANG · Sequential Stochastic Optimization

CHAN · Time Series: Applications to Finance

CHATTERJEE and HADI · Sensitivity Analysis in Linear Regression

CHATTERJEE and PRICE · Regression Analysis by Example, *Third Edition*

CHERNICK · Bootstrap Methods: A Practitioner's Guide

CHERNICK and FRIIS · Introductory Biostatistics for the Health Sciences

CHILÈS and DELFINER · Geostatistics: Modeling Spatial Uncertainty

CHOW and LIU · Design and Analysis of Clinical Trials: Concepts and Methodologies, *Second Edition*

CLARKE and DISNEY · Probability and Random Processes: A First Course with Applications, *Second Edition*

*COCHRAN and COX · Experimental Designs, *Second Edition*

CONGDON · Applied Bayesian Modelling

CONGDON · Bayesian Statistical Modelling

CONGDON · Bayesian Models for Categorical Data

CONOVER · Practical Nonparametric Statistics, *Second Edition*

COOK · Regression Graphics

COOK and WEISBERG · Applied Regression Including Computing and Graphics

COOK and WEISBERG · An Introduction to Regression Graphics

CORNELL · Experiments with Mixtures, Designs, Models, and the Analysis of Mixture Data, *Third Edition*

*Now available in a lower priced paperback edition in the Wiley Classics Library.

COVER and THOMAS · Elements of Information Theory

COX · A Handbook of Introductory Statistical Methods

*COX · Planning of Experiments

CRESSIE · Statistics for Spatial Data, *Revised Edition*

CSÖRGÖ and HORVÁTH · Limit Theorems in Change Point Analysis

DANIEL · Applications of Statistics to Industrial Experimentation

DANIEL · Biostatistics: A Foundation for Analysis in the Health Sciences, *Sixth Edition*

*DANIEL · Fitting Equations to Data: Computer Analysis of Multifactor Data, *Second Edition*

DASU and JOHNSON · Exploratory Data Mining and Data Cleaning

DAVID and NAGARAJA · Order Statistics, *Third Edition*

*DEGROOT, FIENBERG, and KADANE · Statistics and the Law

DEL CASTILLO · Statistical Process Adjustment for Quality Control

DENISON, HOLMES, MALLICK, and SMITH · Bayesian Methods for Nonlinear Classification
    and Regression

DETTE and STUDDEN · The Theory of Canonical Moments with Applications in
    Statistics, Probability, and Analysis

DEY and MUKERJEE · Fractional Factorial Plans

DILLON and GOLDSTEIN · Multivariate Analysis: Methods and Applications

DODGE · Alternative Methods of Regression

*DODGE and ROMIG · Sampling Inspection Tables, *Second Edition*

*DOOB · Stochastic Processes

DOWDY and WEARDEN, and CHILKO · Statistics for Research, *Third Edition*

DRAPER and SMITH · Applied Regression Analysis, *Third Edition*

DRYDEN and MARDIA · Statistical Shape Analysis

DUDEWICZ and MISHRA · Modern Mathematical Statistics

DUNN and CLARK · Applied Statistics: Analysis of Variance and Regression, *Second Edition*

DUNN and CLARK · Basic Statistics: A Primer for the Biomedical Sciences, *Third Edition*

DUPUIS and ELLIS · A Weak Convergence Approach to the Theory of Large Deviations

EDLER and KITSOS (editors) · Recent Advances in Quantitative Methods in Cancer and Human
    Health Risk Assessment

*ELANDT-JOHNSON and JOHNSON · Survival Models and Data Analysis

ENDERS · Applied Econometric Time Series

ETHIER and KURTZ · Markov Processes: Characterization and Convergence

EVANS, HASTINGS, and PEACOCK · Statistical Distributions, *Third Edition*

FELLER · An Introduction to Probability Theory and Its Applications, Volume I, *Third Edition,*
    Revised; Volume II, *Second Edition*

FISHER and VAN BELLE · Biostatistics: A Methodology for the Health Sciences

*FLEISS · The Design and Analysis of Clinical Experiments

FLEISS · Statistical Methods for Rates and Proportions, *Second Edition*

FLEMING and HARRINGTON · Counting Processes and Survival Analysis

FULLER · Introduction to Statistical Time Series, *Second Edition*

FULLER · Measurement Error Models

GALLANT · Nonlinear Statistical Models

GELMAN and MENG (editors): Applied Bayesian Modeling and Casual Inference from
    Incomplete-data Perspectives

GHOSH, MUKHOPADHYAY, and SEN · Sequential Estimation

GIESBRECHT and GUMPERTZ · Planning, Construction, and Statistical Analysis of Comparative
    Experiments

GIFI · Nonlinear Multivariate Analysis

GLASSERMAN and YAO · Monotone Structure in Discrete-Event Systems

GNANADESIKAN · Methods for Statistical Data Analysis of Multivariate Observations,
    Second Edition

*Now available in a lower priced paperback edition in the Wiley Classics Library.

*Now available in a lower priced paperback edition in the Wiley Classics Library.

KADANE and SCHUM · A Probabilistic Analysis of the Sacco and Vanzetti Evidence

KALBFLEISCH and PRENTICE · The Statistical Analysis of Failure Time Data,
    Second Edition

KARIYA and KURATA · Generalized Least Squares

KASS and VOS · Geometrical Foundations of Asymptotic Inference

KAUFMAN and ROUSSEEUW · Finding Groups in Data: An Introduction to Cluster Analysis

KEDEM and FOKIANOS · Regression Models for Time Series Analysis

KENDALL, BARDEN, CARNE, and LE · Shape and Shape Theory

KHURI · Advanced Calculus with Applications in Statistics, *Second Edition*

KHURI, MATHEW, and SINHA · Statistical Tests for Mixed Linear Models

KLEIBER and KOTZ · Statistical Size Distributions in Economics and Actuarial Sciences

KLUGMAN, PANJER, and WILLMOT · Loss Models: From Data to Decisions

KLUGMAN, PANJER, and WILLMOT · Solutions Manual to Accompany Loss Models:
    From Data to Decisions

KOTZ, BALAKRISHNAN, and JOHNSON · Continuous Multivariate Distributions,
    Volume 1, *Second Edition*

KOTZ and JOHNSON (editors) · Encyclopedia of Statistical Sciences: Volumes 1 to 9
    with Index

KOTZ and JOHNSON (editors) · Encyclopedia of Statistical Sciences: Supplement Volume

KOTZ, READ, and BANKS (editors) · Encyclopedia of Statistical Sciences: Update Volume 1

KOTZ, READ, and BANKS (editors) · Encyclopedia of Statistical Sciences: Update Volume 2

KOVALENKO, KUZNETZOV, and PEGG · Mathematical Theory of Reliability of
    Time-Dependent Systems with Practical Applications

LACHIN · Biostatistical Methods: The Assessment of Relative Risks

LAD · Operational Subjective Statistical Methods: A Mathematical, Philosophical, and
    Historical Introduction

LAMPERTI · Probability: A Survey of the Mathematical Theory, *Second Edition*

LANGE, RYAN, BILLARD, BRILLINGER, CONQUEST, and GREENHOUSE ·
    Case Studies in Biometry

LARSON · Introduction to Probability Theory and Statistical Inference, *Third Edition*

LAWLESS · Statistical Models and Methods for Lifetime Data, *Second Edition*

LAWSON · Statistical Methods in Spatial Epidemiology

LE · Applied Categorical Data Analysis

LE · Applied Survival Analysis

LEE and WANG · Statistical Methods for Survival Data Analysis, *Third Edition*

LePAGE and BILLARD · Exploring the Limits of Bootstrap

LEYLAND and GOLDSTEIN (editors) · Multilevel Modelling of Health Statistics

LIAO · Statistical Group Comparison

LINDVALL · Lectures on the Coupling Method

LINHART and ZUCCHINI · Model Selection

LITTLE and RUBIN · Statistical Analysis with Missing Data, *Second Edition*

LLOYD · The Statistical Analysis of Categorical Data

MAGNUS and NEUDECKER · Matrix Differential Calculus with Applications in
    Statistics and Econometrics, *Revised Edition*

MALLER and ZHOU · Survival Analysis with Long Term Survivors

MALLOWS · Design, Data, and Analysis by Some Friends of Cuthbert Daniel

MANN, SCHAFER, and SINGPURWALLA · Methods for Statistical Analysis of
    Reliability and Life Data

MANTON, WOODBURY, and TOLLEY · Statistical Applications Using Fuzzy Sets

MARCHETTE · Random Graphs for Statistical Pattern Recognition

MARDIA and JUPP · Directional Statistics

*Now available in a lower priced paperback edition in the Wiley Classics Library.

*Now available in a lower priced paperback edition in the Wiley Classics Library.

RIPLEY · Spatial Statistics

RIPLEY · Stochastic Simulation

ROBINSON · Practical Strategies for Experimenting

ROHATGI and SALEH · An Introduction to Probability and Statistics, *Second Edition*

ROLSKI, SCHMIDLI, SCHMIDT, and TEUGELS · Stochastic Processes for Insurance and Finance

ROSENBERGER and LACHIN · Randomization in Clinical Trials: Theory and Practice

ROSS · Introduction to Probability and Statistics for Engineers and Scientists

ROUSSEEUW and LEROY · Robust Regression and Outlier Detection

RUBIN · Multiple Imputation for Nonresponse in Surveys

RUBINSTEIN · Simulation and the Monte Carlo Method

RUBINSTEIN and MELAMED · Modern Simulation and Modeling

RYAN · Modern Regression Methods

RYAN · Statistical Methods for Quality Improvement, *Second Edition*

SALTELLI, CHAN, and SCOTT (editors) · Sensitivity Analysis

*SCHEFFE · The Analysis of Variance

SCHIMEK · Smoothing and Regression: Approaches, Computation, and Application

SCHOTT · Matrix Analysis for Statistics

SCHOUTENS · Levy Processes in Finance: Pricing Financial Derivatives

SCHUSS · Theory and Applications of Stochastic Differential Equations

SCOTT · Multivariate Density Estimation: Theory, Practice, and Visualization

*SEARLE · Linear Models

SEARLE · Linear Models for Unbalanced Data

SEARLE · Matrix Algebra Useful for Statistics

SEARLE, CASELLA, and McCULLOCH · Variance Components

SEARLE and WILLETT · Matrix Algebra for Applied Economics

SEBER · Multivariate Observations

SEBER and LEE · Linear Regression Analysis, *Second Edition*

SEBER and WILD · Nonlinear Regression

SENNOTT · Stochastic Dynamic Programming and the Control of Queueing Systems

*SERFLING · Approximation Theorems of Mathematical Statistics

SHAFER and VOVK · Probability and Finance: Its Only a Game!

SMALL and McLEISH · Hilbert Space Methods in Probability and Statistical Inference

SRIVASTAVA · Methods of Multivariate Statistics

STAPLETON · Linear Statistical Models

STAUDTE and SHEATHER · Robust Estimation and Testing

STOYAN, KENDALL, and MECKE · Stochastic Geometry and Its Applications,
    Second Edition

STOYAN and STOYAN · Fractals, Random Shapes and Point Fields: Methods of
    Geometrical Statistics

STYAN · The Collected Papers of T. W. Anderson: 1943–1985

SUTTON, ABRAMS, JONES, SHELDON, and SONG · Methods for Meta-Analysis in
    Medical Research

TANAKA · Time Series Analysis: Nonstationary and Noninvertible Distribution Theory

THOMPSON · Empirical Model Building

THOMPSON · Sampling, *Second Edition*

THOMPSON · Simulation: A Modeler's Approach

THOMPSON and SEBER · Adaptive Sampling

THOMPSON, WILLIAMS, and FINDLAY · Models for Investors in Real World Markets

TIAO, BISGAARD, HILL, PEÑA, and STIGLER (editors) · Box on Quality and
    Discovery: with Design, Control, and Robustness

*Now available in a lower priced paperback edition in the Wiley Classics Library.

Printed and bound by CPI Group (UK) Ltd, Croydon, CR0 4YY

16/04/2025

14658551-0004